T0180952

Lecture Notes of
the Unione Matematica Italiana

26

More information about this series at http://www.springer.com/series/7172

Editorial Board

Andreas Hochenegger • Manfred Lehn •
Paolo Stellari

Editors

Birational Geometry of Hypersurfaces

Gargnano del Garda, Italy, 2018

 Springer

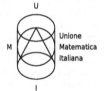

U

M

Unione
Matematica
Italiana

I

Editors

Andreas Hochenegger
Dipartimento di Matematica
Università degli Studi di Milano
Milano, Italy

Manfred Lehn
Institut für Mathematik
Johannes Gutenberg Universität Mainz
Mainz, Germany

Paolo Stellari
Dipartimento di Matematica
Università degli Studi di Milano
Milano, Italy

ISSN 1862-9113   ISSN 1862-9121   (electronic)
Lecture Notes of the Unione Matematica Italiana
ISBN 978-3-030-18637-1   ISBN 978-3-030-18638-8   (eBook)
https://doi.org/10.1007/978-3-030-18638-8

Mathematics Subject Classification (2010): 14-06, 14E08, 16E35, 14D22, 14C30

This Springer imprint is published by the registered company Springer Nature Switzerland AG.
The registered company address is: Gewerbestrasse 11, 6330 Cham, Switzerland

Preface

This volume originates from the "School on Birational Geometry of Hypersurfaces," which took place in the Palazzo Feltrinelli in Gargnano del Garda in March 2018.

The school was about a large number of open questions, techniques, and spectacular results, all surrounding the notion of (stable) rationality of projective varieties and, more specifically, hypersurfaces in projective spaces. Many of these questions look easy to formulate, for example, the (ir)rationality of cubic fourfolds. But after many attempts to address these questions, they proved to be so hard and striking to become among the most challenging open problems in the mainstream of algebraic geometry.

The school aimed at shedding some light on this vast research area by focusing on the two main aspects:

1. Approaches focusing on the (stable) rationality using deformation theory and Chow-theoretic tools like decomposition of the diagonal
2. The connection between K3 surfaces, hyperkähler geometry, and cubic fourfolds, which has a Hodge-theoretic and a homological side

This volume entirely reflects this twofold approach to the subject.

The school has benefitted from the beautiful lectures of Jean-Louis Colliot-Thélène, Daniel Huybrechts, Emanuele Macrì, and Claire Voisin. The contents of their talks appear now in the contributions, forming this volume with the addition of lecture notes by János Kollár and an appendix by Andreas Hochenegger.

We should finally point out that the school (and thus this volume) was made possible by the generous financial support of various institutions and research grants among them the Dipartimento di Matematica "Federigo Enriques" of the Università degli Studi di Milano, and the Foundation Compositio Mathematica, the GNSAGA Indam group, and research projects, namely, ERC-2017-CoG-771507 StabCondEn, FIRB 2012 "Moduli Spaces and Their Applications", PRIN 2015 "Geometria delle Varietà Algebriche," and Sonderforschungsbereich/Transregio 45 "Periods, Moduli Spaces and Arithmetic of Algebraic Varieties."

Finally, it is our great pleasure to thank all the participants of the school. Their intriguing questions and active participation created the stimulating atmosphere that gave origin to the idea of writing this volume.

Milano, Italy   Andreas Hochenegger
Mainz, Germany   Manfred Lehn
Milano, Italy   Paolo Stellari
24 January 2019

Contents

Part II Hypersurfaces

Contributors

Jean-Louis Colliot-Thélène CNRS, Université Paris-Sud, Université Paris-Saclay, Mathématiques, Orsay, France

Andreas Hochenegger Dipartimento di Matematica "Federigo Enriques", Università degli Studi di Milano, Milano, Italy

Daniel Huybrechts Mathematisches Institut, Universität Bonn, Bonn, Germany

János Kollár Princeton University, Princeton, NJ, USA

Emanuele Macrì Department of Mathematics, Northeastern University, Boston, MA, USA

Paolo Stellari Dipartimento di Matematica "F. Enriques", Università degli Studi di Milano, Milano, Italy

Claire Voisin Collège de France, Paris, France

Part I
Birational Invariants and (Stable) Rationality

Birational Invariants and Decomposition of the Diagonal

Claire Voisin

Abstract We give a rather detailed account of cohomological and Chow-theoretic methods in the study of the stable version of the Lüroth problem, which ask how to distinguish (stably) rational varieties from general unirational varieties. In particular, we study the notion of Chow or cohomological decomposition of the diagonal, which is a necessary criterion for stable rationality. Having better stability properties than the previously known obstructions under specialization with mildly singular central fibers, it has been very useful in the recent study of rationality questions.

1 Introduction

This paper is a set of expanded notes for lectures I gave in Miami, Sienne, Udine and Gargnano. The Lüroth problem is very simple to state, namely can one distinguish rational varieties from unirational ones? Here the definitions are the following:

Definition 1.1 A smooth projective variety X over a field K is unirational if there exist an integer N and a dominant rational map $\Phi : \mathbb{P}^N_K \dashrightarrow X$.

Note that one can always (at least if K is infinite) reduce to the case $N = n = \dim X$ by restricting Φ to a general linear subspace $\mathbb{P}^n_K \subset \mathbb{P}^N_K$.

Definition 1.2 A smooth projective variety X over a field K is rational if there exists a birational map $\mathbb{P}^n_K \dashrightarrow X$. The variety X is stably rational if $X \times \mathbb{P}^r$ is rational for some integer r.

More generally, we will say that X and Y are stably birational if $X \times \mathbb{P}^r \overset{birat}{\cong} Y \times \mathbb{P}^s$ for some integers r, s. This is an equivalence relation on the set of irreducible algebraic varieties over K. Of course, all these notions can be reformulated using

C. Voisin (✉)
Collège de France, Paris, France
e-mail: claire.voisin@imj-prg.fr

© Springer Nature Switzerland AG 2019
A. Hochenegger et al. (eds.), *Birational Geometry of Hypersurfaces*,
Lecture Notes of the Unione Matematica Italiana 26,
https://doi.org/10.1007/978-3-030-18638-8_1

only the function field $K(X)$ of X, so that the smoothness or projectivity of X is not important here. However, it is very important in practice to work with smooth projective models in order to exhibit stable birational invariants. The simplest example is the case of algebraic differential forms (see Sect. 2.1): For the space of algebraic differential forms on X of a given degree to be a stable birational invariant of X, one needs to take X to be smooth and projective (or at least complete).

The above mentioned problem had a classical satisfactory solution for curves and surfaces over an algebraically closed field of characteristic 0, namely they are rational once they are rationally connected, that is contain plenty of rational curves. However, after some delicate episodes (we refer to [9] for a precise history of the subject), it was found that in dimension 3, these two notions do not coincide. The three contributions leading to this conclusion were very different. We refer to Kollár's paper in this book for an account of one of the methods, namely "birational rigidity" which in its simple form proposed by Iskovskikh and Manin [28], consisted in proving that the considered variety (they were considering smooth quartic hypersurfaces in $\mathbb{P}^4$) has a very small birational automorphisms group, unlike projective space which has a huge group of birational automorphisms, called the Cremona group. The other approach, proposed by Clemens and Griffiths, has been extremely efficient in dimension 3, starting with the celebrated example of the cubic threefold hypersurface that they had solved. It involves the geometry of the intermediate Jacobian and its theta divisor. The relationships with birational geometry in dimension 3 is the fact that under the blow-up of a smooth curve, this Jacobian gets an extra summand added, which is the Jacobian of a curve.

The Clemens–Griffiths method works a priori only in dimension 3, although the developments of categorical methods might lead to higher dimensional variants. We refer for such developments to the notes of Macrì and Stellari in this volume. Both the Iskovskikh–Manin method and the Clemens–Griffiths method deal with rationality but not stable rationality for which we need to analyze the rationality not only of X but also of all the products $X \times \mathbb{P}^r$. For the Clemens–Griffiths method, this limitation is due to the fact that the rationality criterion they use works only in dimension 3. For the Iskovskikh–Manin method, the limitation is due to the fact that analyzing the birational automorphisms of $X \times \mathbb{P}^r$ seems to be very hard.

The third method due to Artin and Mumford not only works in any dimension, but also it rests on the introduction of invariants that have higher degree versions which are more and more subtle as the degree increases. A last crucial point is the fact that these invariants are stable birational invariants. They were the first to prove the following result:

Theorem 1.3 ([3]) *There exist unirational threefolds X which are not stably rational.*

The invariant used by Artin–Mumford is the torsion in Betti cohomology of degree 3 of a smooth projective model of X. We will describe this example in Sect. 2.1.1. The Artin–Mumford method has been further developed by Colliot-Thélène and Ojanguren [16] who used *higher degree unramified cohomology* groups with torsion coefficients as stable birational invariants in order to construct new examples of this phenomenon but having trivial Artin–Mumford invariants. We will introduce unramified cohomology in Sect. 2.2.2. We will also describe its main properties, and compute it in small degree. The degree 2 case is in fact the Brauer group and it is immediately related to the Artin–Mumford invariant which is the topological version of it. The degree 3 case was shown in [18] to measure the defect of the Hodge conjecture in degree 4 with integral coefficients. These developments build on one hand on Bloch–Ogus theory [11] that we will survey in Sect. 2.2.1, and on the other hand on the Bloch–Kato conjecture proved by Voevodsky [53], that is the main recent new ingredient in the theory of unramified cohomology, together with Kerz' work [30].

We now explain our input to the subject. The theory of algebraic cycles of complex algebraic varieties received a great impulse from Bloch–Srinivas contribution [11] who gave an elegant proof and various generalizations of Mumford's theorem [41] saying that a smooth projective complex variety with trivial CH_0 group (in the sense that all points are rationally equivalent) has no nonzero algebraic differential form of degree > 0. Their approach used the "decomposition of the diagonal" which we will describe in Sect. 3.1. The decomposition of the diagonal is the beginning of a Künneth decomposition. It says that after removing the first term $X \times x$ of the diagonal Δ_X, the remaining cycle is supported on $D \times X$, where $D \subset X$ is a proper closed algebraic subset. We will show that for quantities with enough functoriality under correspondences, such a decomposition allows to show that they are supported on D in a strong sense. The first instance of this phenomenon was of course the Bloch–Srinivas improvement of Mumford's theorem saying that if X has $CH_0(X) = \mathbb{Z}$, the positive degree rational cohomology of X has coniveau ≥ 1: more precisely, it is supported on the divisor D appearing in the decomposition of the diagonal.

The Bloch–Srinivas decomposition of the diagonal is with $\mathbb{Q}$-coefficients, and as we will see, there are many further obstructions to get a decomposition of the diagonal (Chow-theoretic or cohomological) with *integral* coefficients. We will discuss many of them in Sects. 3.3.2 and 3.3. Actually, in the cohomological setting, we have a complete understanding of this condition at least in dimension 3. The relevance of this study for rationality questions is the fact that *the existence of such a decomposition is a necessary criterion for stable rationality*. In the Chow setting, this property governs all the invariants of a Chow-theoretic/cohomological nature that we mentioned previously, including unramified cohomology (see Sect. 3.3.2), in the sense that they vanish if the variety has a Chow decomposition of the diagonal with integral coefficients.

What we realized in [60] is the fact that the existence of a Chow decomposition (and with some care, also of a cohomological decomposition) is stable under the following operation: degenerate (or specialize) a smooth general fiber X_t to a mildly singular special fiber X_0 and then desingularize X_0 to $\widetilde{X}_0$. (This statement is the degeneration theorem 4.4.) The paper [60] had considered only the simplest such mild singularities, namely nodal singularities in dimension at least 2. This already led us to the following conclusion:

Theorem 1.4 *There exist unirational threefolds which are not stably rational although all their unramified cohomology groups are trivial. The very general quartic double solids are such examples.*

Note that the only possibly nontrivial unramified cohomology groups for rationally connected threefolds are in fact the group $H^2_{nr}(X, \mathbb{Q}/\mathbb{Z})$, that is, the Artin–Mumford invariant. The gain over Theorem 1.3 is the fact that these varieties are very simple to construct and exhibit (in fact they are general hypersurfaces in a toric fourfold), while Fano threefolds with a nontrivial Artin–Mumford invariant as exhibited in [3] are hard to construct.

The quartic double solids appearing in Theorem 1.4 are Fano threefolds which specialize to Artin–Mumford double solids X_0, which are nodal. The situation is thus the following: The desingularized Artin–Mumford double solid $\widetilde{X}_0$ does not admit a decomposition of the diagonal because it has a nontrivial Artin–Mumford invariant. This implies that a general deformation X_t of X_0 neither admits a decomposition of the diagonal by the specialization theorem mentioned above. However for all deformations smoothifying a node, the Artin–Mumford invariant disappears. One can summarize the above argument in the following statement which does not involve explicitly the decomposition of the diagonal:

Proposition 1.5 *Let $\pi : X \to C$ be a flat projective morphism of relative dimension $n \geq 2$, where C is a smooth curve. Assume that the fiber X_t is smooth for $t \neq 0$, and has at worst ordinary quadratic singularities for $t = 0$. Then if* Tors $H^3_B(\widetilde{X}_0, \mathbb{Z}) \neq 0$, *the general fiber X_t is not stably rational.*

The paper [17] describes the exact conditions on the singularities which make the specialization theorem (hence Proposition 1.5) work. Colliot-Thélène and Pirutka applied their method to the case of the general quartic hypersurface in $\mathbb{P}^4$ for which they proved the analogue of Theorem 1.4. Although the precise nature of the allowed singularities is not known because it is related to the rationality of the exceptional divisors, their criterion was explicit enough to allow many other applications that we will try to survey in Sect. 4.3. The most striking and important consequence is the following result obtained in [26]:

Theorem 1.6 *Stable rationality is not deformation invariant. There exist families of smooth projective varieties such that the fiber X_t is rational over a dense set (a countable union of algebraic subsets) of the base, but the very general fiber is not stably rational.*

We will also describe in that section the Totaro method [51], which combines the specialization method with the Kollár argument in [31] of reduction to nonzero characteristic and analysis of algebraic differential forms on the central fiber. Finally we will explain Schreieder's further improvement (see [45, 46]) of Proposition 1.5.

2 Birational Invariants

We will say that a property or a quantity is birationally invariant, resp. stably birationally invariant, if it is constant on any birational equivalence class of varieties, resp. on any stable birational equivalence class of varieties.

2.1 Classical Birational Invariants and Functoriality

The following obvious lemma allows to produce stable birational invariants:

Lemma 2.1 *Let $X \mapsto I(X)$ be a group defined for smooth varieties over a given field K. Assume that $I(X)$ is covariant for morphisms of K-varieties and has the property that:*

(i) $I(U) \to I(X)$ is surjective for $U \subset X$ a dense Zariski open set, and
(ii) it is an isomorphism if $\operatorname{codim}(X \setminus U \subset X) \geq 2$.
  Then:

(a) $I(X)$ is a birational invariant for smooth projective varieties over K.
(b) If furthermore

(iii) $I(X) \cong I(X \times \mathbb{A}^1)$ (by push-forward) for any X, then $I(X)$ is a stable birational invariant for smooth projective varieties X over K.

Proof Let $\phi : X \dashrightarrow Y$ be a birational map between smooth and projective varieties over K. Then there is an open set $U \subset X$ such that $\operatorname{codim}(X \setminus U \subset X) \geq 2$ and $\phi_{|U}$ is a morphism. Then we have $I(X) \cong I(U)$ by (ii) and by covariant functoriality a morphism $\phi_{U*} : I(U) \to I(Y)$, hence a morphism $\phi_* : I(X) \to I(Y)$. It remains to see that ϕ_* is an isomorphism. Replacing ϕ by ϕ^{-1}, we get $\phi_{V*}^{-1} : I(V) \to I(X)$ for some Zariski open set V of Y such that $I(V) \cong I(Y)$. Let $U' \subset U$ be defined as $\phi^{-1}(V)$. Then $\phi^{-1} \circ \phi$ is the identity on U', hence $(\phi^{-1})_* \circ \phi_* : I(U') \to I(U')$ is the identity. As $I(U') \to I(X)$ is surjective by (i), we conclude that $(\phi^{-1})_* \circ \phi_* : I(X) \to I(X)$ is the identity which proves the first statement after exchanging ϕ and ϕ^{-1}. This proves (a).

For statement (b), we observe that (iii) implies that $I(X) \cong I(X \times \mathbb{A}^l)$ for any X and any l, and then that $I(X) \cong I(X \times \mathbb{P}^l)$ for any X and any l. Indeed we have

$$I(X) \cong I(X \times \mathbb{A}^l) \twoheadrightarrow I(X \times \mathbb{P}^l) \to I(X),$$

where the second arrow is surjective by (i) and the composite is the identity. Together with (a) (proved for smooth *projective* varieties), this implies (b). $\square$

The obvious application is the case of the fundamental group $\pi_1(X_{an})$ when $K = \mathbb{C}$, where X_{an} is $X(\mathbb{C})$ endowed with the Euclidean topology. It clearly satisfies properties (i) and (ii). Unfortunately, the birational invariant so constructed is trivial for rationally connected varieties by the following result which is due to Serre [47] in the case of unirational varieties.

Theorem 2.2 *Let X be a smooth projective rationally connected variety over the complex numbers. Then $\pi_1(X_{an}) = \{e\}$.*

Proof As X is rationally connected, there exist a smooth projective variety B, and a rational map

$$\phi : B \times \mathbb{P}^1 \dashrightarrow X \tag{2.1}$$

which has the property that

(i) $\phi(B \times 0) = \{x\}$ for a fixed point $x \in X(\mathbb{C})$ and
(ii) $\phi_\infty := \phi_{|B \times \infty} : B \dashrightarrow X$ is dominant (say generically finite).

Using the same arguments as above, there is an induced morphism $\phi_* : \pi_1(B_{an} \times \mathbb{CP}^1) \to \pi_1(X_{an})$. This morphism is trivial by (i) and its image is of finite index by (ii). This implies that $\pi_1(X_{an})$ is finite. The end of the proof is an argument of Serre : Consider the universal cover $\widetilde{X}_{an} \to X_{an}$. Then $\widetilde{X}_{an}$ is the analytic space of an algebraic variety $\widetilde{X}$ which is rationally connected because all rational curves contained in X lift to $\widetilde{X}$. This implies that the degree of the covering map $\widetilde{X} \to X$ is 1 (and thus that X is in fact simply connected) by the following Euler-Poincaré characteristic argument: When X is rationally connected over a field of characteristic 0, one has $H^i(X, \mathcal{O}_X) = 0$ for $i > 0$ (see Proposition 2.4 below). This implies that $\chi(X, \mathcal{O}_X) = 1$. This equality has to hold for both X and $\widetilde{X}$, giving

$$\chi(X, \mathcal{O}_X) = 1, \quad \chi(\widetilde{X}, \mathcal{O}_{\widetilde{X}}) = 1. \tag{2.2}$$

However, for a proper étale cover, one has

$$\chi(X, \mathcal{O}_X) = \deg(\widetilde{X}/X)\chi(\widetilde{X}, \mathcal{O}_{\widetilde{X}}).$$

Comparing with (2.2), we get that $\deg(\widetilde{X}/X) = 1$. $\qquad\qquad\square$

The contravariant version of Lemma 2.1 is the following:

Lemma 2.3 *Let $X \mapsto I(X)$ be a group defined for smooth varieties over a given field k. Assume that $I(X)$ is contravariant and has the property that:*

(i) *$I(X) \to I(U)$ is injective for $U \subset X$ a Zariski open set, and*
(ii) *it is an isomorphism if $\mathrm{codim}\,(X \setminus U \subset X) \geq 2$.*

Then:

(a) *$I(X)$ is a birational invariant for smooth projective varieties over K.*
(b) *If furthermore (iii) $I(X) = I(X \times \mathbb{A}^1)$ (by pull-back) for any X, then $I(X)$ is a stable birational invariant for smooth projective varieties over K.*

This lemma applies to closed differential forms of fixed positive degree. In characteristic 0, algebraic differential forms on smooth projective varieties are closed, but of course this is not true on nonprojective varieties and in fact condition (b) is not satisfied for algebraic differential forms, while it is for closed differential forms. The stable birational invariant that we get is trivial for rationally connected varieties in characteristic 0 by the following proposition:

Proposition 2.4 *Let X be smooth projective rationally connected over a field of characteristic 0. Then $H^0(X, \Omega_X^{\otimes l}) = 0$ and $H^l(X, \mathcal{O}_X) = 0$ for any $l > 0$.*

Proof The second statement is a consequence of the first by Hodge symmetry, which gives that over $\mathbb{C}$, $H^l(X, \mathcal{O}_X)$ is canonically isomorphic to the complex conjugate of $H^0(X, \Omega_X^l)$, which is naturally a subspace of $H^0(X, \Omega_X^{\otimes l})$. Consider a rational map $\phi : B \times \mathbb{P}^1 \dashrightarrow X$ as in (2.1). As X is projective, ϕ is well defined along a generic fiber $\mathbb{P}_b^1 := b \times \mathbb{P}^1$. As $\phi_{B \times \infty}$ is dominant, we can assume that ϕ is a submersion at (b, ∞). The differential ϕ_{b*} of ϕ along $\mathbb{P}_b^1$ gives a morphism

$$\phi_{b*} : T_{\mathbb{P}_b^1} \oplus T_{B,b} \otimes \mathcal{O}_{\mathbb{P}_b^1} \to \phi_b^* T_X$$

of vector bundles along $\mathbb{P}_b^1$. As $\phi(B \times 0)$ reduces to one point, ϕ_{b*} vanishes at 0. On the other hand, ϕ_{b*} is by assumption surjective at ∞. We thus conclude that the vector bundle $(\phi^* T_X)_{|\mathbb{P}_b^1}(-0)$ is generically generated by sections, hence that it is a direct sum of line bundles $\mathcal{O}_{\mathbb{P}_b^1}(a_i)$ with $a_i \geq 0$ on $\mathbb{P}_b^1$. Hence $(\phi^* T_X)_{|\mathbb{P}_b^1}$ is a direct sum of line bundles $\mathcal{O}_{\mathbb{P}_b^1}(b_i)$ with $b_i > 0$. It follows that for any $l > 0$, $H^0(\mathbb{P}_b^1, (\phi^* \Omega_X^{\otimes l})_{|\mathbb{P}_b^1}) = 0$. As $b \in B$ is generic, this implies that $H^0(X, \Omega_X^{\otimes l}) = 0$ for $l > 0$. □

Proposition 2.4 is not true in nonzero characteristic. The problem is that the dominant map ϕ could be nonseparable, hence nowhere submersive. Kollár [31] exhibited such a phenomenon for some mildly singular double covers of a hypersurfaces in projective space.

Theorem 2.5 ([31]) *Let $X \subset \mathbb{P}^n$ be a hypersurface of degree $2d$. Then X specializes to a double cover X_0 of a hypersurface of degree d branched along a hypersurface $Y_0 \subset X_0$ of degree $2d$. Assume $\operatorname{char} K = 2$ and $3d > n + 2$. Then X_0 has a desingularization $\widetilde{X}_0$ admitting a nonzero section of $\Omega_{\widetilde{X}_0}^{n-2} \otimes L^{-1}$, where the line bundle L is big and effective.*

We will see in Sect. 4.3 how Totaro uses this construction. Totaro only uses the effectivity of L, while Kollár needs the bigness of L, in order to apply the following result:

Proposition 2.6 *A separably uniruled (in particular, a ruled) variety Y does not admit a nonzero section of $\Omega_Y^l \otimes L^{-1}$ for some $l \leq \dim Y$, where the line bundle L is big.*

Proof If there is a variety Z admitting a morphism $f : Z \to B$ with general fiber $\mathbb{P}^1$ and a separable dominant map $\phi : Z \dashrightarrow Y$ not mapping the fibers of f to points, then we may assume ϕ is generically finite separable and dominant. If there is a nonzero section of $\Omega_Y^{n-2} \otimes L^{-1}$, there is a nonzero section of $\Omega_Z^l \otimes \phi^* L^{-1}$, where the line bundle $\phi^* L$ is also big, and in particular has positive degree along the fibers of f. But this is clearly impossible as $\Omega_{Z|Z_b} = f^*\Omega_B \oplus \Omega_{Z_b}$, and the first term is a trivial vector bundle along the fiber Z_b while the second term is a negative line bundle along the fibers $Z_b \cong \mathbb{P}^1$.   □

One major application obtained by Kollár is:

Theorem 2.7 ([31]) *If $X \subset \mathbb{P}^n_{\mathbb{C}}$ is a very general hypersurface of degree $d \geq 2^{\lceil \frac{n+3}{3} \rceil}$, X is not ruled, hence not rational.*

Proof (We give the argument only for even d) It suffices to exhibit one hypersurface in the above range of degree and dimension which is not ruled. The crucial point is that ruledness is stable under specialization. This result is due to Matsusaka [38]. Consider a hypersurface X defined over $\mathbb{Z}$, which admits a reduction modulo 2 of the form described in Theorem 2.5. If X is ruled, so is the specialization X_0 or rather its desingularization $\widetilde{X}_0$. But Proposition 2.6 precisely says that $\widetilde{X}_0$ is not ruled.   □

Remark 2.8 It is not true that (separable) rational connectedness is stable under specialization. Under specialization to nonzero characteristic, a family of rational curves sweeping-out X can specialize to a family of rational curves sweeping-out X_0 but nonseparably. This problem does not appear for ruledness because in this case the morphism from the family of curves to X has degree 1, hence also its specialization. Hence the specialized morphism is separable.

2.1.1   The Artin–Mumford Invariant

Our last examples of classical birational invariants will need functoriality properties slightly different from what we used in Lemmas 2.1 and 2.3, namely functoriality under correspondences. Let $X \mapsto I(X)$ be an invariant of smooth projective varieties. Assume that any correspondence $\Gamma \subset X \times Y$ with $\dim \Gamma = \dim X = \dim Y$ induces $\Gamma^* : I(Y) \to I(X)$ and that this action is compatible with composition of correspondences. In particular a morphism $\phi : X \to Y$ between smooth projective varieties of the same dimensions induces $\phi^* : I(Y) \to I(X)$ and $\phi_* : I(X) \to I(Y)$. Assume also that the projection formula $\phi_* \circ \phi^* = (\deg \phi) Id : I(Y) \to I(Y)$ holds. Assume the characteristic is 0 or resolution of singularities holds in the following sense: for any rational map $\phi : X \dashrightarrow Y$, with Y projective, there exists a smooth variety $\tau : \widetilde{X} \to X$, obtained from X by a sequence of blow-ups along smooth centers, such that $\phi \circ \tau$ gives a morphism $\widetilde{X} \to Y$.

Lemma 2.9 *Let $X \mapsto I(X)$ be an invariant of smooth projective varieties satisfying the functoriality properties above. Then $I(X)$ is invariant under birational maps of smooth projective varieties if and only if it is invariant under blow-up.*

Proof Let $\phi : X \dashrightarrow Y$ be a birational map. The graph $\Gamma_\phi \subset X \times Y$ induces a morphism $\Gamma_\phi^* : I(Y) \to I(X)$. If

$$\tilde{\phi} : \tilde{X} \to Y, \ \tau : \tilde{X} \to X$$

is a resolution of indeterminacies of ϕ (or singularities of Γ_ϕ), with τ a composition of blow-ups, one has

$$\Gamma_\phi^* = \tau_* \circ \phi^* \tag{2.3}$$

because $\Gamma_\phi = (\tau, Id_Y)(\Gamma_{\tilde{\phi}})$ or equivalently $\Gamma_\phi = {}^t\Gamma_\tau \circ \Gamma_{\tilde{\phi}}$. Invariance of I under blow-ups guarantees that $\tau_* : I(\tilde{X}) \to I(X)$ is an isomorphism. But ϕ^* is injective on $I(Y)$ because $\phi_* \circ \phi^* = Id$ on $I(Y)$. Hence by (2.3), Γ_ϕ^* is injective. In order to prove surjectivity, we now use resolution of singularities for ϕ^{-1}. We thus have a diagram

$$\tilde{\phi}^{-1} : \tilde{Y} \to X, \ \tau' : \tilde{Y} \to Y$$

where τ' is a composition of blow-ups.

As before we have $\Gamma_\phi^* = \tilde{\phi}_*^{-1} \circ \tau'^*$, where now τ'^* is an isomorphism by assumption while $\tilde{\phi}_*^{-1}$ is surjective by the projection formula $\tilde{\phi}_*^{-1} \circ (\tilde{\phi}^{-1})^* = Id_{I(X)}$. Thus Γ_ϕ^* is surjective. $\square$

Remark 2.10 The proposition above becomes a triviality if we use the weak factorization instead of resolutions of singularities.

Let us now introduce the Artin–Mumford invariant which was used in [3]. It will be generalized in the next section but the simplest version of it is the following: X is defined over the complex numbers and

$$I(X) = \text{Tors } H_B^3(X, \mathbb{Z}), \tag{2.4}$$

where $H_B^i(X, A)$ denotes the Betti cohomology group $H^i(X_{an}, A)$.

Proposition 2.11 *The Artin–Mumford invariant is a stable birational invariant of smooth projective varieties.*

Proof As all the Betti cohomology groups with integral coefficients have functoriality under correspondences, the same holds for their torsion subgroups. Similarly for the projection formula. By Lemma 2.9, in order to show birational invariance of Tors $H_B^3(X, \mathbb{Z})$, it thus suffices to show its invariance under blow-up. We now use the blow-up formula

$$H_B^i(\tilde{X}, \mathbb{Z}) = H_B^i(X, \mathbb{Z}) \oplus H_B^{i-2}(Z, \mathbb{Z}) \oplus H_B^{i-4}(Z, \mathbb{Z}) \oplus \dots,$$

where $\tau : \widetilde{X} \to X$ is the blow-up of X along the smooth locus Z with exceptional divisor $\tau_E : E \to Z$ and the first map is τ^* while the other maps are $j_* \circ (e^s \cup) \circ \tau_E^*$, with $e = [E]_{|E} \in H_B^2(E, \mathbb{Z})$. The end of the proof follows from the observation that the blow-up formula remains true if we replace integral cohomology by its torsion and that Tors $H_B^1(W, \mathbb{Z}) = 0$ for any topological space W. This last fact follows indeed from the cohomology long exact sequence associated with the short exact sequence of constant sheaves

$$0 \to \mathbb{Z} \xrightarrow{m} \mathbb{Z} \to \mathbb{Z}/m\mathbb{Z} \to 0$$

on W. The blow-up formula then gives

$$\text{Tors } H_B^3(\widetilde{X}, \mathbb{Z}) = \text{Tors } H_B^3(X, \mathbb{Z}).$$

In order to get stable birational invariance, it remains to see invariance under $X \mapsto X \times \mathbb{P}^r$. This follows from Künneth formula which gives $H_B^3(X \times \mathbb{P}^r, \mathbb{Z}) = H_B^3(X, \mathbb{Z}) \oplus H_B^1(X, \mathbb{Z})$, hence

$$\text{Tors } H_B^3(X \times \mathbb{P}^r, \mathbb{Z}) = \text{Tors } H_B^3(X, \mathbb{Z}) \oplus \text{Tors } H_B^1(X, \mathbb{Z}) = \text{Tors } H_B^3(X, \mathbb{Z}).$$

$\square$

Remark 2.12 The same proof shows as well that Tors $H_B^2(X, \mathbb{Z})$ is also a birational invariant. However, this invariant is trivial for rationally connected varieties, because they are simply connected by Theorem 2.2.

The Artin–Mumford invariant of X has an important interpretation as the topological part of the Brauer group of X, which detects Brauer–Severi varieties on X. These varieties are fibered over X into projective spaces, but are not projective bundles $\mathbb{P}(E)$ for some vector bundle E on X. Given such a fibration $\pi : Z \to X$ with fibers Z_x isomorphic to $\mathbb{P}^r$, $Z \cong \mathbb{P}(E)$ for some vector bundle of rank $r + 1$ if and only if there exists a line bundle L on Z which restricts to $\mathcal{O}(1)$ on each fiber. The topological part of the obstruction to the existence of L is the obstruction to the existence of $\alpha \in H_B^2(Z, \mathbb{Z})$ which restricts to $h_x := c_1(\mathcal{O}_{Z_x}(1)) \in H_B^2(Z_x, \mathbb{Z})$. The relevant piece of the Leray spectral sequence of π gives the exact sequence

$$H_B^2(Z, \mathbb{Z}) \to H^0(X, R^2\pi_*\mathbb{Z}) \xrightarrow{d_2} H^3(X, R^0\pi_*\mathbb{Z}) = H_B^3(Z, \mathbb{Z}),$$

where the second map is 0 with $\mathbb{Q}$-coefficients by the degeneration at E_2 of the Leray spectral sequence (or because there is a line bundle on Z whose restriction to the fibers is $\mathcal{O}_{Z_x}(-r - 1)$, namely the canonical bundle K_Z). The image $d_2(h)$ is thus a torsion class in $H_B^3(Z, \mathbb{Z})$, called the Brauer class. The same argument shows that the order of the Brauer class divides $r + 1$.

The Artin–Mumford invariant was used by Artin and Mumford to exhibit unirational threefolds which are not stably rational. Let $S_f \subset \mathbb{P}^3$ be a quartic surface defined by a degree 4 homogeneous polynomial f. Let $X_f \to \mathbb{P}^3$ be the double cover of $\mathbb{P}^3$ ramified along S_f. It is defined as $\operatorname{Spec}(\mathcal{O}_{\mathbb{P}^3} \oplus \mathcal{O}_{\mathbb{P}^3}(-2))$, where the algebra structure $\mathcal{A} \otimes \mathcal{A} \to \mathcal{A}$ on $\mathcal{A} = \mathcal{O}_{\mathbb{P}^3} \oplus \mathcal{O}_{\mathbb{P}^3}(-2))$ is natural on the summands $\mathcal{O}_{\mathbb{P}^3} \otimes \mathcal{O}_{\mathbb{P}^3}$ and $\mathcal{O}_{\mathbb{P}^3} \otimes \mathcal{O}_{\mathbb{P}^3}(-2)$ and sends $\mathcal{O}_{\mathbb{P}^3}(-2) \otimes \mathcal{O}_{\mathbb{P}^3}(-2)$ to $\mathcal{O}_{\mathbb{P}^3}$ via the composition

$$\mathcal{O}_{\mathbb{P}^3}(-2) \otimes \mathcal{O}_{\mathbb{P}^3}(-2) \to \mathcal{O}_{\mathbb{P}^3}(-4) \xrightarrow{f} \mathcal{O}_{\mathbb{P}^3}.$$

The local equation for $X_f \subset \operatorname{Spec}(\oplus_{l \geq 0} \mathcal{O}_{\mathbb{P}^3}(-2l))$ is thus $u^2 = f$, from which we conclude that X_f has ordinary quadratic singularities if S_f does. When S_f is smooth, X_f has trivial Artin–Mumford invariant. This follows from Lefschetz theorem on hyperplane sections as X_f can be seen as a hypersurface (not ample but positive) in the $\mathbb{P}^1$-bundle $\operatorname{Proj}(\operatorname{Sym}(\mathcal{O}_{\mathbb{P}^3} \oplus \mathcal{O}_{\mathbb{P}^3}(-2)))$ over $\mathbb{P}^3$. Assume now that S_f has ordinary quadratic singularities and let $\widetilde{X}_f$ be the desingularization of X_f by blow-up of the nodes. Note that $\widetilde{X}_f$ is unirational. This is true for all quartic double solids but becomes particularly easy once S_f has a node. Indeed, choose a node $O \in S_f$. The lines in $\mathbb{P}^3$ passing through O intersect S_f in the point O with multiplicity 2 and two other points. The inverse image of such a line Δ in X_f has a singular point at O (that we see now as a point of X_f), and its proper transform C_Δ in $\widetilde{X}_f$ is the double cover of $\Delta \cong \mathbb{P}^1$ ramified over the two remaining intersection points of Δ and S_f. It follows that C_Δ is rational and we thus constructed a conic bundle structure $a : \widetilde{X}_f \to \mathbb{P}^2$ on $\widetilde{X}_f$. On the other hand, if we choose a generic plane P in $\mathbb{P}^3$, its inverse image $\Sigma_{f,P}$ in X_f is a del Pezzo surface, hence is rational, and via a, it is a double cover of $P \cong \mathbb{P}^2$. The double cover $\widetilde{X}_f \times_P \Sigma_{f,P}$ of X_f is then rational, being rational over the function field of $\Sigma_{f,P}$ since it is a conic bundle over $\Sigma_{f,P}$ which has a section. We thus constructed a degree 2 unirational parametrization of X_f:

$$\widetilde{X}_f \times_P \Sigma_{f,P} \overset{birat}{\cong} \mathbb{P}^3 \dashrightarrow X_f.$$

Artin and Mumford construct f in such a way that X_f is nodal and $\widetilde{X}_f$ has a nontrivial Artin–Mumford invariant. Their construction is as follows: Project S_f from one of its nodes O. Then this projection makes the blow-up $\widetilde{S}_f$ of S_f at O a double cover of $\mathbb{P}^2$ ramified along a sextic curve. This sextic curve D is not arbitrary: it has to be tangent to a conic $C \subset \mathbb{P}^2$ at any of their intersection points. This conic indeed corresponds to the exceptional curve of $\widetilde{S}_f$. Another way to see it is to write the equation f as $X_0^2 q + X_0 t + s$, where q, t, s are homogeneous of respective degrees 2, 3, 4 in three variables X_1, X_2, X_3. The ramification curve of the 2:1 map $\widetilde{S}_f \to \mathbb{P}^2$ is defined by the discriminant of f seen as a quadratic polynomial in X_0, that is,

$$g = t^2 - 4qs. \tag{2.5}$$

The conic C is defined by $q = 0$ and (2.5) shows that $g_{|C}$ is a square, and g is otherwise arbitrary. Artin and Mumford choose g to be a product of two degree 3 polynomials, each of which satisfies the tangency condition along C. Note that S_f has then 9 extra nodes coming from the intersection of the two cubics.

Theorem 2.13 ([3]) *If the ramification curve D is the union of two smooth cubics E, F meeting transversally and tangent to C at each of their intersection points, the desingularized quartic double solid $\widetilde{X}_f$ has* Tors $H_B^3(\widetilde{X}_f, \mathbb{Z}) \neq 0$.

Rather than giving the complete proof of this statement, we describe Beauville's construction [9] of the Brauer-Severi variety $Z \to \widetilde{X}_f$ providing a Brauer class which is a 2-torsion class in $H_B^3(\widetilde{X}_f, \mathbb{Z})$ as described previously. The Artin–Mumford condition implies that the polynomial f is the discriminant of a $(4, 4)$-symmetric matrix M whose entries are linear forms in four variables (the quartic surface S_f is then called a quartic symmetroid). This defines a family of quadric surfaces $\mathcal{Q}$ over $\mathbb{P}^3$ if we see M as an equation of type $(2, 1)$ on $\mathbb{P}_1^3 \times \mathbb{P}_2^3$, and the associated double cover of $\mathbb{P}_2^3$ parameterizes the choice of a ruling in the corresponding quadric $\mathcal{Q}_t \subset \mathbb{P}_1^3$. The family of lines in a given ruling on a given fiber is a curve $\Delta \cong \mathbb{P}^1$ but the natural embedding of Δ in $G(2, 4)$ gives Δ as a conic. This way we get a family of rational curves over $\widetilde{X}_f$, smooth away from the surface S_f parameterizing singular quadrics. We refer to [9] and also to [34] for the local analysis which shows how to actually construct a $\mathbb{P}^1$-fibration on the whole of $\widetilde{X}_f$.

The last, less classical, birational invariant that we will mention is defined as follows. For a smooth complex variety X, one has the cycle class map

$$cl : \mathcal{Z}^2(X) \to H_B^4(X, \mathbb{Z})$$

and we will denote by $H_B^4(X, \mathbb{Z})_{alg} \subset H_B^4(X, \mathbb{Z})$ the image of cl. The group $H_B^4(X, \mathbb{Z})_{alg}$ is contained in the group $\mathrm{Hdg}^4(X, \mathbb{Z})$ of integral Hodge classes of degree 4 on X.

Lemma 2.14 *The groups* Tors $(H_B^4(X, \mathbb{Z})/H_B^4(X, \mathbb{Z})_{alg})$ *and* $\mathrm{Hdg}^4(X, \mathbb{Z})/H_B^4(X, \mathbb{Z})_{alg}$ *are birational invariants of the smooth projective variety X.*

Proof The two groups satisfy the functoriality conditions needed to apply Lemma 2.9, hence in order to show birational invariance, it suffices to show their invariance under blow-up. However, for the blow-up $\widetilde{X} \to X$ of $Z \subset X$, one has

$$H_B^4(\widetilde{X}, \mathbb{Z}) = H_B^4(X, \mathbb{Z}) \oplus H_B^2(Z, \mathbb{Z}) \oplus H_B^0(Z, \mathbb{Z}),$$

where the last term appears only if codim $Z \geq 3$. In this decomposition, all the maps are natural and induced by algebraic correspondences. In particular this is a decomposition into a direct sum of Hodge structures. This decomposition thus induces

$$H_B^4(\widetilde{X}, \mathbb{Z})_{alg} = H_B^4(X, \mathbb{Z})_{alg} \oplus H_B^2(Z, \mathbb{Z})_{alg} \oplus H_B^0(Z, \mathbb{Z})_{alg},$$

and

$$\mathrm{Hdg}^4(\widetilde{X}, \mathbb{Z}) = \mathrm{Hdg}^4(X, \mathbb{Z}) \oplus \mathrm{Hdg}^2(Z, \mathbb{Z}) \oplus \mathrm{Hdg}^0(Z, \mathbb{Z}).$$

Using the facts that $H_B^2(Z, \mathbb{Z})/H_B^2(Z, \mathbb{Z})_{alg}$ has no torsion and $\mathrm{Hdg}^2(Z, \mathbb{Z}) = H_B^2(Z, \mathbb{Z})_{alg}$, which both follow from the integral Hodge conjecture in degree 2 (or Lefschetz theorem on $(1, 1)$-classes), we conclude that

$$\mathrm{Tors}\,(H_B^4(\widetilde{X}, \mathbb{Z})/H_B^4(\widetilde{X}, \mathbb{Z})_{alg}) = \mathrm{Tors}\,(H_B^4(X, \mathbb{Z})/H_B^4(X, \mathbb{Z})_{alg}),$$

$$\mathrm{Hdg}^4(\widetilde{X}, \mathbb{Z})/H_B^4(\widetilde{X}, \mathbb{Z})_{alg} = \mathrm{Hdg}^4(X, \mathbb{Z})/H_B^4(X, \mathbb{Z})_{alg},$$

which proves the desired result.

The invariance of these groups under $X \mapsto X \times \mathbb{P}^r$ is proved similarly. $\quad\square$

Note that if the rational Hodge conjecture holds for degree 4 Hodge classes on X, these two groups are naturally isomorphic:

Lemma 2.15 *For any smooth projective variety X,*

$$\mathrm{Tors}\,(H_B^4(X, \mathbb{Z})/H_B^4(X, \mathbb{Z})_{alg}) = \mathrm{Tors}\,(\mathrm{Hdg}^4(X, \mathbb{Z})/H_B^4(X, \mathbb{Z})_{alg}).$$

If X satisfies the rational Hodge conjecture in degree 4, the group $\mathrm{Tors}\,(H_B^4(X, \mathbb{Z})/ H_B^4(X, \mathbb{Z})_{alg}))$ identifies with the group $\mathrm{Hdg}^4(X, \mathbb{Z})/H_B^4(X, \mathbb{Z})_{alg}$ which measures the defect of the Hodge conjecture for integral Hodge classes of degree 4 on X.

Proof Indeed, a torsion element in $H_B^4(X, \mathbb{Z})/H_B^4(X, \mathbb{Z})_{alg}$ is given by a class α on X such that $N\alpha$ is algebraic on X. Then α is an integral Hodge class on X, which proves the first statement. Finally, the rational Hodge conjecture in degree 4 for X is equivalent to the fact that the group $\mathrm{Hdg}^4(X, \mathbb{Z})/H_B^4(X, \mathbb{Z})_{alg}$ is of torsion, which proves the second statement. $\quad\square$

2.2 Unramified Cohomology

2.2.1 The Bloch–Ogus Spectral Sequence

Let X be an algebraic variety (in particular, it is irreducible and we can speak of its function field). If X is defined over $\mathbb{C}$, we can consider two topologies on $X(\mathbb{C})$, namely the Euclidean (or analytic) topology and the Zariski topology. We will denote X_{an}, resp. X_{Zar}, the topological space $X(\mathbb{C})$ equipped with the Euclidean topology, resp. the Zariski topology. As Zariski open sets are open for the Euclidean topology, the identity of $X(\mathbb{C})$ is a continuous map

$$f : X_{an} \to X_{Zar}.$$

Given any abelian group A, the Bloch–Ogus spectral sequence is the Leray spectral sequence of f, abutting to the cohomology $H_B^i(X, A) := H^i(X_{an}, A)$. It starts with

$$E_2^{p,q}(A) = H^p(X_{Zar}, \mathcal{H}^q(A)),$$

where $\mathcal{H}^q(A)$ is the sheaf on X_{Zar} associated with the presheaf $U \mapsto H_B^q(U, A)$. The Betti cohomology groups $H_B^n(X, A) = H^n(X_{an}, A)$ thus have a filtration, (which is in fact when X is smooth the coniveau filtration) namely the Leray filtration for which $Gr_L^p H_B^{p+q}(X_{an}, A) = E_\infty^{p,q}$, the latter group being a subquotient of $E_2^{p,q}$.

A fundamental result of Bloch–Ogus [11] is the Gersten-Quillen resolution for the sheaves $\mathcal{H}^q(A)$. It is constructed as follows: For any variety Y, we denote by $H^i(\mathbb{C}(Y), A)$ the direct limit over all dense Zariski open sets $U \subset Y$ of the groups $H_B^i(U, A)$:

$$H^i(\mathbb{C}(Y), A) = \lim_{\substack{\longrightarrow \\ \emptyset \neq U \subset Y, \text{open}}} H_B^i(U, A). \tag{2.6}$$

Let now Z be a normal irreducible closed algebraic subset of X, and let Z' be an irreducible reduced divisor of Z. At the generic point of Z', both Z' and Z are smooth. There is thus a residue map $\partial : H^i(\mathbb{C}(Z), A) \to H^{i-1}(\mathbb{C}(Z'), A)$. It is defined as the limit over all pairs of dense Zariski open sets $V \subset Z_{reg}$, $U \subset Z'_{reg}$ such that $U \subset V \cap Z'_{reg}$, of the residue maps

$$Res_{Z,Z'} : H^i((V \setminus V \cap Z')_{an}, A) \to H^{i-1}(U_{an}, A).$$

If now $Z' \subset Z$ is a divisor, with Z not necessarily normal along Z', we can introduce the normalization $n : \widetilde{Z} \to Z$ with restriction $n' : Z'' \to Z'$, where $Z'' = n^{-1}(Z')$, and then define $\partial : H^i(\mathbb{C}(Z), A) \to H^{i-1}(\mathbb{C}(Z'), A)$ as the composite

$$H^i(\mathbb{C}(Z), A) \cong H^i(\mathbb{C}(\widetilde{Z}), A) \xrightarrow{\partial} H^{i-1}(\mathbb{C}(Z''), A) \xrightarrow{n'_*} H^{i-1}(\mathbb{C}(Z'), A). \tag{2.7}$$

In (2.7), the pushforward morphism

$$n'_* : H^{i-1}(\mathbb{C}(Z''), A) \to H^{i-1}(\mathbb{C}(Z'), A)$$

is defined by restricting to pairs of Zariski open sets $U \subset Z''_{reg}$, $V \subset Z'_{reg}$ such that n' restricts to a proper (in fact, finite) morphism $U \to V$. More precisely, as Z'' is not necessarily irreducible, we should in the above definition write $Z'' = \cup_j Z''_j$ as a union of irreducible components, and take the sum over j of the morphisms (2.7) defined for each Z''_j.

For each subvariety $j : Z \hookrightarrow X$, we consider the group $H^i(\mathbb{C}(Z), A)$ as a constant sheaf supported on Z and we get the corresponding sheaf $j_* H^i(\mathbb{C}(Z), A)$ on X_{Zar}. Finally, we observe that we have a natural sheaf morphism

$$\mathcal{H}^i(A) \to H^i(\mathbb{C}(X), A)$$

where we recall that the second object is a constant sheaf on X_{Zar}. This sheaf morphism is simply induced by the natural maps $H^i(U_{an}, A) \to H^i(\mathbb{C}(X), A)$ for any Zariski open set $U \subset X$, given by (2.6). The residue maps have the following property: Let D_1, $D_2 \subset Y$ be two smooth divisors in a smooth variety, let Z be a smooth reduced irreducible component of $D_1 \cap D_2$ and let $\alpha \in H^i_B(U, A)$, where $U : Y \setminus (D_1 \cup D_2)$. Then

$$\mathrm{Res}_Z(\mathrm{Res}_{D_1}(\alpha)) = -\mathrm{Res}_Z(\mathrm{Res}_{D_2}(\alpha)), \tag{2.8}$$

where on the left Z is seen as a divisor in D_1, and on the right it is seen as a divisor of D_2. Considering the case where $Y \subset X$ is the regular locus of any subvariety of codimension k of X, D, $D' \subset Y$ are of codimension $k + 1$, and $Z \subset D \cap D' \subset Y$ is of codimension $k + 2$ in X, we conclude from (2.8) that for any i, the two sheaf maps

$$\partial : \oplus_{\mathrm{codim}\, Y=k} H^i(\mathbb{C}(Y), A) \to \oplus_{\mathrm{codim}\, D=k+1} H^{i-1}(\mathbb{C}(D), A)$$

and

$$\partial : \oplus_{\mathrm{codim}\, D=k+1} H^{i-1}(\mathbb{C}(D), A) \to \oplus_{\mathrm{codim}\, Z=k+2} H^{i-2}(\mathbb{C}(Z), A)$$

satisfy $\partial \circ \partial = 0$.

Theorem 2.16 (Bloch–Ogus, [11]) *Let X be smooth. The complex*

$$0 \to \mathcal{H}^i(A) \to H^i(\mathbb{C}(X), A) \to \underset{\substack{D \text{ irred} \\ \mathrm{codim}\, D=1}}{\oplus} H^{i-1}(\mathbb{C}(D), A) \to \ldots \to \underset{\substack{Z \text{ irred} \\ \mathrm{codim}\, Z=i}}{\oplus} H^0(\mathbb{C}(Z), A) \to 0 \tag{2.9}$$

is exact, hence provides an acyclic resolution of $\mathcal{H}^i(A)$.

It is clear that this resolution is acyclic. Indeed, all the sheaves appearing in the resolution are acyclic, being constant sheaves for the Zariski topology on algebraic subvarieties of X. Note that the codimension i subvarieties Z of X appearing above are all irreducible, so that $H^0(\mathbb{C}(Z), A) = A$ and the global sections of the last sheaf appearing in this resolution is the group $\mathcal{Z}^i(X) \otimes A$ of codimension i cycles with coefficients in A.

Theorem 2.16 says first that the sheaf map $\mathcal{H}^i(A) \to H^i(\mathbb{C}(X), A)$ is injective, which is by no means obvious. The meaning of this assertion is that if a class $\alpha \in H^i_B(U, A)$ vanishes on a dense Zariski open set $V \subset U$, then U can be covered by Zariski open sets V_i such that $\alpha_{|V_i} = 0$. This is a moving lemma for the support of cohomology.

We now come back to the Bloch–Ogus spectral sequence and describe the consequences of this theorem, following [11].

Theorem 2.17

(i) For any two integers $p > q$, one has $E_2^{p,q}(A) = H^p(X_{Zar}, \mathcal{H}^q(A)) = 0$.

(ii) For $p \le q$, one has

$$H^p(X_{Zar}, \mathcal{H}^q(A)) = \frac{\mathrm{Ker}\,(\partial : \oplus_{\mathrm{codim}\,Z=p} H^{q-p}(\mathbb{C}(Z), A) \to \oplus_{\mathrm{codim}\,Z=p+1} H^{q-p-1}(\mathbb{C}(Z), A))}{\mathrm{Im}\,(\partial : \oplus_{\mathrm{codim}\,Z=p-1} H^{q-p+1}(\mathbb{C}(Z), A) \to \oplus_{\mathrm{codim}\,Z=p} H^{q-p}(\mathbb{C}(Z), A))}.$$

$$(2.10)$$

(iii) The group $H^p(X, \mathcal{H}^p(\mathbb{Z}))$ is isomorphic to the group $\mathcal{Z}^p(X)/\mathrm{alg}$ of codimension p cycles of X modulo algebraic equivalence.

Proof

(i) Indeed, Theorem 2.16 says that $\mathcal{H}^q(A)$ has an acyclic resolution of length q.

(ii) As (2.9) is an acyclic resolution of $\mathcal{H}^q(A)$, the complex of global sections of (2.9) has degree p cohomology equal to $H^p(X_{Zar}, \mathcal{H}^q(A))$. This is exactly the contents of (2.10).

(iii) We use (ii), which gives in this case

$$H^p(X_{Zar}, \mathcal{H}^p(\mathbb{Z})) = \frac{\oplus_{\mathrm{codim}\,Z=p} H^0(\mathbb{C}(Z), \mathbb{Z})}{\mathrm{Im}\,(\partial : \oplus_{\mathrm{codim}\,Z=p-1} H^1(\mathbb{C}(Z), \mathbb{Z}) \to \oplus_{\mathrm{codim}\,Z=p} H^0(\mathbb{C}(Z), \mathbb{Z}))}.$$

We already mentioned that the numerator is the group $\mathcal{Z}^p(X)$. The proof is concluded by recalling the following two facts :

(1) A cycle Z of codimension p on X is algebraically equivalent to 0 if it belongs to the group generated by divisors homologous to 0 in the (desingularization of a) subvarieties of codimension $p - 1$ of X.

(2) A divisor D in a smooth complex manifold is cohomologous to 0 if and only if there exists a degree 1 integral Betti cohomology class α on $X \setminus |D|$ such that $\mathrm{Res}\,\alpha = D$. Here we denote by $|D|$ the support of D.

□

The vanishing (i) in Theorem 2.17 is very important. Let us give some applications taken from [11]. We will give further applications in Sect. 2.2.3. First of all, by the vanishing (i), we conclude that there is no nonzero Leray differential $d_r, r \ge 2$ starting from $E_2^{p,p}(\mathbb{Z}) = H^p(X_{Zar}, \mathcal{H}^p(\mathbb{Z}))$. It follows that $E_\infty^{p,p}(\mathbb{Z})$ is a quotient of the group $H^p(X_{Zar}, \mathcal{H}^p(\mathbb{Z}))$. Furthermore, by the same vanishing (i) above, the Bloch–Ogus filtration on $H_B^{2p}(X, \mathbb{Z})$ has $L^{p+1} = 0$, and thus $L^p H_B^{2p}(X, \mathbb{Z}) = Gr_L^p H_B^{2p}(X, \mathbb{Z}) = E_\infty^{p,p}(\mathbb{Z})$. We conclude that there is a natural composite map

$$H^p(X_{Zar}, \mathcal{H}^p(\mathbb{Z})) \to E_\infty^{p,p}(\mathbb{Z}) \hookrightarrow H_B^{2p}(X, \mathbb{Z}). \tag{2.11}$$

It is proved in [11] that, via the identification given by Theorem 2.17(iii), this map is the cycle class map in Betti cohomology. Note that by definition, the kernel of the cycle class map $\mathcal{Z}^p(X)/\text{alg} \to H_B^{2p}(X, \mathbb{Z})$ is the Griffiths group $\text{Griff}^p(X)$. We finally have the following result for codimension 2 cycles which describes the kernel of the cycle class map.:

Theorem 2.18 ([11]) *Let X be a smooth variety over $\mathbb{C}$. There is a natural exact sequence*

$$H_B^3(X, \mathbb{Z}) \to H^0(X_{Zar}, \mathcal{H}^3(\mathbb{Z})) \to H^2(X_{Zar}, \mathcal{H}^2(\mathbb{Z})) \to H_B^4(X, \mathbb{Z}).$$

Proof The maps are the natural ones. The first map is given by restriction to Zariski open sets. The second map is the differential d_2 of the Bloch–Ogus spectral sequence and the last map is the one appearing in (2.11) and just identified with the cycle class map. The proof of the exactness follows from inspection of the Bloch–Ogus spectral sequence. The kernel of the map $H^2(X_{Zar}, \mathcal{H}^2(\mathbb{Z})) = E_\infty^{2,2} \to H_B^4(X, \mathbb{Z})$ must be in the image of some d_r and obviously only $r = 2$ is possible. This shows exactness in the third term. Finally, by the vanishing of Theorem 2.17(i), the only nonzero d_r starting from $H^0(X_{Zar}, \mathcal{H}^3(\mathbb{Z}))$ is d_2. It follows that $\text{Ker } d_2 = E_\infty^{0,3}$, and this is a quotient of $H_B^3(X, \mathbb{Z})$. This shows exactness in the second term. $\square$

2.2.2 Unramified Cohomology

The following definition was first introduced in [16] in the setting of étale cohomology.

Definition 2.19 Let X be an algebraic variety over $\mathbb{C}$ and let A be an abelian group. Then $H_{nr}^i(X, A) = H^0(X_{Zar}, \mathcal{H}^i(A))$.

This definition can be made in fact over other fields, with Betti cohomology replaced by étale cohomology. If A is finite, and X is over $\mathbb{C}$, étale and Betti cohomology compare naturally. The advantage of Betti cohomology is that we can consider integral coefficients, while étale cohomology needs coefficients like $\mathbb{Z}_\ell$ which are projective limits of $\mathbb{Z}/l^n\mathbb{Z}$. However a big advantage of étale cohomology is that it fits naturally with Galois cohomology. In fact, we have a natural isomorphism

$$\varinjlim_{U \subset X, \text{open}} H_{et}^i(U, A) = H_{\text{Gal}}^i(\mathbb{C}(X), A), \tag{2.12}$$

where A is finite, and the direct limit is over the dense Zariski open sets of X. The term on the right is the cohomology of the Galois group of the field $\mathbb{C}(X)$ with coefficients in A. The term on the left is the analogue in the étale setting of what we defined to be $H^i(\mathbb{C}(X), A)$ in the Betti context. If A is finite, then

$$H^i_{et}(U, A) \cong H^i_B(U, A)$$

hence $H^i(\mathbb{C}(X), A) = H^i_{Gal}(\mathbb{C}(X), A)$.

One consequence of Theorem 2.16 is the following formula for unramified cohomology: this is actually cohomology without residues.

Proposition 2.20 *Assuming X smooth over $\mathbb{C}$, one has*

$$H^i_{nr}(X, A) = \mathrm{Ker}\,(H^i(\mathbb{C}(X), A) \xrightarrow{\partial} \oplus_{\mathrm{codim}\, Z=1} H^{i-1}(\mathbb{C}(Z), A)). \quad (2.13)$$

In particular, the restriction map $H^i_{nr}(X, A) \to H^i_{nr}(U, A)$ is injective for any Zariski dense open set U of X.

Proof Looking at Definition 2.19, this is a particular case of formula (2.10). □

We now get the following important consequence:

Theorem 2.21 *Unramified cohomology groups $H^i_{nr}(X, A)$ are birational invariants of smooth projective varieties.*

We should make precise here that we consider complex varieties over $\mathbb{C}$ if we want to work with Betti cohomology and any coefficients, and that for more general fields, we use étale cohomology and have to restrict coefficients as mentioned above.

Proof of Theorem 2.21 This is an immediate application of Propositions 2.20 and 2.3, because formula (2.13) shows that the natural restriction map $H^i_{nr}(X, A) \to H^i_{nr}(U, A)$ is injective when U is a dense Zariski open set of X, and that it is an isomorphism if $\mathrm{codim}\,(X \setminus U \subset X) \geq 2$. One uses of course the obvious contravariant functoriality of unramified cohomology. □

We refer to Sect. 3.3.2 for the proof that unramified cohomology is in fact a stable birational invariant. The following example shows that unramified cohomology generalizes Artin–Mumford invariant to higher degree.

Proposition 2.22 *Let X be a smooth projective complex variety. Then*

$$H^2_{nr}(X, \mathbb{Q}/\mathbb{Z}) = \mathrm{Tors}\, H^2(X_{an}, \mathcal{O}^*_{X_{an}}), \quad (2.14)$$

where $\mathcal{O}^*_{X_{an}}$ is the sheaf of invertible holomorphic functions on X_{an}, is the Brauer group of X. In particular, if X is rationally connected, $H^2_{nr}(X, \mathbb{Q}/\mathbb{Z}) \cong$ Tors $H^3_B(X, \mathbb{Z})$ is the Artin–Mumford group of X.

Proof Let us show the following precise version of (2.14):

$$H^2_{nr}(X, \mathbb{Z}/n\mathbb{Z}) = n - \text{Tors}\,(H^2(X_{an}, \mathcal{O}^*_{X_{an}})). \qquad (2.15)$$

Consider the exact sequence

$$0 \to \mathbb{Z}/n\mathbb{Z} \to \mathcal{O}^*_{X_{an}} \to \mathcal{O}^*_{X_{an}} \to 1,$$

where the second map is $x \mapsto x^n$ and $\mathbb{Z}/n\mathbb{Z}$ is identified with the group of n-th roots of unity. The associated long exact sequence shows that

$$n - \text{Tors}\, H^2(X_{an}, \mathcal{O}^*_{X_{an}}) \cong H^2(X_{an}, \mathbb{Z}/n\mathbb{Z})/\text{Im}\, cl_n,$$

where

$$cl_n : H^1(X_{an}, \mathcal{O}^*_{X_{an}}) = H^1(X, \mathcal{O}^*_X) = \text{CH}^1(X) \to H^2(X_{an}, \mathbb{Z}/n\mathbb{Z})$$

is the cycle class modulo n. We consider the Bloch–Ogus exact sequence for the sheaf $\mathbb{Z}/n\mathbb{Z}$ on X_{an}. The $E_2^{p,q}$-terms in degree 2 are, by Theorem 2.17(i)

$$E_2^{0,2} = H^0(X_{Zar}, \mathcal{H}^2(\mathbb{Z}/n\mathbb{Z})) = H^2_{nr}(X, \mathbb{Z}/n\mathbb{Z}), \;\; E_2^{1,1} = H^1(X_{Zar}, \mathcal{H}^1(\mathbb{Z}/n\mathbb{Z})).$$

The last term maps to $H^2(X_{an}, \mathbb{Z}/n\mathbb{Z})$ as all the higher d_r vanish on it, again by Theorem 2.17(i), and one proves that the map is the cycle class cl_n. No d_r for $r \geq 2$ starts from or arrives to $E_2^{0,2}$, by Theorem 2.17(i) again. Hence $E_2^{0,2} = E_\infty^{0,2}$ is the quotient of $H^2(X_{an}, \mathbb{Z}/n\mathbb{Z})$ by the image of $E_2^{1,1}$. $\qquad\square$

2.2.3 Bloch–Kato Conjecture and Applications

Define the Milnor K-theory groups of a field K (or a ring R) as follows

$$K_i^M(K) = (K^*)^{\otimes i}/I,$$

where I is the ideal generated by $x \otimes (1-x)$ for $x \in K^*$, $1 - x \in K^*$. In particular, we have $K_1^M(K) = K^*$. Fix an integer n prime to the characteristic of K. The exact sequence of Galois modules

$$0 \to \mu_n \to \overline{K}^* \to \overline{K}^* \to 1,$$

where $\mu_n \subset \overline{K}^*$ is the group of n-th roots of unity, gives a map

$$\partial : K^*/n \to H^1(K, \mu_n) := H^1(G_K, \mu_n), \qquad (2.16)$$

where $G_K = \mathrm{Gal}\,(\overline{K}/K)$, which is known by Hilbert's Theorem 90 to be an isomorphism (this is equivalent to the vanishing $H^1(G_K, \overline{K}^*) = 0$). More generally, one has a morphism (called the Galois symbol or norm residue map)

$$\partial_i : K_i^M(K)/n \to H^i(G_K, \mu_n^{\otimes i}) \tag{2.17}$$

which to $(x_1, \ldots, x_i)$ associates $\alpha(x_1) \cup \ldots \cup \alpha(x_i)$. The following fundamental result generalizing the isomorphism (2.16) is the Bloch–Kato conjecture solved by Voevodsky [53].

Theorem 2.23 *The map ∂_i is an isomorphism for any i and n prime to char K.*

This result was known for $i = 2$ as the Merkur'ev-Suslin theorem [39]. The following result is proved in [18], and [6] (see also [7]) to be a consequence of the Bloch–Kato conjecture (now Voevodsky's theorem).

Theorem 2.24 *Let X be a smooth complex variety. Then the sheaves $\mathcal{H}^i(\mathbb{Z})$ on X_{Zar} have no torsion.*

In other words, if an integral Betti cohomology class α defined on a Zariski open set U of X is of n-torsion for some integer n, then U is covered by Zariski open sets V such that $\alpha_{|V} = 0$.

Proof of Theorem 2.24 The exact sequence of sheaves on X_{an}

$$0 \to \mathbb{Z} \xrightarrow{n} \mathbb{Z} \to \mathbb{Z}/n\mathbb{Z} \to 0$$

provides an associated long exact sequence of sheaves on X_{Zar}

$$\ldots \mathcal{H}^i(\mathbb{Z}) \to \mathcal{H}^i(\mathbb{Z}/n\mathbb{Z}) \to \mathcal{H}^{i+1}(\mathbb{Z}) \xrightarrow{n} \mathcal{H}^{i+1}(\mathbb{Z}) \ldots$$

from which one concludes that the sheaves $\mathcal{H}^i(\mathbb{Z})$ have no n-torsion (for any n, i) if and only if the natural sheaf maps

$$\mathcal{H}^i(\mathbb{Z}) \to \mathcal{H}^i(\mathbb{Z}/n\mathbb{Z}) \tag{2.18}$$

are surjective for all i, n. This is however implied by Voevodsky's theorem as follows: Voevodsky gives the isomorphisms

$$K_j^M(\mathbb{C}(D))/n \cong H_{Gal}^j(\mathbb{C}(D), \mathbb{Z}/n\mathbb{Z})$$

for all j, n and closed algebraic subsets D of X. If one combines these isomorphisms with the Bloch–Ogus resolution of $\mathcal{H}^i(\mathbb{Z}/n\mathbb{Z})$ on one hand and the Gersten-Quillen resolution of the sheaves $\mathcal{K}_i^M(\mathcal{O}_X)$ established by Kerz [30] on the other hand, one concludes that the natural maps

$$\mathcal{K}_i^M(\mathcal{O}_X) \to \mathcal{H}^i(\mathbb{Z}/n\mathbb{Z}) \tag{2.19}$$

are sheaf isomorphisms. On the other hand, we note that for a Zariski open set $U \subset X$, we have the inclusion

$$\Gamma(\mathcal{O}_U^*) \subset \Gamma(\mathcal{O}_{U_{an}}^*)$$

where on the right we consider the invertible *holomorphic* functions on U. There are natural maps given by the exponential exact sequence on U_{an}

$$c : K_1^M(\Gamma(\mathcal{O}_{U_{an}})) = \Gamma(\mathcal{O}_{U_{an}}^*) \to H^1(U_{an}, \mathbb{Z}), \ c : K_i^M(\Gamma(\mathcal{O}_{U_{an}})) \to H^i(U_{an}, \mathbb{Z})$$

and the maps $K_i^M(\Gamma(\mathcal{O}_U)) \to H^i(U_{an}, \mathbb{Z}/n\mathbb{Z})$ appearing in (2.19) fit in a commutative diagram

$$
\begin{array}{ccccc}
K_i^M(\Gamma(\mathcal{O}_U)) & \longrightarrow & K_i^M(\Gamma(\mathcal{O}_{U_{an}})) & \overset{c}{\longrightarrow} & H^i(U_{an}, \mathbb{Z}) \\
\downarrow{\scriptstyle f} & & \downarrow{\scriptstyle f_{an}} & & \downarrow{\scriptstyle g} \\
K_i^M(\Gamma(\mathcal{O}_U))/n & \longrightarrow & K_i^M(\Gamma(\mathcal{O}_{U_{an}}))/n & \overset{c_n}{\longrightarrow} & H^i(U_{an}, \mathbb{Z}/n\mathbb{Z})
\end{array}
\tag{2.20}
$$

where the first vertical maps f and f_{an} given by reduction mod n are obviously surjective and the vertical map g is the map (2.18), or rather its global sections version over U. Voevodsky's theorem implies a fortiori the surjectivity of the bottom horizontal map c_n at the sheaf level, so by surjectivity of f_{an}, we conclude that $c_n \circ f_{an} = g \circ c$ is surjective at the sheaf level. A fortiori g is surjective at the sheaf level, that is, the sheaf maps (2.18) are surjective. $\qquad\square$

Corollary 2.25 *The groups $H_{nr}^i(X, \mathbb{Z})$ have no torsion, for any smooth algebraic variety over $\mathbb{C}$.*

We will however also see that these groups are trivial for X unirational. (We refer to Theorem 3.22 in Sect. 3.3.2 for details of proof and for a more general statement.) The unirationality assumption guarantees by functoriality considerations that the groups $H_{nr}^i(X, \mathbb{Z})$ are torsion for $i > 0$. The torsion freeness statement then implies that they are trivial. It follows that we cannot use the unramified cohomology groups with integral coefficients to distinguish rational varieties from unirational ones. In fact, unramified cohomology with *torsion coefficients* are the right invariant to use, as it already appeared in Proposition 2.22. The following result proved in [18] uses Theorem 2.24 to describe the next group $H_{nr}^3(X, \mathbb{Q}/\mathbb{Z})$ (or $H_{nr}^3(X, \mathbb{Z}/n\mathbb{Z})$). In fact we relate it to the birationally invariant group we introduced in Lemma 2.14.

Theorem 2.26 ([18])

(i) For any smooth algebraic variety X over $\mathbb{C}$, there is an exact sequence

$$0 \to H_{nr}^3(X, \mathbb{Z}) \otimes \mathbb{Z}/n\mathbb{Z} \to H_{nr}^3(X, \mathbb{Z}/n\mathbb{Z}) \to n - \text{Tors}\,(H_B^4(X, \mathbb{Z})/H_B^4(X, \mathbb{Z})_{alg}) \to 0 \tag{2.21}$$

(ii) *If X is rationally connected, then $H_{nr}^4(X, \mathbb{Z}/n\mathbb{Z}) \cong n - \mathrm{Tors}\,(H_B^4(X, \mathbb{Z})/$*
    $H_B^4(X, \mathbb{Z})_{alg})$ and $H_{nr}^4(X, \mathbb{Q}/\mathbb{Z})$ measures the defect of the Hodge conjecture
    for degree 4 integral Hodge classes on X.

Proof The second statement follows from the first by Theorem 3.22(ii), using
Lemma 2.15 and the fact that if X is rationally connected, the Hodge conjecture
holds for rational Hodge classes of degree 4 on X (see [12, 19], and Sect. 3.3.1).

We now prove (i). The result is obtained by examining the Bloch–Ogus spectral
sequence for degree 4 integral cohomology. Recall from Sect. 2.2.1 that we have
$E_2^{p,q}$-terms $H^p(X_{Zar}, \mathcal{H}^q(\mathbb{Z}))$ with $p + q = 2$ converging to $H_B^4(X, \mathbb{Z})$. By
Theorem 2.17(i), only

$$H^0(X_{Zar}, \mathcal{H}^4(\mathbb{Z})),\ H^1(X_{Zar}, \mathcal{H}^3(\mathbb{Z})),\ H^2(X_{Zar}, \mathcal{H}^2(\mathbb{Z}))$$

appear. Furthermore, as we already saw, the group $H^2(X_{Zar}, \mathcal{H}^2(\mathbb{Z}))$ maps onto its
image $E_\infty^{2,2}$ in $H_B^4(X, \mathbb{Z})$, which identifies with $H_B^4(X, \mathbb{Z})_{alg}$. We conclude that the
Bloch–Ogus filtration on $H_B^4(X, \mathbb{Z})$ induces a filtration on $H_B^4(X, \mathbb{Z})/H_B^4(X, \mathbb{Z})_{alg}$
with two successive quotients, namely $E_\infty^{1,3}$ and $E_\infty^{0,4}$. The space $E_\infty^{0,4}$ is a subspace
of $E_2^{0,4} = H^0(X_{Zar}, \mathcal{H}^4(\mathbb{Z}))$, hence it has no torsion by Theorem 2.24. It thus
follows that

$$\mathrm{Tors}\,(H_B^4(X, \mathbb{Z})/H_B^4(X, \mathbb{Z})_{alg}) = \mathrm{Tors}\,E_\infty^{1,3}.$$

Finally, applying again Theorem 2.17(i), we see that no d_r can start from $E_\infty^{1,3}$, so
that $E_\infty^{1,3} = E_2^{1,3} = H^1(X_{Zar}, \mathcal{H}^3(\mathbb{Z}))$. Finally we have to compute the n-torsion
of the last group and for this we use the short exact sequence of sheaves on X_{Zar}
given by Theorem 2.24:

$$0 \to \mathcal{H}^3(\mathbb{Z}) \xrightarrow{n} \mathcal{H}^3(\mathbb{Z}) \to \mathcal{H}^3(\mathbb{Z}/n\mathbb{Z}) \to 0.$$

Taking the long exact sequence associated to it, we get

$$0 \to H_{nr}^3(X, \mathbb{Z}) \otimes \mathbb{Z}/n\mathbb{Z} \to H_{nr}^3(X, \mathbb{Z}/n\mathbb{Z}) \to n - \mathrm{Tors}\,(H^1(X_{Zar}, \mathcal{H}^3(\mathbb{Z}))) \to 0,$$

that is (2.21).   □

It is well-known that the integral Hodge conjecture is not true in general in degree
4. This was first observed by Atiyah and Hirzebruch [4]; further examples were
found by Kollár [32], and we refer to [48] for a development of Kollár's method.
One question which remained open was whether there are such counterexamples to
the integral Hodge conjecture for degree 4 Hodge classes on a rationally connected
variety (such a variety is then stably irrational by Lemma 2.14). That such examples
exist follows from Theorem 2.26 and earlier work of Colliot-Thélène-Ojanguren
[16].

Theorem 2.27 ([16]) *There exist unirational six-folds* X, *which satisfy* $H^3_{nr}(X, \mathbb{Z}/2\mathbb{Z}) \neq 0$.

The varieties X are constructed as quadric bundles over $\mathbb{P}^3$. No smooth model is provided in [16] but in fact, although the formulas we gave above need a smooth projective model, it is not actually needed to compute unramified cohomology, as this is a birational invariant, hence can be computed using only the function field, which is what the authors do in [16]. By Theorem 2.26 (ii), any smooth projective model of such a variety X has an integral Hodge class of degree 4 which is not algebraic, but it is not obvious to see it geometrically.

2.3 Further Stable Birational Invariants

We work over the complex numbers. We describe in this section a number of interesting birational invariants constructed from the group of 1-cycles. It is not clear whether these invariants can be nontrivial for some rationally connected varieties.

2.3.1 Curve Classes

Let X be a smooth projective rationally connected variety of dimension n over $\mathbb{C}$. As $H^2(X, \mathcal{O}_X) = 0$, the Hodge structure on $H^{2n-2}_B(X, \mathbb{Z}) \cong H^B_2(X, \mathbb{Z})$ is trivial, that is, purely of type $(n-1, n-1)$. For any smooth projective variety X as above, the cycle class map $\mathcal{Z}_1(X) \otimes \mathbb{Q} \to \mathrm{Hdg}^{2n-2}(X, \mathbb{Q}) := H^{2n-2}_B(X, \mathbb{Q}) \cap H^{n-1,n-1}(X)$ is surjective, as follows from the Lefschetz theorem on $(1, 1)$-classes and the hard Lefschetz theorem in degree 2, which provides an isomorphism

$$l^{n-2} : \mathrm{Hdg}^2(X, \mathbb{Q}) \cong \mathrm{Hdg}^{2n-2}(X, \mathbb{Q}).$$

The following was observed in [48]:

Proposition 2.28 *The quotient group*

$$\mathrm{Hdg}^{2n-2}(X, \mathbb{Z})/H^{2n-2}(X, \mathbb{Z})_{alg}, \tag{2.22}$$

where $H^{2n-2}(X, \mathbb{Z})_{alg}$ *denotes the image of the cycle class map* $\mathcal{Z}_1(X) \to \mathrm{Hdg}^{2n-2}(X, \mathbb{Z})$, *is a stable birational invariant.*

Proof Using Lemma 2.9, we only have to prove invariance under blow-up and under $X \mapsto X \times \mathbb{P}^r$. When we blow-up X along a smooth subvariety $Z \subset X$, the extra (Hodge) classes of degree $2n - 2$ are generated by the classes of vertical lines of the exceptional divisor $E_Z \to Z$. They are all algebraic so that the quotient (2.22) remains unchanged. When taking the product with $\mathbb{P}^r$, the extra Hodge homology classes of degree 2 are generated by the class of a line in $\mathbb{P}^r$, hence are algebraic, so that the quotient (2.22) remains unchanged. ☐

We conjectured in [59] that the group (2.22) is trivial for rationally connected varieties. This conjecture is proved in [56] in the case of threefolds. More generally, we prove the following:

Theorem 2.29 ([56]) *Let X be a smooth projective threefold which is uniruled or has trivial canonical bundle. Then the integral Hodge classes of degree 4 on X are algebraic.*

The conjecture is also proved in [27] for Fano fourfolds, building on the K-trivial case in Theorem 2.29. In the paper [56], we also proved that the conjecture would be a consequence of the Tate conjecture for divisor classes on surfaces defined over a finite field.

2.3.2 Griffiths Group

Here is a more refined birational invariant that one can define using 1-cycles. Recall that the Griffiths group $\mathrm{Griff}_k(X)$ (see [24]) is defined as the group of k-cycles of X homologous to 0 modulo algebraic equivalence.

Proposition 2.30 *The group $\mathrm{Griff}_1(X)$ is a stable birational invariant of the smooth projective variety X.*

Proof Using Lemma 2.9, we only have to prove invariance under blow-up and under $X \mapsto X \times \mathbb{P}^r$. When we blow-up X along a smooth subvariety $Z \subset X$, the blow-up formulas show that the extra elements in the group $\mathrm{Griff}_1(\widetilde{X})$ come from $\mathrm{Griff}_0(Z)$ which is 0 as 0-cycles homologous to 0 are algebraically equivalent to 0. When we take the product of X with $\mathbb{P}^r$, the extra 1-cycles in $X \times \mathbb{P}^r$ are coming from 0-cycles of X, and the extra 1-cycles homologous to 0 from 0-cycles homologous to 0 on X, which are algebraically equivalent to 0. ☐

It is not known if the group $\mathrm{Griff}_1(X)$ can be nonzero for a rationally connected variety. It is tempting to conjecture that it is always trivial for X rationally connected. This has been proved by Tian and Zong [50] for Fano complete intersections of index at least 2. For such a variety X, they prove that all rational curves deform to a union of lines.

Remark 2.31 If $\dim X = 3$, then $\mathrm{Griff}_1(X) = \mathrm{Griff}^2(X)$. If furthermore X is rationally connected, then $\mathrm{Griff}^2(X) = 0$ by Theorem 3.21.

2.3.3 Torsion 1-Cycles with Trivial Abel–Jacobi Invariant

It is an important and classical result due to Roitman [44] that the kernel of the Albanese map

$$\mathrm{alb}_X : \mathrm{CH}_0(X)_{hom} \to \mathrm{Alb}(X)$$

has no torsion if X is a smooth projective variety over $\mathbb{C}$ (or any algebraically closed field of characteristic 0). Let $J_3(X) = J^{2n-3}(X)$ be the intermediate Jacobian built from the Hodge structure on $H_B^{2n-3}(X, \mathbb{Z}) \cong H_3^B(X, \mathbb{Z})$. If X is rationally connected, $H^{3,0}(X) = 0 = H^{n,n-3}(X)$ and $J_3(X)$ is an abelian variety which is the target of the Abel–Jacobi map

$$\phi_X^{n-1} : \mathrm{CH}_1(X)_{hom} \to J_3(X). \tag{2.23}$$

The Abel–Jacobi map for three-dimensional varieties played an important role in the study of the rationality problem, thanks to Clemens–Griffiths criterion that we will revisit in Sect. 5.2. The following provides another birationally invariant group:

Proposition 2.32 *The group* $\mathrm{Tors}\,(\mathrm{Ker}\,\phi_X^{n-1})$ *is a stable birational invariant of the smooth projective variety* X.

Proof Using Lemma 2.9, we only have to prove invariance under blow-up and under $X \mapsto X \times \mathbb{P}^r$. When we blow-up X along a smooth subvariety $Z \subset X$, the blow-up formulas for Chow groups and cohomology give

$$\mathrm{CH}_1(\widetilde{X}) = \mathrm{CH}_1(X) \oplus \mathrm{CH}_0(Z)$$

$$H_2^B(\widetilde{X}, \mathbb{Z}) = H_2^B(X, \mathbb{Z}) \oplus H_0(Z, \mathbb{Z}),$$

hence

$$\mathrm{CH}_1(\widetilde{X})_{hom} = \mathrm{CH}_1(X)_{hom} \oplus \mathrm{CH}_0(Z)_{hom}$$

and similarly $J_3(\widetilde{X}) = J_3(X) \oplus J_1(Z)$ where $J_1(Z) = \mathrm{Alb}(Z)$. The Abel–Jacobi map $\phi_{\widetilde{X}}^{n-1}$ is the direct sum of the Abel–Jacobi map ϕ_X^{n-1} and the Albanese map of Z. It follows that

$$\mathrm{Tors}\,(\mathrm{Ker}\,\phi_{\widetilde{X}}^{n-1}) = \mathrm{Tors}\,(\mathrm{Ker}\,\phi_X^{n-1}) \oplus \mathrm{Tors}\,(\mathrm{Ker}\,\mathrm{alb}_Z)$$

and the second group on the right is trivial by Roitman's theorem.

Similarly, for any $r \geq 1$, we have

$$\mathrm{CH}_1(X \times \mathbb{P}^r)_{hom} = \mathrm{CH}_1(X)_{hom} \oplus \mathrm{CH}_0(X)_{hom}$$

and $J_3(X \times \mathbb{P}^r) = J_3(X) \oplus J_1(X)$ where $J_1(X) = \text{Alb}(X)$. The Abel–Jacobi map $\phi_{X \times \mathbb{P}^r}^{n-1+r}$ is the direct sum of the Abel–Jacobi map ϕ_X^{n-1} and the Albanese map of X. It follows that

$$\text{Tors}\,(\text{Ker}\,\phi_{X \times \mathbb{P}^r}^{n-1+r}) = \text{Tors}\,(\text{Ker}\,\phi_X^{n-1}) \oplus \text{Tors}\,(\text{Ker}\,\text{alb}_X)$$

and the second group on the right is trivial by Roitman's theorem.   □

It is again an open question whether a smooth projective rationally connected variety over $\mathbb{C}$ can have some nonzero torsion in $\text{Ker}\,\phi_X^{n-1}$.

Remark 2.33 If dim $X = 3$, then the 1-cycles are codimension 2 cycles and a difficult theorem of Bloch (see Theorem 3.19) thus applies and says that $\text{Ker}\,\phi_X^{n-1} = \text{Ker}\,\phi_X^2$ has no torsion in this case.

3   0-Cycles

3.1   Bloch–Srinivas Principle

The Bloch–Srinivas principle [12] says the following:

Theorem 3.1 *Let $Y \to B$ be a flat morphism of varieties defined over a field k, with B smooth, and let Z be a cycle on Y. Assume that $K \supseteq k$ is an algebraically closed field of infinite transcendence degree over k and that for any point $b \in B(K)$, the restricted cycle $Z_{|Y_b}$ is rationally equivalent to 0. Then there exist an integer $N > 0$ and a dense Zariski open set $U \subset B$ such that $N Z_{|Y_U} = 0$ in $\text{CH}(Y_U)$, where $Y_U := \phi^{-1}(U) \subset Y$.*

The condition on K guarantees that it contains any finitely generated extension of k. The assumptions we imposed on B and ϕ are used to give a meaning to the restricted cycles $Z_{|Y_b}$. As the conclusion concerns only a dense Zariski open set of B, smoothness of B is not restrictive. The theorem is obtained by embedding $k(B)$ into K and by applying the assumption to the generic point η of B, which is defined over $k(B)$ but can be seen as defined over K via $k(B) \subset K$. As Z vanishes in $\text{CH}(Y_{\eta_K})$, one easily concludes by a trace argument that it is torsion in $\text{CH}(Y_\eta)$. Finally, as η is the generic point of B, the vanishing of $N Z$ in $\text{CH}(Y_\eta)$ implies the vanishing of $N Z$ in $\text{CH}(Y_U)$ for some dense Zariski open set U of B, which proves the theorem. Note that the same argument proves as well the following statement:

Proposition 3.2 *Under the same assumptions as in Theorem 3.1, there exist a dense Zariski open set $U \subset B_{reg}$ and a finite cover $U' \to U$ such that $Z_{U'} = 0$ in $\text{CH}(Y_{U'})$, where $Y_{U'} := U' \times_U Y_U$ and $Z_{U'}$ is the pull-back of $Z_{|Y_U}$ to $Y_{U'}$.*

If X is a complex variety, then X is defined over a field k which has finite transcendence degree over $\mathbb{Q}$ and $\mathbb{C}$ satisfies the desired properties with respect to k. We then conclude:

Theorem 3.3 *Let* $\phi : Y \to B$ *be a morphism of complex varieties and let* Z *be a cycle on* Y. *Assume that for any complex point* $b \in B(\mathbb{C})$, *the restricted cycle* $Z_{|Y_b}$ *is rationally equivalent to* 0. *Then there exist an integer* $N > 0$ *and a dense Zariski open set* $U \subset B$ *such that* $NZ_{|Y_U} = 0$ *in* $\mathrm{CH}(Y_U)$, *where* $Y_U := \phi^{-1}(U) \subset Y$.

This theorem leads us to the "decomposition of the diagonal" first introduced by Bloch and Srinivas, some applications of which we will describe below:

Theorem 3.4 ([12]) *Let* X *be a variety of dimension* n *over* $\mathbb{C}$ *and assume that there exists a closed algebraic subset* $W \subset X$ *such that* $\mathrm{CH}_0(W) \to \mathrm{CH}_0(X)$ *is surjective. Then for some integer* $N > 0$, *one has a decomposition*

$$N\Delta_X = Z_W + Z \text{ in } \mathrm{CH}^n(X \times X), \tag{3.24}$$

where Δ_X *is the diagonal of* X, Z_W *is supported on* $X \times W$ *and* Z *is supported on* $D \times X$ *for some proper closed algebraic subset* $D \subset X$.

Proof The assumption is equivalent, by the localization exact sequence, to the vanishing of $\mathrm{CH}_0(X \setminus W)$. We can then apply Theorem 3.3 to $Z = \Delta_{X|X \times (X \setminus W)}$ and conclude that for some Zariski open set $U \subset X$, and for some integer $N > 0$,

$$N\Delta_{X|U \times (X \setminus W)} = 0 \text{ in } \mathrm{CH}^n(U \times (X \setminus W)).$$

By the localization exact sequence, letting $D := X \setminus U$, this is equivalent to the decomposition (3.24). $\qquad\qquad\square$

In these notes, we are interested in rationally connected varieties X, which have "trivial" CH_0 group over an algebraically closed field, as all points of X are rationally equivalent in X. We are thus in the situation of Theorem 3.4, where we can take for W any point $x \in X$. One then gets:

Theorem 3.5 *Let* X *be a complex algebraic variety of dimension* n, *such that all points of* X *are rationally equivalent to any given point* $x \in X$. *Then there is a divisor* $D \subset X$ *and an integer* N *such that*

$$N\Delta_X = N(X \times x) + Z \text{ in } \mathrm{CH}^n(X \times X) \tag{3.25}$$

where Z *is supported on* $D \times X$.

The decomposition (3.25) is a Chow decomposition of the diagonal with rational coefficients, due to the presence of the coefficient N. It was used by Bloch and Srinivas to give a new proof and a generalization of Mumford's theorem [41]. Note conversely that, if X is smooth projective, and admits a decomposition as in (3.25), then $CH_0(X) = \mathbb{Z}$. Indeed, for any $y \in X$, we get by letting act the correspondences appearing in (3.25) on any 0-cycle z (see (3.27)):

$$Nz = N(\deg z)x \text{ in } CH_0(X),$$

This shows that up to torsion, $CH_0(X) = \mathbb{Z}$, and in particular that $\text{Alb } X = 0$. On the other hand Roitman's theorem [44] says that the kernel of the Albanese map has no torsion, hence finally $CH_0(X) = \mathbb{Z}$.

3.2 Universal Chow Group of 0-Cycles

The universal CH_0 group of X is not a group but a functor. If X is a variety defined over a field K, this functor, from the category of fields containing K to the category of abelian groups, associates to any field $L \supseteq K$ the group $CH_0(X_L)$. The crucial point is that it provides much more information on X than the group $CH_0(X)$, even if K is very big like $\mathbb{C}$, because the considered fields L are not algebraically closed. The interest of this notion for rationality questions comes from the following facts:

Lemma 3.6 *One has* $CH_0(\mathbb{P}_K^n) = \mathbb{Z}$ *for any field K. One can take for generator the class of any K-point of* $\mathbb{P}_K^n$.

Proof (See also [21, 1.9].) This follows indeed by induction from the localization exact sequence

$$CH_0(\mathbb{P}_K^{n-1}) \to CH_0(\mathbb{P}_K^n) \to CH_0(\mathbb{A}_K^n) \to 0,$$

where $\mathbb{P}_K^{n-1} \subset \mathbb{P}_K^n$ is any hyperplane, and from $CH_0(\mathbb{A}_K^n) = 0$ which is almost trivial: any effective 0-cycle of $\mathbb{A}_K^1$ is the divisor of a polynomial $P \in K[X]$. □

Note that the same proof shows that $CH_0(X \times \mathbb{P}_K^n) \cong CH_0(X)$, see [21, 3.1].
   The following definition appears in [5]:

Definition 3.7 A variety X over K has universally trivial CH_0-group if X has a 0-cycle z of degree 1 and $CH_0(X_L) = \mathbb{Z}z$ for any field $L \supseteq K$.

We then have

Proposition 3.8 *If X and Y are smooth projective over K and are stably birational over K, then X has universally trivial CH_0-group if and only Y does.*

   In particular, if X is stably rational over K, X has universally triviallly trivial CH_0-group.

Let us first recall the following basic facts that will be used many times in the sequel. Let X be a smooth projective variety of dimension n. Any cycle $\Gamma \in CH^n(X \times X)$

(also called a self-correspondence) acts on Chow groups of X in the following way: the upper-star action $\Gamma^* : CH(X) \to CH(X)$ is defined by

$$\Gamma^*(z) = pr_{1*}(pr_2^* z \cdot \Gamma), \tag{3.26}$$

where $pr_i : X \times X \to X$ are the two projections, and the lower-star action $\Gamma_* : CH(X) \to CH(X)$ is defined by

$$\Gamma_*(z) = pr_{2*}(pr_1^* z \cdot \Gamma). \tag{3.27}$$

Obviously $\Gamma_* = {}^t\Gamma^*$ where ${}^t\Gamma$ is the image of Γ under the involution of $X \times X$ exchanging the factors, but it is important for us in this section to use the two actions.

Proof of Proposition 3.8 The CH_0-group has the functoriality properties needed to apply Lemma 2.9. Hence assuming resolution of singularities, it suffices to show invariance under blow-up and invariance under $X \mapsto X \times \mathbb{P}^r$. The former follows more generally from the blow-up formulas for Chow groups, and the later was noted above. An alternative proof which does not use resolution of singularities is as follows: Let $\phi : X \dashrightarrow Y$ be a birational map. Then the graphs $\Gamma_\phi \subset X \times Y$ and $\Gamma_{\phi^{-1}} \subset Y \times X$ are correspondences which satisfy

$$\Gamma_{\phi^{-1}} \circ \Gamma_\phi = \Delta_X + Z \text{ in } CH^n(X \times X) \tag{3.28}$$

$$\Gamma_\phi \circ \Gamma_{\phi^{-1}} = \Delta_Y + Z' \text{ in } CH^n(Y \times Y)$$

where the self-correspondences Z (resp. Z') have the property of being supported on $D \times X$ (resp. $D' \times Y$) for some proper closed algebraic subset D of X (resp. D' of Y). But a correspondence Z satisfying this property acts trivially on $CH_0(X)$, and similarly for Z'. Thus we conclude that

$$(\Gamma_{\phi^{-1}})_* \circ (\Gamma_\phi)_* = Id_{CH_0(X)}, \ (\Gamma_\phi)_* \circ (\Gamma_{\phi^{-1}})_* = Id_{CH_0(Y)}.$$

$\square$

We now consider the previous situation where X is a smooth complex projective variety. The precise relationship between CH_0-triviality and universal CH_0-triviality is described by the Bloch–Srinivas Theorem 3.5:

Proposition 3.9 *If $CH_0(X) = \mathbb{Z}$, the universal CH_0 group of X is trivial modulo torsion; more precisely, there is an integer $N > 0$ such that $NCH_0(X_L)_0 = 0$ for any field $L \supseteq \mathbb{C}$.*

Here $CH_0(X_L)_0$ is the Chow group of 0-cycles of degree 0.

Proof We use Bloch–Srinivas decomposition of the diagonal of $X/\mathbb{C}$

$$N\Delta_X = N(X \times x) + Z \text{ in } CH^n(X \times X) \tag{3.29}$$

with Z supported on $D \times X$. It clearly remains true for X_L, which gives

$$N\Delta_{X_L} = N(X_L \times x) + Z_L \text{ in } \mathrm{CH}^n(X_L \times X_L), \qquad (3.30)$$

with Z_L supported on $D_L \times X_L$. Both sides of this equality act on $\mathrm{CH}_0(X_L)_0$. The action of Z_{L*} and $N(X_L \times x)_*$ are clearly 0 on $\mathrm{CH}_0(X_L)_0$, while $(N\Delta_{X_L})_* = N\,Id_{\mathrm{CH}_0(X_L)_0}$. $\qquad\square$

The next question (and the central subject of these notes) is whether one can get rid of the coefficient N, that is whether X has universally trivial CH_0 and we will see in the next sections that there are many obstructions to that, which all provide interesting obstructions to stable rationality. We will shift to the language of decomposition of the diagonal, that was studied first in [58] in relation with rationality questions.

Definition 3.10 A n-dimensional variety X over K admitting a K-point x (or a 0-cycle of degree 1) has a Chow decomposition of the diagonal if one can write

$$\Delta_X = X \times x + Z \text{ in } \mathrm{CH}^n(X \times X), \qquad (3.31)$$

where Z is a cycle of $X \times X$ which is supported on $D \times X$, where $D \subset X$ is a proper closed algebraic subset.

It is immediate to see that the definition is independent of the choice of x. The equivalence of the two definitions is contained in the following result proved in [5]:

Proposition 3.11 *A variety X over K admitting a K-point x (or a 0-cycle of degree 1) has a Chow decomposition of the diagonal if and only if X has universally trivial CH_0 group.*

Proof If we look at the proof of Proposition 3.9, and put $N = 1$, we see that it proves the "iff" direction. Conversely, assume X has universally trivial CH_0 group and let $L = K(X)$. The diagonal of X provides a L-point η_X of X_L (namely the generic point). By assumption, we get that

$$\eta_X = x_L \text{ in } \mathrm{CH}_0(X_L). \qquad (3.32)$$

Now we use the fact that

$$\mathrm{CH}_0(X_L) = \mathrm{CH}^n(X_L) = \varinjlim_{U \subset X} \mathrm{CH}^n(U \times X), \quad n = \dim X,$$

where the direct limit is over all the dense Zariski open sets of X. The points η_X and x_L are the limits of the cycles $(\Delta_X)_{|U \times X}$ and $U \times x$ respectively. Formula (3.32) thus says that there exists a Zariski open set $U \subset X$ such that

$$(\Delta_X)_{|U \times X} - U \times x = 0 \text{ in } \mathrm{CH}^n(U \times X),$$

which is equivalent to a decomposition of the diagonal (3.31) with $D = X \setminus U$ by the localization exact sequence.   □

The study of the decomposition (3.31) and its consequences will allow us in next section to exhibit many obstructions, some topological, to the universal triviality of CH_0.

3.3 Decomposition of the Diagonal: Consequences

Here we will work over $\mathbb{C}$ and use integral Betti cohomology classes, but of course $\mathbb{Z}_\ell$-étale cohomology classes could be used in general. A cohomology class $\alpha \in H_B^{2n}(X \times X)$ acts on integral cohomology of X by the same formulas as (3.26) and (3.27). We will denote by α^* and α_* these actions. When α is an integral Hodge class, in particular when $\alpha = [\Gamma]$ is algebraic, the two maps $\alpha^* : H_B^l(X, \mathbb{Z}) \to H_B^l(X, \mathbb{Z})$ are morphisms of Hodge structures. In particular, when $l = 2j+1$ is odd, there are corresponding endomorphisms α^*, α_* of the associated Jacobian $J^l(X) = H_B^l(X, \mathbb{C})/(F^i H_B^l(X) \oplus H_B^l(X, \mathbb{Z})_{tf})$.

We will use freely the fact that the actions of Γ^*, resp. Γ_* on Chow groups are compatible via the cycle class map and Abel–Jacobi map with the action of $[\Gamma]^*$, resp. $[\Gamma]_*$, on cohomology and Jacobians, see [55, 9.2].

3.3.1 Consequences of a Cohomological Decomposition of the Diagonal

Let X be smooth projective of dimension n over $\mathbb{C}$. We will say that X has a cohomological decomposition of the diagonal if one can write

$$[\Delta_X] = [X \times x] + [Z] \text{ in } H_B^{2n}(X \times X, \mathbb{Z}), \tag{3.33}$$

where Z is a cycle of $X \times X$ which is supported over $D \times X$, with $D \subset X$ a proper closed algebraic subset, that we can assume to be a divisor. Clearly, if X has a Chow decomposition of the diagonal as in (3.31), then it has a cohomological decomposition of the diagonal by taking cohomology classes. Note that (3.33) implies that $([\Delta_X] - [X \times x])_{|U \times X} = 0$ in $H_B^{2n}(U \times X, \mathbb{Z})$ but that this is a priori a stronger statement, because the latter is just saying that the homology class $[\Delta_X] - [X \times x]$ comes from an integral homology class β supported on $D \times X$ for some proper closed algebraic subset D, but it is not saying that this β can be taken algebraic on $D \times X$. In order to draw consequences of (3.33), we use it in the following form: We observe that we can choose D to be smooth generically along each component of $pr_1(\text{Supp } Z)$. It then follows that the cycle Z lifts to a codimension $n - 1$ cycle $\widetilde{Z}$ of $\widetilde{D} \times X$, where $\tilde{j} : \widetilde{D} \to X$ is a desingularization of $D \subset X$. Then (3.33) rewrites as

$$[\Delta_X] - [X \times x] = (\tilde{j}, Id_X)_*([\widetilde{Z}]) \text{ in } H_B^{2n}(X \times X, \mathbb{Z}). \qquad (3.34)$$

We now get the following consequence:

Lemma 3.12 *If X has a cohomological decomposition of the diagonal as in (3.34), then for any $\alpha \in H_B^*(X, \mathbb{Z})$ of degree $* > 0$, one has*

$$\alpha = \tilde{j}_*([\widetilde{Z}]^*\alpha) \text{ in } H_B^*(X, \mathbb{Z}). \qquad (3.35)$$

Similarly, for any $\alpha \in H^*(X, \mathbb{Z})$ of degree $* < 2n$, one has

$$\alpha = [\widetilde{Z}]_*(\tilde{j}^*\alpha) \text{ in } H_B^*(X, \mathbb{Z}). \qquad (3.36)$$

Proof For (3.35), we let both sides of (3.34) act on $H_B^*(X, \mathbb{Z})$ by the upper-star action. We observe that $[X \times x]^*\alpha = 0$ if $* = \deg \alpha > 0$ and $[\Delta_X]^*\alpha = \alpha$. Finally we have

$$((\tilde{j}, Id_X)_*([\widetilde{Z}]))^*(\alpha) = \tilde{j}_*([\widetilde{Z}]^*\alpha) \text{ in } H_B^*(X, \mathbb{Z}),$$

which proves (3.35).

For (3.36), we argue similarly but use the lower-star action. We observe that $[X \times x]_*\alpha = 0$ if $* = \deg \alpha < 2n$ and $[\Delta_X]_*\alpha = \alpha$. Finally we have

$$((\tilde{j}, Id_X)_*([\widetilde{Z}]))_*\alpha = [\widetilde{Z}]_*(\tilde{j}^*\alpha) \text{ in } H_B^*(X, \mathbb{Z}),$$

which proves (3.36). $\qquad\qquad\qquad\qquad\qquad\qquad\qquad\qquad\qquad\qquad\qquad\qquad\square$

We now get the following:

Theorem 3.13 *If X has a cohomological decomposition of the diagonal, then the following hold:*

1. $H^{i,0}(X) = 0$ *(hence also $H^{0,i}(X) = 0$ for $i > 0$).*
2. *Tors $H_B^i(X, \mathbb{Z}) = 0$ for $i \leq 3$. Dually Tors $H_B^i(X, \mathbb{Z}) = 0$ for $i \geq 2n - 2$.*
3. *Integral Hodge classes of degree 4 on X are algebraic.*
4. *Integral cohomology classes of degree $2n - 2$ on X are algebraic.*

Remark 3.14 Statement 1 is due to Bloch and Srinivas [12] and uses only the cohomological decomposition of the diagonal with $\mathbb{Q}$-coefficients. Statement 3 is proved by Bloch and Srinivas in [12] with $\mathbb{Q}$-coefficients. Statements 2 and 3 appear in [18] and statement 4 in [58].

Remark 3.15 If X is rationally connected of dimension 3 over $\mathbb{C}$, the only property, among these four properties, which can be violated is the vanishing of

Tors $H_B^3(X, \mathbb{Z})$ and of Tors $H_B^4(X, \mathbb{Z})$. Indeed, by Theorem 2.2, the other cohomology groups have no torsion. Furthermore, by Theorem 2.29, properties 3 and 4, which coincide in this case, are satisfied.

Proof of Theorem 3.13 We use formula (3.35). If $\alpha \in H^{i,0}(X)$ with $i > 0$, then $\tilde{j}_*([\tilde{Z}]^*\alpha) = 0$ in $H^{i,0}(X)$ as this is a holomorphic form on X which vanishes on the dense Zariski open set $X \setminus D$. Thus (3.35) gives $\alpha = 0$, proving 1.

Remark 3.16 To make this argument totally rigorous, we should use the action of classes of correspondences on Dolbeault cohomology, rather than Betti cohomology (they coincide on X but not on U). We refer to the discussion starting the proof of Theorem 3.20 for more detail.

If α is torsion and of degree $* \leq 3$, then $[\tilde{Z}]^*\alpha$ is torsion and of degree $* - 2 \leq 1$, hence vanishes in $H^{*-2}(\tilde{D}, \mathbb{Z})$. Hence (3.35) gives $\alpha = \tilde{j}_*([\tilde{Z}]^*\alpha) = 0$. The other statements are obtained by duality or can be obtained directly by using formula (3.36). This proves 2.

If α is an integral Hodge class of degree 4, then $[\tilde{Z}]^*\alpha$ is an integral Hodge class of degree 2 on $\tilde{D}$, hence is algebraic by the Lefschetz $(1, 1)$-theorem. Thus $\alpha = \tilde{j}_*([\tilde{Z}]^*\alpha)$ is algebraic and 3 holds.

For the remaining statement, we use (3.36). If α is an integral cohomology class of degree $2n - 2$ on X, then $\tilde{j}^*\alpha \in H^{2n-2}(\tilde{D}, \mathbb{Z})$ is algebraic on $\tilde{D}$ which is smooth of dimension $n - 1$. Thus $\alpha = [\tilde{Z}]_*(\tilde{j}^*\alpha)$ is algebraic on X. □

3.3.2 Consequences of a Chow Decomposition of the Diagonal

We now describe consequences of a Chow decomposition of the diagonal that a priori cannot be obtained from a cohomological decomposition of the diagonal, for which we refer to Theorem 3.13 .

Theorem 3.17 *If X has a Chow decomposition of the diagonal, then*

1. *The Griffiths group $\mathrm{Griff}_1(X)$ is trivial.*
2. *The kernel of the Abel–Jacobi map $\phi_X^{2n-3} : \mathrm{CH}^{n-1}(X)_{hom} \to J^{2n-3}(X)$ has no torsion.*
3. *The kernel of the Abel–Jacobi map $\phi_X^3 : \mathrm{CH}^3(X)_{hom} \to J^5(X)$ has no torsion.*

We start the proof by redoing in the Chow setting the analysis done previously in the cohomological setting. A Chow decomposition of the diagonal $\Delta_X = X \times x + Z$ in $\mathrm{CH}^n(X \times X)$ rewrites by desingularization in the form

$$\Delta_X - X \times x = (\tilde{j}, Id_X)_*(\tilde{Z}) \text{ in } \mathrm{CH}^n(X \times X) \tag{3.37}$$

where $\widetilde{D}$ is smooth of dimension $n - 1$ and maps to X via $\widetilde{j}$. We get the following consequence by letting both sides of (3.37) act on $CH(X)$, either by the lower star or by the upper star action:

Lemma 3.18 *If X has a Chow decomposition of the diagonal, then for any $z \in CH^*(X)$ of codimension $* > 0$, one has*

$$z = \widetilde{j}_*(\widetilde{Z}^*z) \text{ in } CH^*(X). \tag{3.38}$$

Similarly, for any $z \in CH^*(X)$ of codimension $* < n$, one has

$$z = \widetilde{Z}_*(\widetilde{j}^*z) \text{ in } CH^*(X). \tag{3.39}$$

Proof of Theorem 3.17 By assumption, X has a Chow decomposition of the diagonal that we write as in (3.37). If $z \in CH_1(X)$, we get $z = \widetilde{j}_*(\widetilde{Z}^*z)$ in $CH_1(X)$ by Lemma 3.18, and if z is homologous to 0, $\widetilde{Z}^*z$ is a 0-cycle homologous to 0 on $\widetilde{D}$. It is thus algebraically equivalent to 0 and so $z = \widetilde{j}_*(\widetilde{Z}^*z)$ is also algebraically equivalent to 0. This proves 1.

Assume now that z is of torsion and annihilated by the Abel–Jacobi map. Then $\widetilde{Z}^*z$ is a torsion 0-cycle on $\widetilde{D}$ which is annihilated by the Albanese map and Roitman's theorem gives that $\widetilde{Z}^*z = 0$. Thus $z = 0$, which proves 2.

If $\dim X \geq 4$ and $z \in CH^3(X)$, we have by Lemma 3.18, $z = \widetilde{j}_*(\widetilde{Z}^*z)$ in $CH^3(X)$, where $\widetilde{Z}^*z$ is a codimension 2 cycle on $\widetilde{D}$. If now z is of torsion and annihilated by the Abel–Jacobi map, $\widetilde{Z}^*z$ is of torsion and annihilated by the Abel–Jacobi map, hence it vanishes in $CH^2(\widetilde{D})$ by the following result of Bloch:

Theorem 3.19 (Bloch) *The kernel of the Abel–Jacobi map for codimension 2 cycles homologous to zero on complex projective manifolds has no torsion.*

It then follows that $z = \widetilde{j}_*(\widetilde{Z}^*z) = 0$, which proves 3.   □

We conclude this section with an implication of cohomological type which is due to Totaro [51] and will be used in Sect. 4.3. The statement is due to Bloch and Srinivas when char $K = 0$.

Theorem 3.20 *Let X be a smooth projective variety of dimension n defined over a field K of any characteristic. Assume X has a Chow decomposition of the diagonal. Then $H^0(X, \Omega_X^i) = 0$ for $i > 0$.*

Proof We use the algebraic de Rham cycle class for cycles in any smooth variety Y over K. For any cycle $Z \in CH^k(Y)$, we get a class $[Z] \in H^k(Y, \Omega_Y^k)$. Furthermore, if X is smooth projective of dimension n, a class $\alpha \in H^n(Y \times X, \Omega_{Y \times X}^n)$, with Y smooth but not necessarily projective, induces a morphism

$$\alpha^* : H^p(X, \Omega_X^q) \to H^p(Y, \Omega_Y^q)$$

$$\alpha^*(a) = pr_{1*}(\alpha \cup pr_2^*a).$$

We now start from our Chow decomposition of the diagonal in the form

$$(\Delta_X)_{|U \times X} = U \times x \text{ in } CH^n(U \times X) \tag{3.40}$$

for some Zariski dense open set of X. Taking de Rham cycle classes, we get

$$[\Delta_X]_{|U \times X} = [U \times x] \text{ in } H^n(U \times X, \Omega^n_{U \times X}). \tag{3.41}$$

We let both sides act on elements $a \in H^{i,0}(X)$ for $i > 0$. The right hand side acts by 0 and the left hand side acts by restriction of forms to U. We thus conclude that for any $a \in H^0(X, \Omega^i_X)$ with $i > 0$, $a_{|U} = 0$, hence $a = 0$ because $U \subset X$ is a dense Zariski open set. □

We finish this section with an important result due to Bloch and Srinivas [12] and uses in fact only the Chow decomposition of the diagonal with $\mathbb{Q}$-coefficients.

Theorem 3.21 *Let X be a smooth projective complex variety admitting a Chow decomposition of the diagonal with $\mathbb{Q}$-coefficients (equivalently, by Theorem 3.4, $CH_0(X) = \mathbb{Z}$). Then the Griffiths group $\mathrm{Griff}^2(X)$ is trivial and the Abel–Jacobi map*

$$\phi^2_X : CH^2(X)_{hom} \to J^3(X)$$

is an isomorphism.

Proof We write the decomposition of the diagonal as

$$N\Delta_X = N(X \times x) + (\tilde{j}, Id_X)_*(\tilde{Z}) \text{ in } CH^n(X \times X), \tag{3.42}$$

where $\tilde{j} : \tilde{D} \to X$ s a morphism from a smooth variety of dimension $n - 1$. This provides for any $z \in CH^2(X)$ the equality

$$Nz = \tilde{j}_*(\tilde{Z}^*z), \tag{3.43}$$

where $\tilde{Z}^*z$ is a codimension 1 cycle on $\tilde{D}$. If z is cohomologous to 0, so is $\tilde{Z}^*z$, hence $\tilde{Z}^*z$ is algebraically equivalent to 0 and Nz is algebraically equivalent to 0 by (3.43). We thus proved that $\mathrm{Griff}^2(X)$ is a torsion group. On the other hand, using the cohomological version of (3.42), we conclude that for any $\alpha \in H^3(X, \mathbb{Z})$, $N\alpha = \tilde{j}_*(\tilde{Z}^* * \alpha)$, hence vanishes on $U = X \setminus D$. Using notation of Sect. 2.2, this implies that the map $H^3(X, \mathbb{Z}) \to H^0(X_{Zar}, \mathcal{H}^3(\mathbb{Z}))$ is of N-torsion, hence is trivial as the second group has no torsion by Theorem 2.24. Recalling the Bloch–Ogus exact sequence

$$H^3_B(X, \mathbb{Z}) \to H^0(X_{Zar}, \mathcal{H}^3(\mathbb{Z})) \to \mathrm{Griff}^2(X) \to 0$$

from Theorem 2.18, we conclude that in this case $\text{Griff}^2(X) = H^0(X_{Zar}, \mathcal{H}^3(\mathbb{Z}))$ has no torsion. Hence it is in fact trivial, which proves the first statement.

The second statement is obtained as follows: we use again (3.43). If now z is homologous to 0 and annihilated by ϕ_X^2, then $\widetilde{Z}^* z$ is a codimension 1 cycle on $\widetilde{D}$ which is homologous to 0 and annihilated by $\phi_{\widetilde{D}}^1$, hence is trivial. Thus $Nz = 0$ in $\text{CH}^2(X)$ by (3.43). We then conclude that $z = 0$ using Theorem 3.19.   □

We finally turn to unramified cohomology. The following result was proved in [18]:

Theorem 3.22 *Let X be a smooth projective complex variety. (i) If $N\Delta_X$ decomposes as in (3.25), $H_{nr}^i(X, A)$ is of N-torsion for any $i > 0$ and any coefficients A. In particular, if X has a Chow decomposition of the diagonal, the unramified cohomology groups $H_{nr}^i(X, A)$ vanish for any $i > 0$ and any coefficients A.*

(ii) If X satisfies $\text{CH}_0(X) = \mathbb{Z}$, $H_{nr}^i(X, \mathbb{Z})$ vanishes for any $i > 0$.

Proof Statement (ii) follows from (i), using Theorem 3.5, which guarantees the existence of a decomposition of $N\Delta_X$ assuming $\text{CH}_0(X) = \mathbb{Z}$, and Corollary 2.25, which tells that $H_{nr}^i(X, \mathbb{Z})$ has no torsion.

The proof of (i) uses the fact that Chow correspondences $\Gamma \in \text{CH}^n(X \times Y)$ with X, Y smooth and Y projective of dimension n act on unramified cohomology providing

$$\Gamma^* : H_{nr}^l(Y, A) \to H_{nr}^l(X, A). \tag{3.44}$$

We refer to the appendix of [18] for a precise construction of this action. It factors through the cycle class $[\Gamma]_{mot} \in H^n((X \times Y)_{Zar}, \mathcal{H}^n(\mathbb{Z})) = \mathcal{Z}^n(X \times Y)/\text{alg}$ introduced in Theorem 2.17(iii). The construction of the action then rests on the basic functoriality properties of the groups $H^p(W_{Zar}, \mathcal{H}^q(A))$, for W smooth, under pullback, and push-forward under proper maps and the existence of a cup-product. Having this action, we simply let act on $H_{nr}^i(X, A)$ both sides of the decomposition

$$N[\Delta_X]_{mot} = N[X \times x]_{mot} + [Z]_{mot}$$

with Z supported on $D \times X$ for some proper closed algebraic subset $D \subset X$. The left hand side acts as $N Id$. The term $N[X \times x]_{mot}$ acts trivially on $H_{nr}^i(X, A)$ for $i > 0$. The fact that $[Z]_{mot}^* = 0$ on $H_{nr}^i(X, A)$ follows from the fact that, denoting $U := X \setminus D$ and $j_U : U \to X$ the inclusion, we clearly have

$$j_U^* \circ [Z]_{mot}^* = 0 : H_{nr}^i(X, A) \to H_{nr}^i(U, A)$$

for any i since Z is supported on $D \times X$. On the other hand, the restriction map j_U^* is injective on $H_{nr}^i(X, A)$ by Proposition 2.20.   □

Corollary 3.23

(i) The unramified cohomology of $\mathbb{P}^n$ with any coefficients vanishes in degree > 0.

(ii) Unramified cohomology with any coefficients is a stable birational invariant.

Proof Clearly $\mathbb{P}^n$ admits a decomposition of the diagonal. This follows from the computation of $\mathrm{CH}(\mathbb{P}^n \times \mathbb{P}^n)$ as the free abelian group with basis $h_1^i \cdot h_2^j$, $0 \leq i, j \leq n$, where $h_1 = pr_1^* c_1(\mathcal{O}_{\mathbb{P}^n}(1))$, $h_2 = pr_2^* c_1(\mathcal{O}_{\mathbb{P}^n}(1)) \in \mathrm{CH}^1(\mathbb{P}^n \times \mathbb{P}^n)$. Thus (i) follows from Theorem 3.22(i).

For the proof of (ii), as we already proved birational invariance of unramified cohomology in Theorem 2.21, it suffices to show invariance under $X \mapsto X \times \mathbb{P}^r$. We use for this the following partial or relative decomposition of the diagonal of $X \times \mathbb{P}^r$:

$$\Delta_{X \times \mathbb{P}^r} = \sum_{i=0}^{r} p_{13}^* \Delta_X \cdot p_{24}^* (h_1^i \cdot h_2^{r-i}), \qquad (3.45)$$

where $p_{13} : X \times \mathbb{P}^r \times X \times \mathbb{P}^r \to X \times X$ and $p_{24} : X \times \mathbb{P}^r \times X \times \mathbb{P}^r \to \mathbb{P}^r \times \mathbb{P}^r$ are the obvious projections, and the h_1, h_2 are as above codimension 1 cycles on $\mathbb{P}^r \times \mathbb{P}^r$. We let act both sides of the decomposition above on $H_{nr}^i(X \times \mathbb{P}^r, A)$, say by the upper-star action. The left hand side acts trivially, and a term $p_{13}^* \Delta_X \cdot p_{24}^* (h_1^i \cdot h_2^{r-i})$ acts nontrivially on $H_{nr}^i(X \times \mathbb{P}^r, A)$ only if it dominates $X \times \mathbb{P}^r$ by the projection p_{12}, as follows from the argument already given above and using the injectivity of the restriction map to an open set. The only term which dominates $X \times \mathbb{P}^r$ by the projection p_{12} is

$$W := p_{13}^* \Delta_X \cdot p_4^*(h_2^r),$$

which acts on $H_{nr}^i(X \times \mathbb{P}^r, A)$ by the composite map:

$$H_{nr}^i(X \times \mathbb{P}^r, A) \overset{rest}{\to} H_{nr}^i(X \times pt, A) = H_{nr}^i(X, A) \overset{p_X^*}{\to} H_{nr}^i(X \times \mathbb{P}^r, A).$$

It follows from the above arguments that $W^* = Id$ on $H_{nr}^i(X \times \mathbb{P}^r, A)$, from which we conclude immediately that the pull-back map $p_X^* : H_{nr}^i(X, A) \to H_{nr}^i(X \times \mathbb{P}^r, A)$ is an isomorphism. $\qquad\square$

3.3.3 Cohomological Versus Chow Decomposition

We explained above that the existence of a Chow decomposition of the diagonal has a priori stronger consequences than the existence of a cohomological decomposition of the diagonal. We are going to discuss here how the two properties relate.

Note first that, by Theorem 3.13, 1, a smooth complex projective variety admitting a cohomological decomposition of the diagonal has $h^{i,0}(X) = 0$ for $i > 0$, hence Bloch's conjecture predicts that $\mathrm{CH}_0(X) = \mathbb{Z}$. The Bloch–Srinivas theorem 3.5 then shows that X admits a Chow decomposition of the diagonal with $\mathbb{Q}$-coefficients. In conclusion, when working with $\mathbb{Q}$-coefficients, having a cohomological and a Chow decomposition of the diagonal should be equivalent.

Turning to integral coefficients, the following result appears in [61].

Proposition 3.24 *A smooth projective variety defined over an algebraically closed field admits a Chow decomposition of the diagonal if and only if it admits a decomposition of the diagonal modulo algebraic equivalence.*

Proof We use the following result of [52] and [54].

Theorem 3.25 *Let* $\Gamma \in CH^*(X \times X)$ *be a self-correspondence which is algebraically equivalent to* 0. *Then* Γ *is nilpotent in the ring* $CH^*(X \times X)$ *of self-correspondences of* X.

Starting from our decomposition

$$\Delta_X = X \times x + Z$$

modulo algebraic equivalence, with Z supported on $D \times X$, let $\Gamma = \Delta_X - X \times x - Z \in CH^n(X \times X)$. Theorem 3.25 implies that $\Gamma^{\circ N} = 0$ in $CH^n(X \times X)$ for some $N > 0$. We finally observe that $\Gamma^{\circ N} = \Delta_X - X \times x - Z'$ in $CH^n(X \times X)$ for some Z' supported on $D' \times X$, for some proper closed algebraic subset $D' \subset X$. The equality

$$\Gamma^{\circ N} = 0 = \Delta_X - X \times x - Z' \text{ in } CH^n(X \times X)$$

thus gives a Chow decomposition of the diagonal for X.   □

In the surface case, we have the following result (proved in [61], and reproved in [29]).

Theorem 3.26 *Let* X *be a smooth complex projective surface with* $CH_0(X) = \mathbb{Z}$. *Then the following are equivalent:*

1. *X admits a Chow decomposition of the diagonal.*
2. *X admits a cohomological decomposition of the diagonal.*
3. *Tors* $H_B^*(X, \mathbb{Z}) = 0$.

Proof The implications $1 \Rightarrow 2 \Rightarrow 3$ are clear (the second one is Theorem 3.13, 2). Let us prove $3 \Rightarrow 1$. The condition Tors $H^*(X, \mathbb{Z}) = 0$ implies that X admits a Künneth decomposition with integral coefficients, so that we can write

$$[\Delta_X] = \sum_i \alpha_i \otimes \beta_i \text{ in } H_B^4(X \times X, \mathbb{Z}) \tag{3.46}$$

for some integral cohomology classes α_i, β_i. As $CH_0(X) = \mathbb{Z}$ we have by Mumford's theorem [41] or Bloch–Srinivas that $H^{i,0}(X) = 0$ for $i > 0$, which in our case implies that the whole cohomology of X is algebraic. (In particular X has no odd degree cohomology.) In formula (3.46), the classes α_i, β_i are classes of algebraic cycles (points, curves, or X itself), which gives a cohomological decomposition of the diagonal that takes the form

$$[\Delta_X - X \times x - Z] = 0, \tag{3.47}$$

where Z is a cycle supported on $D \times X$ for some curve $D \subset X$. We then apply Theorem 3.21 to $Y = X \times X$ which has $\mathrm{CH}_0(Y) = \mathbb{Z}$. This theorem tells us that the group $\mathrm{Griff}^2(Y)$ is trivial. It follows that the cycle $\Delta_X - X \times x - Z$ homologous to 0 is algebraically equivalent to 0. We conclude that X has a decomposition of the diagonal modulo algebraic equivalence, hence admits a Chow decomposition of the diagonal by Proposition 3.24. □

The following question is open:

Question 3.27 *Do there exist smooth projective complex varieties which admit a cohomological decomposition of the diagonal, but no Chow decomposition of the diagonal?*

The answer might be affirmative in view of the discussion made in the previous sections concerning what is controlled by the Chow, resp. cohomological decompositions of the diagonal. If we look at the proof of Proposition 3.24, we see that the key point is the nilpotence of self-correspondences algebraically equivalent to 0 (Theorem 3.25). A big conjecture in the theory of algebraic cycles is the following nilpotence conjecture:

Conjecture 3.28 *For any smooth projective variety X over $\mathbb{C}$, self-correspondences $\Gamma \in \mathrm{CH}(X \times X)_{\mathbb{Q}}$ with $\mathbb{Q}$-coefficients and homologous to 0 are nilpotent, that is, $\Gamma^{\circ N} = 0$ in $\mathrm{CH}(X \times X)_{\mathbb{Q}}$ for some $N > 0$.*

This conjecture is not formulated for self-correspondences $\Gamma \in \mathrm{CH}(X \times X)$, that is with $\mathbb{Z}$-coefficients, and is presumably false, although we are not aware of an explicit counterexample. In fact, there is a different and more general nilpotence conjecture by Voevodsky [52] which predicts the following:

Conjecture 3.29 *For any smooth projective variety Y, any cycle $Z \in \mathrm{CH}(Y)_{\mathbb{Q}}$ with $\mathbb{Q}$-coefficients and homologous to 0 is smash-nilpotent, namely $Z^N = 0$ in $\mathrm{CH}(X^N)_{\mathbb{Q}}$ for some $N > 0$.*

This conjecture implies Conjecture 3.28 by putting $Y = X \times X$ and realizing that $\Gamma^{\circ N}$ is obtained from $\Gamma^N \in \mathrm{CH}((X \times X)^N)_{\mathbb{Q}}$ by a natural correspondence. However, Conjecture 3.29 is shown not to be true with integral coefficients in [48, Theorem 5].

4 The Degeneration Method

4.1 A Specialization Result

First of all, let us explain a version of Fulton's specialization map [21, 20.3].

Proposition 4.1 *Let* $\pi : Y \to C$ *be a flat morphism to a smooth curve over* $\mathbb{C}$. *Let* Z *be a cycle on* Y *such that for the very general complex point* $t \in C$, $Z_{|Y_t}$ *is rationally equivalent to* 0, *where* $Y_t := \pi^{-1}(t) \subset Y$. *Then for any* $t \in C$, $Z_{|Y_t}$ *is rationally equivalent to* 0.

Remark 4.2 There is no smoothness assumption in this statement, neither for Y, nor for the morphism π. Indeed, by flatness of π and smoothness of C, the fibers Y_t are Cartier divisors, so the restricted cycle $Z_{|Y_t}$ is well-defined.

Proof We apply Proposition 3.2. It says that there exist a base change $C' \to C$, where we obviously can assume that C' is smooth, and a Zariski open set $U' \subset C'$, such that $Z_{U'} = 0$ in $\mathrm{CH}(Y_{U'})$. The cycle $Z_{C'} \in \mathrm{CH}(Y_{C'})$ thus vanishes on the Zariski open set $Y_{U'} \subset Y_{C'}$, and it follows from the localization exact sequence that there are finitely many fibers $Y_{t'_i} \subset Y_{C'}$ such that $Z_{C'}$ is supported on the union of the $Y_{t'_i}$'s. Clearly the restriction $Z_{|Y_{t'}}$ vanishes for any $t \neq t'_i$ for all i, but in fact this is also true for $t' = t'_i$. Indeed, let $j_i : Y_{t'_i} \to Y_{C'}$ be the inclusion. Then $Y_{t'_i}$ is a Cartier divisor, which furthermore has the property that $\mathcal{O}_{Y_{t'_i}}(Y_{t'_i})$ is trivial. It follows that $j_i^* \circ j_{i*} : \mathrm{CH}(Y_{t'_i}) \to \mathrm{CH}(Y_{t'_i})$ is 0. This proves the result also for the special points t'_i since we know that $Z_{C'} = \sum_i j_{i*}(Z_i)$. (One uses here the fact that the fibers of $Y_{C'} \to C'$ and $Y \to C$ are the same.) $\quad\square$

Corollary 4.3 *Let* $X \to C$ *be a flat morphism over* $\mathbb{C}$ *where* C *is a smooth curve and* X *is irreducible. Assume that for a very general point* $t \in C$, *the fiber* X_t *has a Chow decomposition of the diagonal. Then any fiber* X_t *has a Chow decomposition of the diagonal.*

Proof Consider the flat morphism $Y := X \times_C X \to C$. By assumption, for a very general point $t \in C$, there exist a divisor $D_t \subset X_t$ and a point $x_t \in X_t$ such that

$$\Delta_{X_t} = X_t \times x_t + Z_t \text{ in } \mathrm{CH}(X_t \times X_t),$$

where the cycle Z_t is supported on $D_t \times X_t$. The data such as D_t, Z_t or x_t are parameterized by a countable union of Chow varieties which are proper over C', and we conclude that, after base change $C' \to C$, there exists a divisor $D \subset X_{C'}$ which does not contain any fiber, there exist a section $\sigma : C' \to X_{C'}$ and a cycle Z supported on $D \times_{C'} X_{C'}$ such that the cycle

$$\Gamma := \Delta_{X'/C'} - X' \times_{C'} \sigma(C') - Z \in \mathrm{CH}(Y_{C'}) \tag{4.48}$$

has the property that for a very general point $t \in C'$, $\Gamma_{|Y_t} = 0$. Note that we can assume that D is the Zariski closure of its generic fiber over C', as the only constraint it has to satisfy, namely (4.48), concerns its generic fiber. By Proposition 4.1, this remains true for any $t \in C'$. As X is irreducible and flat over C, the fibers of $X \to C$ are equidimensional of dimension n, hence we can assume that D does not contain any component of any fiber of $X_{C'} \to C'$ (such a fiber would form an irreducible component of D that does not dominate C', hence would not be in the Zariski closure of the generic fiber of D). Thus $D \cap X_t$ is a proper divisor for any point $t \in C'$, and the condition $\Gamma_{|Y_t} = 0$ for any t thus says that X_t has a Chow decomposition of the diagonal.   □

The following result is proved in [60].

Theorem 4.4 *Let $\pi : X \to C$ be a flat projective morphism of relative dimension $n \geq 2$, where C is a smooth curve. Assume that the fiber X_t is smooth for $t \neq 0$, and has at worst isolated ordinary quadratic singularities for $t = 0$. Then*

(i) If for general $t \in B$, X_t admits a Chow theoretic decomposition of the diagonal (equivalently, $CH_0(X_t)$ is universally trivial), the same is true for any smooth projective model $\widetilde{X}_0$ of X_0.

(ii) If for general $t \in B$, X_t admits a cohomological decomposition of the diagonal, and the even degree integral homology of a smooth projective model $\widetilde{X}_0$ of X_0 is algebraic (i.e. generated over $\mathbb{Z}$ by classes of subvarieties), $\widetilde{X}_0$ also admits a cohomological decomposition of the diagonal.

In order to prove (ii), we will need an intermediate step involving the notion of a homological decomposition of the diagonal for singular projective varieties: to make sense of this, we just need to know that cycles Z have a homology class $[Z]_{hom}$ in Betti integral *homology*, which is standard. Then a homological decomposition of the diagonal of a singular but projective X of pure dimension n is an equality

$$[\Delta_X]_{hom} = [X \times x]_{hom} + [Z]_{hom} \text{ in } H_{2n}(X \times X, \mathbb{Z}),$$

where as usual Z is a cycle supported on $D \times X$ for some nowhere dense closed algebraic subset D of X.

Proof of Theorem 4.4 By Corollary 4.3 and under the assumptions made on the general fibers in (i), the central fiber admits a Chow decomposition of the diagonal. This step does not need any assumption on the singularities of the fibers. Similarly, under the assumptions made on the general fibers in (ii), the central fiber admits a homological decomposition of the diagonal. The proof here uses the fact that as we are over $\mathbb{C}$, for any proper flat analytic morphism $X' \to \Delta$, after shrinking Δ if necessary, there is a continuous retraction $X' \to X_0$. Passing to $X' \times_\Delta X'$, this retraction maps the diagonal $\Delta_{X'_t}$ to the diagonal $\Delta_{X'_0}$. This implies that a homological relation $[\Gamma_t] = 0$ in $H_{2n}(X'_t \times X'_t, \mathbb{Z})$, where Γ_t is as in (4.48) implies a homological relation $[\Gamma_0] = 0$ in $H_{2n}(X'_0 \times X'_0, \mathbb{Z})$, which provides a homological decomposition of the diagonal for $X'_0 = X_0$.

The second step is passing from X_0 to $\widetilde{X}_0$ and this is here that we use the assumption on the singularities. Let us first concentrate on (i). From the decomposition

$$\Delta_{X_0} = X_0 \times x + Z \text{ in } CH_n(X_0 \times X_0),$$

where Z is supported on $D \times X_0$, we deduce by restriction to

$$U \times U, \quad U := X_0 \setminus \operatorname{Sing} X_0 = \widetilde{X}_0 \setminus E,$$

where E is the exceptional divisor of the resolution of singularities of X_0 obtained by blowing-up the singular points:

$$\Delta_U = U \times x + Z_{|U \times U} \text{ in } CH_n(U \times U).$$

By the localization exact sequence, we get a decomposition on $\widetilde{X}_0$ which takes the following form:

$$\Delta_{\widetilde{X}_0} = \widetilde{X}_0 \times x_0 + \widetilde{Z} + \Gamma_1 + \Gamma_2 \text{ in } CH_n(\widetilde{X}_0 \times \widetilde{X}_0),$$

where $\widetilde{Z}$ is supported on $D' \times \widetilde{X}_0$ for some $D' \subsetneq \widetilde{X}_0$, and Γ_1 is supported on $E \times \widetilde{X}_0$, Γ_2 is supported on $\widetilde{X}_0 \times E$. Of course the cycle Γ_1 does not dominate $\widetilde{X}_0$ by the first projection, so we need only to understand Γ_2. But E is a disjoint union of smooth quadrics Q_i of dimension ≥ 1, and for each of them, n-dimensional cycles in $\widetilde{X}_0 \times Q_i$ decompose as $n_i \widetilde{X}_0 \times x_i + Z_i$, where Z_i does not dominate $\widetilde{X}_0$ by the first projection, n_i is an integer, and x_i is any point of Q_i. At this point, we obtained a decomposition of the form

$$\Delta_{\widetilde{X}_0} = \widetilde{X}_0 \times x_0 + \sum_i n_i \widetilde{X}_0 \times x_i + Z \text{ in } CH(\widetilde{X}_0 \times \widetilde{X}_0). \tag{4.49}$$

In order to conclude, we have to use the assumption $n \geq 2$. It implies that $\widetilde{X}_0$ is irreducible or equivalently, connected. Indeed the general fiber X_t is connected as this is a consequence of the existence of a decomposition of the diagonal for X_t. Formula (4.49) tells us by letting both sides act on $CH_0(\widetilde{X}_0)$ that $CH_0(\widetilde{X}_0)$ is generated over $\mathbb{Z}$ by x_0 and the x_i. By Roitman's theorem [44], this implies that $CH_0(\widetilde{X}_0) = \mathbb{Z}$, so that all the x_i's are rationally equivalent to x_0 in $\widetilde{X}_0$. Then (4.49) gives a Chow decomposition of the diagonal for $\widetilde{X}_0$.

The proof of (ii) is quite similar although the tools are slightly different. It is important here to realize that homology and algebraic cycles do not work completely in the same way. For example, we do not have in homology the localization exact sequence.

We know that the central fiber has a homological decomposition of the diagonal in $H_*(X_0 \times X_0, \mathbb{Z})$. A fortiori it has a homological decomposition in the relative homology $H_*(X_0 \times X_0, B, \mathbb{Z})$ where $B = \mathrm{Sing}(X_0) \times X_0 \cup X_0 \times \mathrm{Sing}(X_0)$ As $\widetilde{X}_0 \setminus E \cong X_0 \setminus \mathrm{Sing}\, X_0$, it follows that we get for $\widetilde{X}_0$ a homological decomposition of the diagonal modulo $E \times \widetilde{X}_0 \cup \widetilde{X}_0 \times E$. This shows that we have a relation

$$[\Delta_{\widetilde{X}_0}]_{hom} = [\widetilde{X}_0 \times x_0] + [\widetilde{Z}] + \alpha \quad \text{in } H_{2n}(\widetilde{X}_0 \times \widetilde{X}_0, \mathbb{Z}), \qquad (4.50)$$

where $\alpha \in H_{2n}(E \times \widetilde{X}_0 \cup \widetilde{X}_0 \times E, \mathbb{Z})$. We use now the fact that $E = \sqcup Q_i$ so that the union above is the union of the $Q_i \times \widetilde{X}_0$ and $\widetilde{X}_0 \times Q_j$ intersecting along the union of the $Q_i \times Q_j$. As $Q_i \times Q_j$ has trivial odd degree cohomology, it follows that $H_{2n}(E \times \widetilde{X}_0 \cup \widetilde{X}_0 \times E, \mathbb{Z})$ is generated by the subgroups $H_{2n}(Q_i \times \widetilde{X}_0, \mathbb{Z})$ and $H_{2n}(\widetilde{X}_0 \times Q_i, \mathbb{Z})$. Hence $\alpha = \sum_i \alpha_i + \beta_i$ with $\alpha_i \in H_{2n}(Q_i \times \widetilde{X}_0, \mathbb{Z})$, $\beta_i \in H_{2n}(\widetilde{X}_0 \times Q_i, \mathbb{Z})$.

We assume for simplicity that $H^{2*}(\widetilde{X}_0, \mathbb{Z})$ is algebraic (we only assumed that this assumption holds for some variety birationally equivalent to $\widetilde{X}_0$). We then get using the Künneth decomposition of the even degree cohomology (or homology) of $\widetilde{X}_0 \times Q_i$ and $Q_i \times \widetilde{X}_0$ that each α_i is algebraic and each β_i is algebraically decomposable, that is of the form $\sum_l [Z_l \times Z_l']$ for some algebraic cycles on each summand. Clearly α_i is then the class of a cycle z_i in $\widetilde{X}_0 \times \widetilde{X}_0$ which does not dominate $\widetilde{X}_0$ by the first projection. For the $\beta_i = \sum_l [Z_l \times Z_l']$, if $\dim Z_l' > 0$, then $\dim Z_l < n$ and Z_l does not dominate $\widetilde{X}_0$ by the first projection. Finally, if $\dim Z_l' = 0$, then one gets a contribution $[\widetilde{X}_0 \times x_i]$. Putting this decomposition in (4.50) and using the fact that $[x_i] = [x_0]$ in $H^{2n}(\widetilde{X}_0, \mathbb{Z})$, this clearly provides a homological (or equivalently cohomological as $\widetilde{X}_0$ is smooth) decomposition of the diagonal of $\widetilde{X}_0$. We used in the last step the fact that $n \geq 2$ to guarantee that $H_B^{2n}(\widetilde{X}_0, \mathbb{Z}) = \mathbb{Z}$ is generated by the class of the point x_0. $\qquad \square$

Remark 4.5 The assumptions on the singularities in Theorem 4.4 are too strong and this will be discussed in next section, but some assumptions on the singularities are necessary. Consider the case of the cubic surface degenerating to a cone over an elliptic curve. The general fiber is rational hence has a Chow decomposition of the diagonal, but the desingularization $\widetilde{S}_0$ of the central fiber has nonzero holomorphic forms, so it does not admit a decomposition of the diagonal by Theorem 3.13.

As a first application, let us prove Proposition 1.5 stated in the introduction:

Proof of Proposition 1.5 Indeed, if X_t was stably rational, it would admit a Chow decomposition of the diagonal. Then $\widetilde{X}_0$ would also admit a Chow decomposition of the diagonal by Theorem 4.4, because clearly the fiber dimension has to be ≥ 2. By Theorem 3.13, this contradicts the fact that $\mathrm{Tors}\, H^3(\widetilde{X}_0, \mathbb{Z}) \neq 0$. $\qquad \square$

4.1.1 The Very General Quartic Double Solid Is Not Stably Rational

Recall that a quartic double solid is a hypersurface X in $\mathbb{L} := \mathrm{Spec}\,(\mathrm{Sym}\, \mathcal{O}_{\mathbb{P}^3}(-2))$ $\overset{\pi}{\to} \mathbb{P}^3$ defined by the equation $u^2 = p^* f$, where u is the canonical extra section of $\pi^* \mathcal{O}_{\mathbb{P}^3}(2)$ on $\mathbb{L}$ and $f \in H^0(\mathbb{P}^3, \mathcal{O}_{\mathbb{P}^3}(4))$. Thus quartic double solids are parameterized by $\mathbb{P}(H^0(\mathbb{P}^3, \mathcal{O}_{\mathbb{P}^3}(4)))$. We described in Sect. 2.1.1 the Artin–Mumford double solid X_0 which is nodal, with the property that $\widetilde{X}_0$ has a nontrivial Artin–Mumford invariant.

Theorem 4.6 *The very general quartic double solid X does not admit a cohomological (hence a fortiori Chow-theoretic) decomposition of the diagonal. Similarly, the desingularization of the very general quartic double solid X with $k \leq 7$ nodes in general position does not admit a cohomological decomposition of the diagonal.*

Here we observe that given $k \leq 7$ general points in $\mathbb{P}^3$, there is a linear space of dimension $34 - 4k > 0$ of quartic homogeneous polynomials f having multiplicity ≥ 2 at these k points. There is thus an irreducible variety parameterizing quartic double solids with k nodes in general position. As usual, "very general" in Theorem 4.6 means that the statement is true for a parameter f in the complement of a countable union of proper closed algebraic subsets of this variety.

Theorem 4.6 immediately follows from Theorem 4.4 by degeneration to the Artin–Mumford double solid. Indeed, if X_0 is the Artin–Mumford double solid, $\widetilde{X}_0$ does not admit a cohomological decomposition of the diagonal by Theorem 3.13, because the Artin–Mumford invariant of $\widetilde{X}_0$ is not trivial. Furthermore, the even degree integral Betti cohomology of $\widetilde{X}_0$ is algebraic by Theorem 2.29 because $\widetilde{X}_0$ is a rationally connected threefold. For the nodal case, one needs to check that the Artin–Mumford double solid smoothifies partially to the k-nodal quartic double solid with k nodes in general position, for $k \leq 7$.

As a consequence of Theorem 4.6, one gets the following

Corollary 4.7 *The desingularization of the very general quartic double solid with $k \leq 7$ nodes in general position is not stably rational.*

Note that by Endrass [20], if $\widetilde{X}$ is as in Theorem 4.7, $\widetilde{X}$ has trivial Artin–Mumford invariant. In fact Endrass proves that the desingularization of a quartic double solid with less than 10 points has no torsion in its third Betti cohomology. To our knowledge, the only criterion for stable irrationality of rationally connected threefolds used previously was the Artin–Mumford invariant.

4.2   Colliot-Thélène-Pirutka and Schreieder's Work

It was noticed in [60] that the assumptions on the singularities in Theorem 4.4 were too strong, even if, according to Remark 4.5, some assumptions are necessary. The paper by Colliot-Thélène and Pirutka [17], written in the equivalent language of universally CH_0-trivial varieties (see Sect. 3.2), provides a similar specialization result under weaker assumptions. They prove the following theorem that we in turn reformulate below in the language of decomposition of the diagonal. We will state the result over any algebraically closed field k. The only difference when working over $\mathbb{C}$ is the fact that, $\mathbb{C}$ being a large field, we can use the assumption on the very general fiber of a morphism as being equivalent to the similar assumption on the geometric generic fiber (see the discussion in Sect. 3.1). Note however that the setting of Colliot-Thélène-Pirutka's work is that of a scheme over a DVR, which includes specialization from varieties defined over a number field to varieties defined over a finite field. This is very important in Totaro's work (Theorem 4.15) that we will review later on.

The Colliot-Thélène and Pirutka's condition in [17] asks that the resolution map $\tau : \widetilde{X}_0 \to X_0$ is universally Chow-trivial, which means that for any field L containing the base field k, the morphism $\tau_* : CH_0(\widetilde{X}_{0,L}) \to CH_0(X_{0,L})$ is an isomorphism. This condition is rather strong and needs to be carefully checked geometrically. It says that for each subvariety $M \subset X_0$, the generic fiber $\widetilde{X}_{0,M}$ of the induced morphism $\tau^{-1}(M) \to M$, which is a variety over $k(M)$, has CH_0 universally trivial.

Let us first consider the following condition (*) that is slightly stronger than the Colliot-Thélène-Pirutka condition but is explicit geometrically:

(*) *For any irreducible subvariety $Y \subset X_0$, the map $\tau_Y : E_Y := \tau^{-1}(Y) \to Y$ has a rational section (or a 0-cycle of degree 1) and its generic fiber is smooth geometrically irreducible and has a decomposition of the diagonal over $k(Y)$.*

Remark 4.8 In practice, condition (*) is proved by checking that each generic fiber $E_{Y,\eta}$ is smooth rational over $k(Y)$.

Here the decomposition of the diagonal for $E_{Y,\eta}$ is supposed to hold with respect to the given point or 0-cycle $y_\eta \in E_{Y,\eta}(k(Y))$. Note that we use here the Chow decomposition of the diagonal for any variety defined over any field, in particular not algebraically closed.

Theorem 4.9 *Let $X \to C$ be a flat morphism, where C is a smooth curve over $\mathbb{C}$. Assume the very general fiber X_t is smooth and has a Chow decomposition of the diagonal. Then if the central fiber has a desingularization $\tau : \widetilde{X}_0 \to X_0$ satisfying (*), $\widetilde{X}_0$ has a Chow decomposition of the diagonal.*

Remark 4.10 We recover the case of nodal singularities by considering the standard resolution by blow-up. The condition that the fibers have dimension at least 2 is hidden in condition (*), because in dimension 2, the exceptional fiber of the resolution over a singular point consists of two points, which does not satisfy (*).

Proof of Theorem 4.9 The proof starts as the proof of Theorem 4.4: we thus conclude that X_0 has a Chow decomposition of the diagonal and we want to deduce that $\widetilde{X}_0$ also has one, so that we get by lifting the cycles to $\widetilde{X}_0 \times \widetilde{X}_0$:

$$\Delta_{\widetilde{X}_0} = \widetilde{X}_0 \times x_0 + \widetilde{Z} + \Gamma \text{ in } CH_n(\widetilde{X}_0 \times \widetilde{X}_0), \qquad (4.51)$$

where $\widetilde{Z}$ is supported on $D' \times \widetilde{X}_0$ for some proper closed algebraic subset $D' \subset \widetilde{X}_0$, and Γ is supported on $E \times \widetilde{X}_0 \cup \widetilde{X}_0 \times E$. Here E is the exceptional locus of the considered desingularization of X_0. A key point observed by Colliot-Thélène and Pirutka is the fact that for some dense Zariski open set U of X_0, the cycle Γ satisfies

$$(\tau, \tau)_* \Gamma_{|U \times X_0} = 0 \text{ in } CH_n(U \times X_0).$$

The end of the proof then rests on the following statement:

Lemma 4.11 *Let $\phi : W \to V$ be a proper dominant morphism with irreducible smooth generic fiber. Assume there is a generic relative decomposition of the diagonal for ϕ, namely there exist a rational section ψ of ϕ with image $S \subset W$, a proper closed algebraic subset $W_1 \subset W$ and a cycle $T \subset W \times_V W \to V$ which is supported over $W_1 \subset W$, such that*

$$\Delta_W = W \times_V S + T \text{ in } CH(W \times_V W). \tag{4.52}$$

Then for any smooth projective variety Y of dimension n and any cycle $\Gamma_1 \in CH_n(Y \times W)$ such that $(Id_Y, \phi)_* \Gamma_1$ vanishes in $CH(U \times V)$ for some dense Zariski open set U of Y, there exist a proper closed algebraic subset $V' \subset V$ and a cycle $\Gamma_1' \in CH_n(Y \times W')$, where $W' := \phi^{-1}(V')$, such that $\phi_*' \Gamma_1' = 0$ in $CH_n(U \times V')$ and $\Gamma_1' = \Gamma_1$ in $CH_n(U \times W)$ for some dense Zariski open set U of Y.

Proof As $(Id_Y, \phi)_* \Gamma_1$ vanishes in $CH(U \times V)$ for some dense Zariski open set U of Y, and there is a rational section of ϕ, we can assume that $(Id_Y, \phi)_* \Gamma_1$ actually vanishes as a n-cycle of $U \times V$ by replacing the cycle $\Gamma_1 \in \mathcal{Z}_n(Y \times W)$ by $\Gamma_1 - (Id, \psi)_*(Id, \phi)_*(\Gamma_1)$ which is rationally equivalent to it. Moving cycles, we can assume that the support $\text{Supp}\,\Gamma_1$ of Γ_1 does not have its image in W contained in W_1. As $\dim \text{Supp}\,\Gamma_1 = n = \dim Y$, there exists a dense Zariski open set of Y (that we can assume to be U), such that, over U, $\text{Supp}\,\Gamma_1$ and $pr_W^{-1}(W_1)$ do not intersect. Let $m = \dim W$ and let V^0 be a dense Zariski open set over which $\phi : W \to V$ is smooth and let $W^0 := \phi^{-1}(V^0)$. The group $CH_m(W^0 \times_{V^0} W^0)$ acts on $CH_n(Y \times W^0)$ by composition over V^0. The diagonal Δ_{W^0} acts as the identity and $W^0 \times_{V^0} S$ acts as $(Id, \psi)_* \circ (Id, \phi)_*$. It thus follows from (4.52) and from the vanishing of $(Id, \phi)_* \Gamma_1$ in $CH_n(U \times V)$, that

$$\Gamma_{1|U \times W^0} = T \circ \Gamma_1 \text{ in } CH_n(U \times W^0). \tag{4.53}$$

As T is supported over W_1 and $\text{Supp}\,\Gamma_1$ does not meet $pr_W^{-1}(W_1)$ over U, the cycle $T \circ \Gamma_1$ vanishes over the Zariski open set $U \times W^0$ of $Y \times W$. By (4.53) and the localization exact sequence, $\Gamma_{1|U \times W}$ is rationally equivalent to a cycle Γ_1' supported over a proper closed algebraic subset $V' \subset V$. Denoting by $\phi' : W' := \phi^{-1}(V') \to V$, it remains to see that $(Id, \phi')_*(\Gamma_1') = 0$ in $CH(U \times V')$ if U is small enough. This is because, taking the limit over the Zariski open sets U of Y, Γ_1 can be seen as a 0-cycle of W_K, with $K = k(Y)$, which vanishes in $\mathcal{Z}_0(V_K)$. When we apply the map T_*, to it, the resulting cycle also vanishes as a 0-cycle of V_K, and at the same time it is supported on V_K'. Hence it vanishes in $\mathcal{Z}_0(V_K')$. □

We apply Lemma 4.11 in an iterated way, with $Y = \widetilde{X}_0$, starting from the situation where $W = \widetilde{X}_0$, $V = X_0$, $\phi = \tau$ and Γ_1 is the component of the cycle Γ appearing in (4.51) which is supported on $\widetilde{X}_0 \times E$. (We do not care about the component supported on $E \times \widetilde{X}_0$ as it does not dominate $Y = \widetilde{X}_0$ by the first projection.) We then conclude using the condition (*) and Lemma 4.11 that we can decrease step by step the dimension of $\tau(\text{Supp }\Gamma_{k(Y)})$ until finally we conclude that the cycle Γ_1 vanishes in $U \times \widetilde{X}_0$, for a small enough dense Zariski open set $U \subset \widetilde{X}_0$, and equivalently, Γ_1 is supported on $D \times \widetilde{X}_0$ for some proper closed algebraic subset D of $\widetilde{X}_0$. Formula (4.51) then provides a Chow decomposition of the diagonal for $\widetilde{X}_0$. The proof of the theorem is thus finished. □

Combining Theorems 4.9, 3.22 and Remark 4.8, one gets the following improvement of Proposition 1.5:

Proposition 4.12 *Let $\phi : X \to C$ be a flat morphism, where C is a smooth curve over an algebraically closed field k. Then if the central fiber has a desingularization $\tau : \widetilde{X}_0 \to X_0$ satisfying assumption (*) (for example, if τ has rational generic fibers E_{Y_η} over $k(Y)$ for any $Y \subset \widetilde{X}_0$), and $\widetilde{X}_0$ has a nontrivial Brauer group, the geometric generic fiber $X_{\overline{\eta_C}}$ of ϕ is not stably rational. If $k = \mathbb{C}$, the very general fiber X_t is not stably rational.*

We now come to Schreieder's improvement of Propositions 1.5 and 4.12. This is a very simple observation but it is very useful in practice because it does not need any control of the singularities of the special fiber X_0. The statement is as follows:

Theorem 4.13 (See [46, Proposition 26], [45, Proposition 3.1]) *Let $\phi : X \to C$ be a flat morphism where C is a smooth curve over an algebraically closed field k. Assume that the central fiber has a desingularization $\tau : \widetilde{X}_0 \to X_0$ satisfying the following property: There exists a nontrivial unramified cohomology class η of positive degree on $\widetilde{X}_0$ such that any component E_i of the exceptional divisor is smooth and satisfies $\eta_{|E_i} = 0$. Then the geometric generic fiber $X_{\overline{\eta_C}}$ of ϕ is not stably rational. If $k = \mathbb{C}$, the very general fiber X_t is not stably rational.*

Proof By the first step in the proofs of Theorems 4.4 and 4.9, it suffices to show that the central fiber X_0 itself does not admit a Chow decomposition of the diagonal. Lifting such a decomposition to $\widetilde{X}_0$ would provide as before an equality

$$\Delta_{\widetilde{X}_0} = \widetilde{X}_0 \times x_0 + Z + \Gamma \text{ in } \text{CH}_n(\widetilde{X}_0 \times \widetilde{X}_0), \qquad (4.54)$$

where Z is supported on $D \times \widetilde{X}_0$ for some $D \subsetneq \widetilde{X}_0$, and Γ is supported on $E \times \widetilde{X}_0 \cup \widetilde{X}_0 \times E$. Here $E = \cup_i E_i$ is the exceptional locus of the considered desingularization of X_0.

Now we write $\Gamma = \sum_i \Gamma_i + \sum_i \Gamma_i'$ where Γ_i is supported on $E_i \times \widetilde{X}_0 \cup \widetilde{X}_0 \times E_i$, and we let both sides of the equality (4.54) act on η by the upper-star action. We observe here that $\Gamma_i^* \eta = 0$ by Proposition 2.20, because this is a unramified cohomology class which vanishes on the dense Zariski open set $\widetilde{X}_0 \setminus \cup_i E_i$. Next

$(\Gamma_i')^*\eta = 0$, because $\eta_{|E_i} = 0$. As $\deg \eta > 0$, $\widetilde{X}_0 \times x_0$ acts trivially on η, and thus (4.54) provides

$$\eta = Z^*\eta$$

where the right hand side is 0 again by Proposition 2.20 because $Z^*\eta$ vanishes on $\widetilde{X}_0 \setminus D$.   □

Remark 4.14 We used here unramified cohomology but other invariants as discussed in Sect. 2 can be used as well, for example the differential forms in nonzero characteristic.

4.3 Further Developments and Consequences

In this section, we will describe further variants of Proposition 4.12, and some applications. In the paper [51], Totaro uses a version of the specialization theorem where the geometric degeneration is replaced by specialization mod p of a variety defined over a number field. This generalization is already present in Colliot-Thélène-Pirutka's paper. The second important ingredient is the fact that he uses as an obstruction to the Chow decomposition of the diagonal (or universal triviality of CH_0) for the desingularized central fiber $\widetilde{X}_0$ the space of algebraic differential forms of positive degree as discussed in Sect. 3.3.2 (Theorem 3.20).

Finally, the specialization he uses is the same as in [31], although the degree range in the final statement is slightly different. Kollár's specialization produces in characteristic 2 limits of hypersurfaces in $\mathbb{P}^{n+1}$, of any even degree $\geq 2^{\lceil \frac{n+2}{3} \rceil}$ which admit nonzero algebraic differential forms of degree $n - 1$. The important point to be discussed here is the nature of the singularities : not only the forms have to extend on the desingularization, a point which is discussed in Kollár's paper, but the singularities have to satisfy the condition (*) of Colliot-Thélène and Pirutka. This is done in [51].

Combining all these ingredients, Totaro finally proves the following theorem, where the ground field is assumed to be uncountable of characteristic 0 or 2:

Theorem 4.15 ([51]) *A very general hypersurface of degree $\geq 2^{\lceil \frac{n+2}{3} \rceil}$ in $\mathbb{P}^{n+1}$, $n \geq 3$, is not stably rational.*

The method of the proof actually shows that such hypersurfaces defined over a number field exist, and not only they are not stably rational, but they in fact do not have universally trivial CH_0 group.

Let us finally state the following spectacular asymptotic improvement of Totaro's Theorem. This result by Schreieder [45] uses in an essential way Theorem 4.13.

Theorem 4.16 (Schreieder [45]) *A very general complex projective hypersurface of dimension n and degree at least $\log_2 n + 2$, $n \geq 3$, is not stably rational.*

We conclude this section with some hints on the following theorem solving a longstanding question:

Theorem 4.17 ([26]) *Let $Y \subset \mathbb{P}^2 \times \mathbb{P}^3$ be a very general hypersurface of bidegree $(2, 2)$. Then Y is not stably rational.*

On the other hand, there is a dense set of points b in the parameter space for these Y's, such that Y_b is rational. In particular, rationality and stable rationality are not invariant under deformation.

The proof uses the specialization method described above. Let us give a complete proof for the density statement which is not hard but useful. We will use the following fact from [15, Section 3] (see also [18, Section 8]):

Proposition 4.18 *Let Y be a smooth fourfold fibered in two-dimensional quadrics over a surface. Then integral Hodge classes of degree 4 on Y are algebraic.*

We also have the following standard lemma due to Springer [49] (it is in fact true in any dimension and over any field).

Lemma 4.19 *Let Q be a smooth quadric surface over a field k of characteristic 0. Then Q has a k-point, hence is rational over k, if and only if Q has a 0-cycle z of odd degree.*

Proof Indeed, let C be the family of lines in Q. The curve $C_{\bar{k}}$ is the disjoint union of two copies of $\mathbb{P}^1_{\bar{k}}$. Let $k \subset k'$ be the degree 2 (or 1) extension on which the two geometric components of $C_{\bar{k}}$ are defined. Then $C_{k'}$ is the disjoint union of two curves C_1, C_2 which become isomorphic to $\mathbb{P}^1_{\bar{k}'}$ over $\bar{k}'$. But each of these curves C_i has a divisor of odd degree defined over k', namely the incidence divisor $P^*z \in \mathrm{CH}^1(C_i)$, where $P \subset C_i \times Q$ is the universal correspondence. It follows that each component C_i is isomorphic to $\mathbb{P}^1_{k'}$, and has a k'-point l, providing a line $l \subset Q$ defined over k'. Let i be the Galois involution acting on $C(k')$. Then if $i(l) = l$, (so that in fact $k = k'$ and $i = Id$), l is defined over k and Q has a k-point. Otherwise we get two different conjugate lines l and $i(l)$ in Q which belong to different rulings of Q, and their intersection point is defined over k. □

Corollary 4.20 *Let Y be a fourfold as in Theorem 4.17. Then Y is rational if Y has an integral Hodge class α of degree 4 which has odd intersection number with the fibers Q_s of the morphism $pr_1 : Y \to \mathbb{P}^2$.*

Proof Indeed, Y is fibered via pr_1 into quadric surfaces over $\mathbb{P}^2$. Proposition 4.18 thus applies to Y and α is the class of a codimension algebraic cycle Z on Y. Restricting Z to the generic fiber Y_η of pr_1, we get a 0-cycle of odd degree on Y_η defined over the function field $\mathbb{C}(\eta)$ of $\mathbb{P}^2$ and Lemma 4.19 then tells that Y_η is rational over $\mathbb{C}(\eta)$. A fortiori, Y is rational. □

Corollary 4.20 reduces the proof of the density statement to the following proposition:

Proposition 4.21 *Let B be the family of all smooth fourfolds Y described in Theorem 4.17. Then the set of points $b \in B$ such that Y_b has an integral Hodge class α of degree 4 which has odd intersection number with the fibers of pr_1 is dense in B for the usual topology.*

Proof This will follow by applying the following infinitesimal criterion (Proposition 4.22) below: Consider our family of fourfolds $\mathcal{Y} \to B$. We have an associated infinitesimal variation of Hodge structures (see [55, 5.1.2]) at any point $t \in B$

$$H^{2,2}(\mathcal{Y}_t) \to \mathrm{Hom}\,(T_{B,t}, H^{1,3}(\mathcal{Y}_t)),$$

$$\alpha \mapsto \overline{\nabla}(\alpha) : T_{B,t} \to H^{1,3}(\mathcal{Y}_t).$$

Using the fact that the Hodge structure on $H^4(\mathcal{Y}_t, \mathbb{Q})$ is of Hodge niveau 2, that is, $H^{4,0}(\mathcal{Y}_t) = 0$, we have (see [55, 5.3.4]):

Proposition 4.22 *If there exist $t_0 \in B$ and $\alpha \in H^{2,2}(\mathcal{Y}_{t_0})$ such that $\overline{\nabla}(\alpha) : T_{B,t_0} \to H^{1,3}(\mathcal{Y}_{t_0})$ is surjective, then for any Euclidean open set $U \subset B$ containing t_0, the image of the natural map*

$$T_{t_0} : \mathcal{H}^{2,2}_{\mathcal{Y}_U,\mathbb{R}} \to H^4(\mathcal{Y}_{t_0}, \mathbb{R})$$

defined by composing the inclusion $\mathcal{H}^{2,2}_{\mathcal{Y},\mathbb{R}} \to \mathcal{H}^4_{\mathcal{Y},\mathbb{R}}$ with a local flat trivialization over U of $\mathcal{H}^4_{\mathcal{Y},\mathbb{R}}$, contains an open subset V_U of $H^4(\mathcal{Y}_{t_0}, \mathbb{R})$.

Here $\mathcal{H}^4_{\mathcal{Y},\mathbb{R}}$ is the flat real vector bundle with fiber $H^4(\mathcal{Y}_b, \mathbb{R})$ over any $b \in B$ and $\mathcal{H}^{2,2}_{\mathcal{Y},\mathbb{R}}$ is the real vector bundle over B with fiber over $t \in B$ the space $H^{2,2}(\mathcal{Y}_t)_{\mathbb{R}}$ of real cohomology classes of type $(2, 2)$ on $\mathcal{Y}_t$. Note that the image of T_{t_0} is by definition the set of real degree 4 cohomology classes on $\mathcal{Y}_t$ which are of type $(2, 2)$ at some point $t' \in U$.

Corollary 4.23 *Under the same assumption, for any $t \in B$, and any Euclidean open set $U \subset B$ containing t, there exists $t' \in U$ and $\alpha_{t'} \in H^{2,2}(\mathcal{Y}_{t'}) \cap H^4(\mathcal{Y}_{t'}, \mathbb{Z})$ such that the degree of α on the fibers Q_s of $pr_1 : \mathcal{Y}_{t'} \dashrightarrow \mathbb{P}^2$ is odd.*

Proof We observe first that the condition on t_0 in Proposition 4.22 is Zariski open, hence is satisfied on a dense open set. We also note that the open subset V_U of $H^4(\mathcal{Y}_{t_0}, \mathbb{R})$ appearing in Proposition 4.22 is in fact a subcone. It is then immediate to prove that a non-empty open subcone of $H^4(\mathcal{Y}_{t_0}, \mathbb{R}) = H^4(\mathcal{Y}_{t_0}, \mathbb{Z}) \otimes \mathbb{R}$ has to contain an integral class which has odd degree on the fibers Q_s. □

What remains to be done is to check the infinitesimal criterion, which is quite well-understood thanks to the Carlson-Griffiths theory of variation of Hodge structures of hypersurfaces (see [55, 5.3.4]). □

5 Cohomological Decomposition of the Diagonal and the Abel–Jacobi Map

5.1 Intermediate Jacobians, Abel–Jacobi Map and Universal Cycle

We already encountered in the previous sections the Abel–Jacobi map

$$\phi_X^2 : CH^2(X)_{hom} \to J^3(X) \tag{5.55}$$

which is an isomorphism by Theorem 3.21 when X is a smooth projective complex manifold with $CH_0(X) = \mathbb{Z}$. The right hand side is an abelian variety but the left hand side is not an algebraic variety, even if it is more than an abstract group. Namely, we can use the families of codimension 2 algebraic cycles on X given by codimension 2 cycles $\mathcal{Z} \in CH^2(B \times X)$ parameterized by smooth connected varieties B and the associated maps $\mathcal{Z}_* : B \to CH^2(X)_{alg}$, $b \mapsto \mathcal{Z}_b - \mathcal{Z}_{b_0}$, where $b_0 \in B$ is a fixed reference point, to say that ϕ_X^2 is a "regular homomorphism". This notion was introduced by Murre [43] and it says that for any cycle $\mathcal{Z}$ as above, the map

$$\phi_{\mathcal{Z}} := \phi_X^2 \circ \mathcal{Z}_* : B \to J^3(X)$$

is a morphism of algebraic varieties.

The question left open concerning the isomorphism (5.55) is the existence of a *universal codimension* 2 *cycle*, which was first asked in [58]:

Definition 5.1 A universal codimension 2 cycle for X is a codimension 2 cycle $\mathcal{Z} \in CH^2(J^3(X) \times X)$ such that $\mathcal{Z}_0 = 0$ and the associated map

$$\phi_{\mathcal{Z}} : J^3(X) \to J^3(X)$$

is the identity.

For codimension 1 cycles, the universal cycle exists and is called the Poincaré divisor. Its existence in this case can be proved using the fact that the complete family of sufficiently ample divisors of given cohomology class on X is via ϕ_X^1 a honest projective bundle on $J^1(X)$. Indeed, the fiber over a divisor class L is the projective space $|L|$ and a point $x \in X$ determines for any L the hyperplane $|L|_x \subset |L|$ of divisors in $|L|$ passing through x.

We will see in next section (see Corollary 5.9) that, as a consequence of the degeneration method, there are Fano threefolds which do not admit a universal codimension 2 cycle. Note that, once one knows that the Abel–Jacobi map ϕ_X^2 is surjective, there exists a codimension 2 cycle $\mathcal{Z} \in CH^2(J^3(X) \times X)$ such that $\mathcal{Z}_0 = 0$ and the associated map

$$\phi_\mathcal{Z} : J^3(X) \to J^3(X)$$

is N times the identity for a certain integer $N > 0$. Indeed, we use for this the fact that ϕ_X^2 is regular. There are countably many complete families of codimension 2 algebraic cycles on X, so the surjectivity of the Abel–Jacobi map implies that there exist a smooth projective variety B and a cycle $\mathcal{Z} \in CH_2(B \times X)$, such that the morphism

$$\phi_\mathcal{Z} : B \to J^3(X)$$

is surjective. We can replace B by a subvariety B' containing the reference point b_0 such that the restriction ϕ' of $\phi_\mathcal{Z}$ to B' is a generically finite map. Then we consider the cycle

$$\mathcal{Z}_J = (\phi', Id_X)_*(\mathcal{Z}') \in CH^2(J^3(X) \times X),$$

where $\mathcal{Z}' := \mathcal{Z}'_{|B' \times X}$. The integer N obtained by this construction is $\deg \phi'$.

The existence of a universal cycle for codimension 1 cycles allows us to prove the following result:

Proposition 5.2 ([58]) *If X has a cohomological decomposition of the diagonal, X has a universal codimension 2 cycle.*

Proof We write the decomposition in the form

$$[\Delta_X] = [X \times x] + (\tilde{j}, Id_X)_*[\tilde{Z}] \text{ in } H_B^{2n}(X \times X, \mathbb{Z}), \tag{5.56}$$

where $\tilde{j} : \tilde{D} \to X$ is a morphism from a smooth projective variety of dimension $n - 1$. As we used several times, this implies that for any $\alpha \in H_B^3(X, \mathbb{Z})$

$$\alpha = \tilde{j}_*([\tilde{Z}]^*\alpha). \tag{5.57}$$

The considered morphisms are morphisms of Hodge structures of odd weight and they induce as well morphisms between the associated intermediate Jacobians. Equation (5.57) then says that

$$\tilde{j}_* \circ [\tilde{Z}]^* = Id_{J^3(X)} : J^3(X) \to J^3(X). \tag{5.58}$$

Let now $\mathcal{D} \in \mathrm{CH}^1(J^1(\widetilde{D}) \times \widetilde{D})$ be a universal codimension 1-cycle. By pull-back to $J^3(X)$ it provides a codimension 1-cycle on $J^3(X) \times \widetilde{D}$ and by push-forward to X, we get finally a codimension 2 cycle on $J^3(X) \times X$ defined by the formula

$$\mathcal{Z} = (Id_{J^3(X)}, \tilde{j})_*(([\widetilde{Z}]^*, Id_{\widetilde{D}})^* \mathcal{D}).$$

The map $\phi_{\mathcal{Z}} : J^3(X) \to J^3(X)$ equals by construction $\tilde{j}_* \circ [\widetilde{Z}]^*$, hence it is the identity of $J^3(X)$ by (5.58). $\qquad\square$

5.2 Extending Clemens–Griffiths Criterion

The discussion in this section is specific to dimension 3, although it concerns stable rationality for them. The stable rationality of X says that $X \times \mathbb{P}^r$ is rational for some r, hence it involves birational geometry of higher dimensional varieties. Because of this, the Clemens–Griffiths criterion that we now describe concerns only rationality of threefolds and not stable rationality.

Let X be a smooth complex projective threefold. Let us assume that $H_B^1(X, \mathbb{Z}) = 0$ and $H^{3,0}(X) = 0$, which will be the case if X is rationally connected. Consider the intermediate Jacobian $J^3(X) = H_B^3(X, \mathbb{C})/(F^2 H_B^3(X, \mathbb{C}) \oplus H_B^3(X, \mathbb{Z})_{tf})$, which in this case equals $H^{1,2}(X)/H_B^3(X, \mathbb{Z})_{tf}$. Here $H_B^3(X, \mathbb{Z})_{tf}$ denotes the abelian group $H_B^3(X, \mathbb{Z})$ modulo torsion. By definition, one has a canonical isomorphism

$$H_1^B(J^3(X), \mathbb{Z}) \cong H_B^3(X, \mathbb{Z})_{tf}. \tag{5.59}$$

The unimodular intersection pairing $\langle \,,\, \rangle_X$ on $H_B^3(X, \mathbb{Z})_{tf}$ provides, thanks to the Hodge-Riemann relations, a principal polarization on $J^3(X)$ of class $\theta_X \in H_B^2(J^3(X), \mathbb{Z})$. If $g = \dim J^3(X)$, the integral degree $2g - 2$ cohomology class (or degree 2 homology class) $\frac{\theta_X^{g-1}}{(g-1)!}$ on $J^3(X)$ is called the minimal class. It is not known if it is algebraic for a general principally polarized abelian variety (A, θ_A), although it is when $(A, \theta_A) = (J^1(C), \theta_C)$ is the Jacobian of a smooth projective curve, or a product of them. The celebrated Clemens–Griffiths criterion [14] says the following:

Theorem 5.3 *If a smooth projective threefold X is rational, then $(J^3(X), \theta_X)$ is the direct product of Jacobians $(J^1(C_i), \theta_{C_i})$ of curves.*

This theorem follows indeed from the fact that a principally polarized abelian variety splits uniquely into a direct sum of simple principally polarized abelian varieties. Furthermore the Jacobian of a smooth projective curve is indecomposable as a ppav, by Riemann's theorem which implies that its Theta divisor is irreducible. This decomposition of $(J^3(X), \theta_X)$ changes under blow-up of a curve $C \subset X$ by the addition of an orthogonal direct summand which is the Jacobian of C. We then conclude that the Griffiths component of X, namely the sum in the decomposition above of all summands not isomorphic as ppav's to Jacobians of curves, does not change under blow-up and thus is a birational invariant (one also uses the fact that if $\phi : X \to Y$ is a morphism which is birational, that is, of degree 1, the morphism $\phi^* : H_B^3(Y, \mathbb{Z}) \to H_B^3(X, \mathbb{Z})$ is compatible with polarizations and thus makes $H^3(Y, \mathbb{Z})$ an orthogonal direct summand of $H_B^3(X, \mathbb{Z})$).

In the Jacobian $J^1(C)$ of a curve C, the image of C by the Albanese map gives an effective 1-cycle Z whose class is the minimal class. The Matsusaka criterion [37] says the following:

Theorem 5.4 *A principally polarized abelian variety (A, θ_A) is a product of Jacobians of curves if and only if it carries an effective 1-cycle $Z = \sum_i n_i C_i$, $n_i > 0$ whose class $[Z] \in H_2^B(A, \mathbb{Z})$ is the minimal class.*

The following result proved in [61] is thus a version of Clemens–Griffiths theorem for stable rationality.

Theorem 5.5 *Let X be a smooth projective threefold. If X has a cohomological decomposition of the diagonal, the minimal class $\frac{\theta_X^{g-1}}{(g-1)!}$ of $J^3(X)$ is algebraic. In particular, if X is stably rational, the minimal class $\frac{\theta_X^{g-1}}{(g-1)!}$ of $J^3(X)$ is algebraic.*

This condition says that there is a 1-cycle $Z = \sum_i n_i C_i$ whose class $[Z] \in H_2^B(J^3(X), \mathbb{Z})$ is the minimal class. The difference with Clemens–Griffiths criterion is that we do not ask it to be effective.

Proof of Theorem 5.5 Recalling the isomorphism (5.59), or rather its dual

$$i : H_B^1(J^3(X), \mathbb{Z}) \cong H_B^3(X, \mathbb{Z})_{tf}, \qquad (5.60)$$

the minimal class $\gamma \in H_2^B(J^3(X), \mathbb{Z})$ is characterized by the fact that

$$\int_\gamma \alpha \cup \beta = \langle i(\alpha), i(\beta) \rangle_X \qquad (5.61)$$

for any $\alpha, \beta \in H_B^1(J^3(X), \mathbb{Z})$. We now assume that X has a cohomological decomposition of the diagonal

$$[\Delta_X] = \widetilde{j}_*([\widetilde{Z}]) + [X \times x] \text{ in } H_B^6(X \times X, \mathbb{Z}) \qquad (5.62)$$

for some cycle $\widetilde{Z} \in CH^2(\widetilde{D} \times X)$. Recalling that $\widetilde{D}$ is the desingularization of a divisor in X, we can assume after blowing-up X that $\widetilde{D} = \sqcup D_i$, where each $j_i = \widetilde{j}_{|D_i}$ is an embedding and that $\widetilde{j}(\widetilde{D})$ has normal crossings. We denote by Z_i the restriction of $\widetilde{Z}$ to $D_i \times X$, and we denote by W_{il} the curve which is the intersection

$j_i(D_i) \cap j_l(D_l)$, that we can see as a divisor in either surface D_i or D_l. Formula (5.62) gives for any $\alpha \in H_B^3(X, \mathbb{Z})_{tf}$

$$\alpha = \sum_i j_{i*}([Z_i]^*\alpha).$$

It follows that, for any $\alpha \in H_B^3(X, \mathbb{Z})_{tf}$,

$$\langle \alpha, \beta \rangle_X = \sum_{il} \langle j_{i*}([Z_i]^*\alpha), j_{l*}([Z_l]^*\beta) \rangle_X \qquad (5.63)$$

$$= \sum_{i \neq l} \int_{W_{il}} [Z_i]^*\alpha \cup [Z_l]^*\beta + \sum_i \langle j_{i*}([Z_i]^*\alpha), j_{i*}([Z_i]^*\beta) \rangle_X$$

$$= \sum_{i<l} \int_{W_{il}} ([Z_i]^*\alpha \cup [Z_l]^*\beta + [Z_l]^*\alpha \cup [Z_i]^*\beta) + \sum_i \int_{D_i} j_i^*[D_i] \cup [Z_i]^*\alpha \cup [Z_i]^*\beta$$

$$= \sum_{i<l} \int_{W_{il}} ([Z_i] + [Z_l])^*\alpha \cup ([Z_i] + [Z_l])^*\beta$$

$$- \sum_{i<l} \int_{W_{il}} ([Z_i]^*\alpha \cup [Z_i]^*\beta + [Z_l]^*\alpha \cup [Z_l]^*\beta) + \sum_i \int_{W_i} [Z_i]^*\alpha \cup [Z_i]^*\beta,$$

where in the last term W_i is the 1-cycle $j_i^*D_i$ of D_i. The conclusion of (5.63) is that we found smooth projective curves C_s (namely the W_{il}'s and the supports of the W_i's), integers n_s (namely the coefficients of the 1-cycle W_i) and codimension 2 cycles $Z_s' \in \mathrm{CH}^2(C_s \times X)$, such that

$$\langle \alpha, \beta \rangle_X = \sum_s n_s \langle [Z_s']^*\alpha), [Z_s']^*\beta \rangle_{C_s}. \qquad (5.64)$$

Let $\phi_{Z_s'} : C_s \to J^3(X)$ be the associated Abel–Jacobi map. For any class $\eta \in H_B^1(J^3(X), \mathbb{Z})$ the class $\alpha = i(\eta)$ satisfies by definition of the isomorphism i:

$$\phi_{Z_s'}^*\eta = [Z_s']^*\alpha \quad \text{in } H_B^1(C_s, \mathbb{Z}). \qquad (5.65)$$

Thus (5.64) rewrites as

$$\langle \alpha, \alpha' \rangle_X = \sum_s n_s \int_{C_s} \phi_{Z_s'}^*\eta \cup \phi_{Z_s'}^*\eta' \qquad (5.66)$$

for any $\eta, \eta' \in H_B^1(J^3(X), \mathbb{Z})$. Comparing with (5.61), we conclude that $\sum_s n_s[\phi_{Z_s'}(C_s)] = \gamma$ is the minimal class. $\quad\square$

We now give a necessary and sufficient set of conditions for a smooth projective threefold to admit a cohomological decomposition of the diagonal. Part of these results were obtained in [58], and they were finally completed in [61]. We assume that X has $H^{3,0}(X) = 0$ and $H^1_B(X, \mathbb{Z}) = 0$ because this is necessary for X to have a cohomological decomposition with rational coefficients. These vanishing conditions allow us to speak of the ppav $(J^3(X), \theta_X)$.

Theorem 5.6 *A smooth complex projective threefold with $H^{3,0}(X) = 0$ and $H^1_B(X, \mathbb{Z}) = 0$ admits a cohomological decomposition of the diagonal if and only if the following conditions are satisfied:*

1. *$H^*_B(X, \mathbb{Z})$ has no torsion.*
2. *$H^4_B(X, \mathbb{Z})$ is algebraic.*
3. *X admits a universal codimension 2 cycle.*
4. *The minimal class of $(J^3(X), \theta_X)$ is algebraic.*

Proof We already proved that these conditions are necessary: 1 and 2 were proved to be necessary in Theorem 3.13. Condition 3 is necessary by Proposition 5.2 and 4 is necessary by Proposition 5.5.

We now prove that these conditions are sufficient. If X satisfies these conditions, then by 1, X has a Künneth decomposition of the diagonal

$$[\Delta_X] = \delta_{6,0} + \delta_{5,1} + \delta_{4,2} + \delta_{3,3} + \delta_{2,4} + \delta_{1,5} + \delta_{0,6}$$

where $\delta_{i,j} \in H^i_B(X, \mathbb{Z}) \otimes H^j_B(X, \mathbb{Z})$ and acts as the projector on $H^j_B(X, \mathbb{Z})$.

As we assumed $H^1_B(X, \mathbb{Z}) = 0$, $\delta_{1,5}$ and $\delta_{5,1}$ are zero. Assuming 2, the even degree cohomology of X is algebraic, since this implies $H^{2,0}(X) = 0$, so that $H^2_B(X, \mathbb{Z})$ is also algebraic by Lefschetz. It follows that the terms $\delta_{6,0}, \delta_{4,2}, \delta_{2,4}$ can be written as $\sum_i n_i [Z_i \times Z'_i]$, where codim $Z_i > 0$. They are thus contained in $D \times X$ for some closed proper algebraic subset D of X. Finally the term $\delta_{0,6}$ is the class of $X \times x$.

It thus remains to show that the class $\delta_{3,3}$ is the class of a cycle supported on $D \times X$ for some closed proper algebraic subset D of X, and by the previous analysis of the other Künneth terms, it suffices in fact that there is a cycle supported on $D \times X$ for some closed proper algebraic subset D of X whose Künneth component of type $(3, 3)$ is $\delta_{3,3}$.

Let $\Gamma = \sum_i n_i C_i$ be a 1-cycle of $J^3(X)$ representing the minimal class. Let $Z \in \mathrm{CH}^2(J^3(X) \times X)$ be a universal codimension 2 cycle and let $Z_i \in \mathrm{CH}^2(\widetilde{C}_i \times X)$ be its pull-back to $\widetilde{C}_i \times X$, where $\widetilde{C}_i$ is the normalization of C_i. We consider the following cycle

$$T := \sum_i n_i (Z_i, Z_i)_* \Delta_{\widetilde{C}_i} \quad \text{in } \mathrm{CH}^3(X \times X), \qquad (5.67)$$

where as usual $\Delta_{\widetilde{C}_i}$ is the diagonal of $\widetilde{C}_i$ and

$$(Z_i, Z_i) := pr^*_{13} Z_i \cdot pr^*_{24} Z_i \in \mathrm{CH}^4(\widetilde{C}_i \times \widetilde{C}_i \times X \times X).$$

Observe that each Z_i is of dimension 2, so that the support of Z_i does not dominate X by the second projection. It follows that T is supported on $D \times X$ for some closed proper algebraic subset D of X. We now have:

Lemma 5.7 *The $(3, 3)$-Künneth component of T is equal to $\delta_{3,3}$.*

Proof We have to show that for any $\alpha,\ \beta \in H_B^3(X, \mathbb{Z})$

$$\langle [T]^*\alpha, \beta \rangle_X = \langle \alpha, \beta \rangle_X. \tag{5.68}$$

We claim that for any curve C and any codimension 2 cycle Z in $C \times X$, one has, denoting $Z' := (Z, Z)_* \Delta_C$,

$$\langle [Z']^*\alpha, \beta \rangle_X = \langle [Z]^*\alpha, [Z]^*\beta \rangle_C. \tag{5.69}$$

Assuming this equality, we get

$$\langle [T]^*\alpha, \beta \rangle_X = \sum_i n_i \langle [Z_i]^*\alpha, [Z_i]^*\beta \rangle_{\widetilde{C}_i}$$

$$= \sum_i n_i \langle j_i^*([Z]^*\alpha), j_i^*([Z]^*\beta) \rangle_{\widetilde{C}_i},$$

where $j_i : \widetilde{C}_i \to J^3(X)$ is the natural map. As Z is a universal cycle, one has $[Z]^* = i^{-1}$ and thus, as $\sum_i n_i j_{i*}([\widetilde{C}_i])$ is the minimal class, the last term is $\langle \alpha, \beta \rangle_X$ by (5.61).

It remains to prove (5.69). This follows from the fact that

$$[Z, Z]^*(\alpha \otimes \beta) = [Z]^*\alpha \otimes [Z]^*\beta \text{ in } H_B^1(C \times C, \mathbb{Z}), \tag{5.70}$$

where $\alpha \otimes \beta := pr_1^*\alpha \cup pr_2^*\beta$ for both X and C.

It follows from (5.70) that

$$\langle [\Delta_C], [Z]^*\alpha \otimes [Z]^*\beta \rangle_{C \times C} = \langle [\Delta_C], [Z, Z]^*(\alpha \otimes \beta) \rangle_{C \times C}$$

$$= \langle [Z, Z]_*([\Delta_C]), \alpha \otimes \beta \rangle_{X \times X} = \langle [Z'], \alpha \otimes \beta \rangle_{X \times X}.$$

The last term is easily seen to be $\langle [Z']^*\alpha, \beta \rangle_X$.   □

The proof of Theorem 5.6 is finished.   □

In the case of rationally connected threefolds, we know that $H_B^2(X, \mathbb{Z})$ and $H_B^5(X, \mathbb{Z})$ have no torsion by Theorem 2.2. We also know that $H_B^4(X, \mathbb{Z})$ is algebraic by Theorem 2.29. We thus get in this case

Theorem 5.8 *A smooth complex projective rationally connected threefold admits a cohomological decomposition of the diagonal if and only if the following conditions are satisfied:*

1. $H^3_B(X, \mathbb{Z})$ *has no torsion.*
2. *X admits a universal codimension 2 cycle.*
3. *The minimal class of $(J^3(X), \theta_X)$ is algebraic.*

It is interesting to note that 1 is the Artin–Mumford invariant, while 3 is our generalization of Clemens–Griffiths criterion that works for stable rationality. The condition 2 has no obvious classical analogue but in [60], we compute it as "universal degree 3 unramified cohomology" of X. In fact this condition is related to the integral Hodge conjecture for $X \times J^3(X)$ or rather its (3, 1)-Künneth component.

We now deduce the following consequence:

Corollary 5.9 *There are rationally connected threefolds not admitting a universal codimension 2 cycle.*

Proof The example is the desingularization of a very general quartic double solid with 7 nodes. It does not admit a cohomological decomposition of the diagonal by Theorem 4.6. On the other hand, its intermediate Jacobian has dimension 3, so it is a Jacobian and the minimal class is algebraic. Finally it has trivial Artin–Mumford invariant by work of Endrass [20]. The condition that fails in Theorem 5.8 must thus be Condition 2.   □

5.3   The Case of Cubic Hypersurfaces

The rationality or stable rationality of cubic hypersurfaces is an almost completely open problem. The results available are:

- *A smooth plane cubic is not rational* as it has $H^{1,0} \neq 0$.
- *A smooth cubic surface X over an algebraically closed field is rational*: This is a particular case of Castelnuovo theorem but it can be proved explicitly in this case: take any two not intersecting lines Δ, Δ' in X. Then for $x \in \Delta$, $x' \in \Delta'$, the line $\langle x, x' \rangle$ in $\mathbb{P}^3$ meets X in a third point $\phi(x, x')$. This defines a birational map

$$\phi : \Delta \times \Delta' \dashrightarrow X.$$

The inverse map is constructed as follows: start from a general point $y \in X$, and let $Q_y := \langle y, \Delta \rangle$, $Q'_y := \langle y, \Delta' \rangle$. Then $Q_y \cap \Delta' = \{x'\}$, $Q'_y \cap \Delta = \{x\}$ and $\phi(x, x') = y$. This construction shows more generally:

- *Any smooth cubic hypersurface of dimension $2m$ containing two m-planes P, P' which do not meet is rational.*

- *A smooth cubic threefold is not rational.* This is the celebrated Clemens–Griffiths theorem, proved in [14] for cubics defined over $\mathbb{C}$, and by Murre (see [42]) in any nonzero characteristic different from 2.

This is essentially all that we know about cubic hypersurfaces. Let us state a few open questions:

Question 5.10 *Does there exist a smooth cubic hypersurface of odd dimension which is rational or stably rational?*

Question 5.11 *Is a smooth cubic threefold stably irrational? Does there exist a stably rational smooth cubic threefold?*

Question 5.12 *Is a very general smooth cubic hypersurface of even dimension ≥ 4 irrational?*

In this section, we are going to study the weaker question whether a smooth cubic hypersurface has a decomposition of the diagonal. We will see that even for cubic threefolds, this already rises serious difficulties.

5.3.1 General Cubic Hypersurfaces

The following construction is certainly classical. In dimension 1, it allows to construct the group structure on a plane cubic curve. Let X be a smooth cubic hypersurface in $\mathbb{P}^n$. The variety $F(X)$ of lines in X is smooth of dimension $2n - 6$. If $x \in X$ and l is a line in $\mathbb{P}^n$ passing through x, then $l \cap X$ contains x and two residual points $y, z \in X$. Conversely, starting from two points y, z in X, the line $l_{y,z} := \langle y, z \rangle$ intersects X in a third point $x \in X$. This shows that there is a birational map

$$\Phi : X^{[2]} \dashrightarrow Q_X,$$

where $Q_X \to X$ is the projective bundle with fiber over x the $\mathbb{P}^{n-1}$ parameterizing lines in $\mathbb{P}^n$ passing through x. The following is proved in [61], see also [22]:

Proposition 5.13 *The map Φ induces an isomorphism between the blow-up of $X^{[2]}$ along $C(X)$ and the blow-up of Q_X along Q_{XX}.*

Here the loci $C(X)$ and Q_{XX} are defined as follows:

- $C(X) \subset X^{[2]}$ is the locus of length 2 subschemes of X that are contained in a line contained in X. Thus $C(X)$ is a $\mathbb{P}^2$-bundle over $F(X)$.
- The locus $Q_{XX} \subset Q_X$ is the set of pairs $(x, [l])$, such that $x \in l$ and the line l is contained in X. It is thus naturally isomorphic to the universal $\mathbb{P}^1$-bundle over $F(X)$.

We now explain two consequences of this proposition. The first one is due to Galkin and Shinder [22] and gives a beautiful evidence for the link between rationality of cubic fourfolds and existence of associated K3 surfaces which has been proposed and studied in [25] and more recently explicitly conjectured and studied in [1, 2, 33]. Let $K_0(\mathrm{Var}_K)$ be the Grothendieck ring whose generators are isomorphism classes of algebraic varieties defined over K, with relation

$$[U] + [Z] = [X] \tag{5.71}$$

whenever $X = U \sqcup Z$, with Z closed, U open. The ring structure is given by product. Denote by $\mathbb{L} \in K_0(\mathrm{Var}_K)$ the class of the affine line and $\langle \mathbb{L} \rangle$ the ideal of $K_0(\mathrm{Var}_K)$ generated by $\mathbb{L}$. The following result is proved in [35].

Theorem 5.14 ([35]) *Let K be a field of characteristic zero. The quotient-ring $K_0(\mathrm{Var}_K)/\langle \mathbb{L} \rangle$ is naturally isomorphic to the free abelian group generated by stable birational equivalence classes of smooth projective connected varieties over K together with its natural ring structure. In particular, if X and $Y_1, \ldots, Y_m$ are smooth projective connected varieties and*

$$[X] = \sum_i n_i [Y_i] \text{ in } K_0(\mathrm{Var}_K)/\langle \mathbb{L} \rangle$$

for some $n_i \in \mathbb{Z}$, then X is stably birationally equivalent to one of the Y_i's.

The class of $\mathbb{P}^n$ is equal to $\sum_{i=0}^n \mathbb{L}^i$, as one argue by induction using (5.71) and

$$\mathbb{P}^n \setminus \mathbb{P}^{n-1} = \mathbb{A}^n = (\mathbb{A}^1)^n.$$

Similarly, one gets that the class of a projective bundle $\mathbb{P}(E) \to X$ with rank $E = r$ is given by

$$[\mathbb{P}(E)] = [\mathbb{P}^{r-1}][X] = (\sum_{i=0}^{r-1} \mathbb{L}^i)[X]. \tag{5.72}$$

For the blow-up $\widetilde{X}$ of a smooth variety X along a smooth subvariety Z of codimension r, the isomorphism $\widetilde{X} \setminus E \cong X \setminus Z$ gives by (5.71)

$$[\widetilde{X}] - (\sum_{i=0}^{r-1} \mathbb{L}^i)[Z] = [X] - [Z]$$

or equivalently

$$[\widetilde{X}] = [X] + (\sum_{i=1}^{r-1} \mathbb{L}^i)[Z]. \tag{5.73}$$

The following result is due to Galkin and Shinder [22].

Theorem 5.15 *Let X be a smooth cubic hypersurface in $\mathbb{P}_K^n$. Then the following equality holds in $K_0(\mathrm{Var}_K)$:*

$$[X^{(2)}] - \mathbb{L}^2[F(X)] = [X](1 + \mathbb{L}^{n-1}). \tag{5.74}$$

If char $K =$, and X is rational of dimension 4, then either there is an explicit nonzero element in $K_0(\mathrm{Var}_K)$ annihilated by $\mathbb{L}^2$, or the Fano variety of lines is birational to $S^{[2]}$, where S is a K3 surface.

Proof Applying Proposition 5.13 and the projective bundle and blow-up formulas (5.72) and (5.73), we get

$$[X^{[2]}] + \mathbb{L}(1 + \mathbb{L} + \mathbb{L}^2)[F(X)] = [X](\sum_{i=0}^{n-1} \mathbb{L}^i) + (\mathbb{L} + \mathbb{L}^2)(1 + \mathbb{L})[F(X)]. \tag{5.75}$$

Now we notice that the difference $X^{[2]} \backslash E_X$ is isomorphic to $X^{(2)} \backslash X$, where $X \subset X^{(2)}$ is the diagonal and $E_X \to X$, $E_X \subset X^{[2]}$, is the exceptional divisor over the diagonal. Plugging again (5.71) and the projective bundle formula for E_X in formula (5.75), we get (5.74).

We now turn to the proof of the second statement. We observe that the symmetric product operation $s^{(2)} : [Y] \mapsto [Y^{(2)}]$ satisfies the following property

$$s^{(2)}([Y] + [Y']) = s^{(2)}([Y]) + s^{(2)}([Y']) + [Y] \cdot [Y'], \tag{5.76}$$

as this is the case for disjoint unions. We also have $s^{(2)}(\mathbb{L}[Y]) = \mathbb{L}^2 s^{(2)}([Y])$ as follows from the fact that $\mathbb{A}^{(2)} = \mathbb{A}^2$ in $K_0(\mathrm{Var}_K)$.

Suppose now that a smooth cubic fourfold X is rational. Then by a sequence of smooth blow-ups starting from X, one gets something isomorphic to a variety Y which is also obtained from $\mathbb{P}^4$ by a sequence of smooth blow-ups. Let us assume for simplicity that we blew-up only surfaces S_i on the X side and T_j on the $\mathbb{P}^4$ side. Then in $K_0(\mathrm{Var}_K)$, we get using (5.73)

$$[X] + \sum_i \mathbb{L}[S_i] = [\mathbb{P}^4] + \sum_j \mathbb{L}[T_j]. \tag{5.77}$$

Taking symmetric products and applying (5.76), we get

$$[X^{(2)}] + \sum_i \mathbb{L}^2[S_i^{(2)}] + \sum_{i \neq i'} \mathbb{L}^2[S_i][S_{i'}] + \sum_i \mathbb{L}[X][S_i]$$

$$= [(\mathbb{P}^4)^{(2)}] + \sum_j \mathbb{L}^2[T_j^{(2)}] + \sum_{j \neq j'} \mathbb{L}^2[T_j][T_{j'}] + \sum_j \mathbb{L}[\mathbb{P}^4][T_j].$$

We now replace in this formula $[X^{(2)}]$ by its expression given in formula (5.74) and get

$$\mathbb{L}^2[F(X)] + [X](1 + \mathbb{L}^4) + \sum_i \mathbb{L}^2[S_i^{(2)}] + \sum_{i \neq i'} \mathbb{L}^2[S_i][S_{i'}] + \sum_i \mathbb{L}[X][S_i]$$

$$= [(\mathbb{P}^4)^{(2)}] + \sum_j \mathbb{L}^2[T_j^{(2)}] + \sum_{j \neq j'} \mathbb{L}^2[T_j][T_{j'}] + \sum_j \mathbb{L}[\mathbb{P}^4][T_j].$$

Using again (5.77) in the form

$$[X] = 1 + \mathbb{L} + \sum_i \mathbb{L}[S_i] - \sum_j \mathbb{L}[T_j] \text{ modulo } \mathbb{L}^2,$$

and the relation

$$[(\mathbb{P}^4)^{(2)}] = [\mathbb{P}^4] + \mathbb{L}^2 \text{ modulo } \mathbb{L}^3,$$

we get after simplification

$$\mathbb{L}^2([F(X)] - \sum_{i<i'}[S_i][S_{i'}] - \sum_j[T_j^{(2)}] - \sum_{j \neq j'}[T_j][T_{j'}] - 1 \qquad (5.78)$$

$$- \sum_i[S_i] + \sum_{i,j}[S_i][T_j] - \sum_j[T_j] + \mathbb{L}\alpha) = 0$$

for some $\alpha \in K_0(\text{Var}_K)$. We thus conclude that either the class $[F(X)] - \sum_{i<i'}[S_i][S_{i'}] - \sum_j[T_j^{(2)}] - \sum_{j \neq j'}[T_j][T_{j'}] - 1 - \sum_i[S_i] + \sum_{i,j}[S_i][T_j] - \sum_j[T_j] + \mathbb{L}\alpha$ is nonzero in $K_0(Var)$ but annihilated by $\mathbb{L}^2$, or the following relation holds in $K_0(\text{Var}_K)/\langle \mathbb{L} \rangle$:

$$[F(X)] - \sum_{i<i'}[S_i][S_{i'}] - \sum_j[T_j^{(2)}] - \sum_{j \neq j'}[T_j][T_{j'}] - 1 \qquad (5.79)$$

$$- \sum_i[S_i] + \sum_{i,j}[S_i][T_j] - \sum_j[T_j] = 0.$$

In the latter case, as this provides an equality

$$[F(X)] = \sum_{i<i'}[S_i][S_{i'}] + \sum_j[T_j^{(2)}] + \sum_{j \neq j'}[T_j][T_{j'}] + [\mathbb{P}^4] + \sum_i[S_i \times \mathbb{P}^2] - \sum_{i,j}[S_i][T_j] + \sum_j[T_j \times \mathbb{P}^2]$$

in $K_0(Var)/\langle \mathbb{L} \rangle$ of combinations of classes of four-dimensional varieties, and that $F(X)$ is irreducible, we conclude by Theorem 5.14 that $F(X)$ is stably birational to one of the terms appearing on the right. Using the fact that $F(X)$ has a unique nondegenerate holomorphic 2-form (see [10]), hence is not rationally connected, we easily conclude that $F(X)$ is in fact birational to $[T_j^{(2)}]$ for some smooth surface T_j with $h^{2,0}(T_j) = 1$. Finally one concludes by surface classification that T_j is birational to a $K3$ surface.   □

Theorem 5.15 does not allow to conclude that a very general smooth cubic fourfold is not rational because multiplication by $\mathbb{L}$ is not injective on $K_0(\mathrm{Var}_K)$ as shown by Borisov [13]. Still it gives a beautiful evidence for the relationship between rationality and the existence of an associated $K3$ surface.

We now turn to another application of Proposition 5.13, which can be found in [61].

Theorem 5.16 *Let X be a smooth cubic hypersurface. Assume that X satisfies the Hodge conjecture for integral Hodge classes modulo 2. Then X has a Chow decomposition of the diagonal if and only if it has a cohomological decomposition of the diagonal.*

Note that the assumption is satisfied by odd dimensional cubic hypersurfaces, because three times their even degree cohomology comes from projective space by Lefschetz theorem on hyperplane sections, hence is algebraic. It is also true for cubic fourfolds by [57] (the later result has been reproved recently by Mongardi and Ottem in [40]).

Sketch of Proof of Theorem 5.16 First of all we show the following result, which uses our assumption on integral Hodge classes and also the fact that the integral cohomology of X has no torsion. We will denote by $\Gamma \subset X \times X \times X^{[2]}$ the graph of the natural rational map $X^2 \dashrightarrow X^{[2]}$.

Proposition 5.17 *If a smooth cubic hypersurface X has a cohomological decomposition of the diagonal, there exists a cycle W cohomologous to 0 in $X^{[2]}$ such that*

$$\Delta_X - X \times x - Z = \Gamma^* W \text{ in } \mathrm{CH}(X \times X), \qquad (5.80)$$

where as usual Z is supported on $D \times X$, with $D \subset X$ proper closed algebraic.

The difficulty to achieve (5.80) is the following: our assumption is that $\Delta_X - X \times x - Z$ is cohomologous to 0, and we can also arrange to make this cycle symmetric, hence coming from a cycle on $X^{[2]}$. The point of (5.80) is that we want it to come from a cycle which is also cohomologous to 0 on $X^{[2]}$.

Having Proposition 5.17, we now use Proposition 5.13 which allows us to analyze the cycle W. In the case of the cubic threefold, the proof is very short and as follows: We know that after blow-up, $X^{[2]}$ becomes isomorphic to the blow-up of a projective bundle over X along a subvariety which is a projective bundle over the surface $F(X)$. Both X and $F(X)$ have trivial Griffiths groups in all dimensions, hence it follows from the blow-up formulas that $X^{[2]}$ also has trivial Griffiths groups. Hence the cycle W which is given by Proposition 5.80 is algebraically equivalent to

0 on $X^{[2]}$. This means that the equality $\Delta_X - X \times x - Z = 0$ holds modulo algebraic equivalence. We can then apply Proposition 3.24 and conclude that X has a Chow decomposition of the diagonal. □

Let us mention the following application of Theorem 5.16 (see [61] for the proof).

Proposition 5.18 *A cubic fourfold X such that $\mathrm{Hdg}^4(X, \mathbb{Z})$ has rank 2 and discriminant not divisible by 4 has universally trivial CH_0-group.*

Here the discriminant is the discriminant of the restricted intersection pairing $\langle \, , \, \rangle_X$ on the rank 2 lattice $\mathrm{Hdg}^4(X, \mathbb{Z})$. Such cubics are said special and were studied first by Hassett [25].

5.3.2 The Case of the Cubic Threefold

Recall from Sect. 5.2 that a smooth projective threefold X with $h^{1,0}(X) = h^{3,0}(X) = 0$ has an associated principally polarized abelian variety $(J^3(X), \theta_X)$ of dimension $g = b_3(X)/2$. The minimal class $\theta_X^{g-1}/(g-1)! \in \mathrm{Hdg}^{2g-2}(J^3(X), \mathbb{Z})$ is an integral Hodge class of degree $2g - 2$ (or homology class of degree 2). For $g \geq 4$, it is not known to be algebraic. Note that Mongardi and Ottem made recent progress [40] on the similar problem for hyper-Kähler manifolds, which in some sense are close to abelian varieties.

In the case where $g = 4, 5$, it is known that the generic principally polarized abelian variety (A, θ_A) is a Prym variety $P(\widetilde{C}/C)$, and there is then a copy of the curve $\widetilde{C}$ in A, whose class is twice the minimal class. Many interesting rationally connected threefolds appear as conic bundles over a rational surface, for which the intermediate Jacobian is a Prym variety (see [8]). This applies particularly to cubic threefolds, whose intermediate Jacobian is well-known to be a Prym variety, thanks to the representation of X as a conic bundle.

Theorem 5.19 ([61]) *A smooth cubic threefold X has a decomposition of the diagonal (or has universally trivial CH_0 group) if and only if the minimal class of $J^3(X)$ is algebraic.*

Proof By Theorem 5.16, it suffices to prove the result for "cohomological decomposition" instead of "Chow decomposition". We now use Theorem 5.6. It says in particular that the algebraicity of the minimal class is a necessary condition for the existence of a cohomological decomposition of the diagonal. It remains to show that it is also sufficient. We know that $H_B^*(X, \mathbb{Z})$ has no torsion and that $H_B^4(X, \mathbb{Z})$ is algebraic, being generated by the class of a line, so the only condition to check is the existence of a universal codimension 2 cycle on X. This follows from the following statement which is taken from [58]:

Proposition 5.20 *Let X be a smooth projective threefold with $h^{1,0}(X) = h^{2,0}(X) = 0$ and such that the minimal class of $J^3(X)$ is algebraic. Then if* *furthermore there exist a smooth projective variety B and a codimension 2 cycle $Z \in CH^2(B \times X)$ such that*

$$\phi_Z : B \to J^3(X)$$

is surjective with rationally connected fibers, X has a universal codimension 2 cycle.

Proof Let $\Gamma = \sum_i n_i C_i$ be a 1-cycle in the minimal class, where $C_i \subset J^3(X)$ are curves, that we can even assume to be smooth. By the Graber-Harris-Starr theorem [23], the map ϕ_Z with rationally connected fibers has sections over each C_i (or a general translate of it). This provides lift $s_i : C_i \to B$, and thus for each C_i, we get a codimension 2 cycle $Z_i \in CH^2(C_i \times X)$ with the property that

$$\phi_{Z_i} : C_i \to J^3(X) \qquad (5.81)$$

is the natural inclusion of C_i into $J^3(X)$.

Let $g := \dim J^3(X)$. As a consequence of the fact that the class of $\Gamma = \sum_i n_i C_i$ is the minimal class, we have

$$\Gamma^{*g} = g! J^3(X). \qquad (5.82)$$

Here $*$ is the Pontryagin product on cycles of $J^3(X)$, which is defined by

$$\gamma * \gamma' = \mu_*(\gamma \times \gamma'),$$

where $\mu : J^3(X) \times J^3(X) \to J^3(X)$ is the sum map.

The meaning of Eq. (5.82) is clear if we assume that $\Gamma = C_i$ is a single curve : it says then that the sum map induces a birational map

$$j_g : C_i^{(g)} \to J^3(X)$$

is birational. In this case, constructing a universal codimension 2 cycle on $J^3(X)$ is easy: namely, starting from Z_i, we construct a codimension 2-cycle on $C_i^g \times X$ defined as $\sum_{i=1}^g ((p_i, p_X)^* Z_i$, which is clearly symmetric, hence descend to a codimension 2 cycle $Z_i^{(g)}$ on $C_i^{(g)} \times X$. One has

$$\phi_{Z_i^{(g)}} = j_g$$

which is birational by assumption, so that via (j_g, Id_X), $Z_i^{(g)}$ descends to a universal codimension 2 cycle on $J^3(X) \times X$.

The general case works similarly, using the curve $C = \sqcup_i C_i$, the cycle Z which is Z_i on the component C_i, and viewing Γ as a 1-cycle on C. $\qquad \square$

The cubic threefold satisfies the assumptions of the proposition by work of Markushevich–Tikhomirov [36]. Indeed they show that the Abel–Jacobi map on the family of elliptic curves of degree 5 in X has rationally connected fibers. What they prove is that a general such elliptic curve $E \subset X$ determines a rank 2 vector bundle $\mathcal{E}$ on X with 6 sections. The fiber of the Abel–Jacobi map passing through $[E]$ identifies to $\mathbb{P}(H^0(X, \mathcal{E}))$. Proposition 5.20 thus applies and the theorem is proved. $\square$

Let us mention the following consequence: We already mentioned that twice the minimal class is algebraic on $J^3(X)$. So if $(J^3(X), \theta_X)$ has an odd degree isogeny to $(J(C), \theta_C)$ for some genus 5 curve C, an odd multiple of the minimal class of $J^3(X)$ is algebraic, hence the minimal class itself is algebraic. This condition happens along a sublocus of codimension ≤ 3 in the moduli space of cubic threefolds, because the locus of Jacobians in $\mathcal{A}_5$ has codimension 3. Playing on this observation, we get the following:

Theorem 5.21 *There is a non-empty codimension ≤ 3 locus in the moduli space of cubic threefolds parameterizing cubic threefolds with universally trivial* CH_0*-group.*

The following questions remain open:

Question 5.22 *Does a general cubic threefold admit a universal codimension 2-cycle?*

This problem has been rephrased in [60] as computing universal unramified degree 3 cohomology of X with torsion coefficients.

Remark 5.23 The Markushevich–Tikhomirov parameterization of $J^3(X)$ with generic fiber isomorphic to $\mathbb{P}(H^0(X, \mathcal{E}))$ does not solve this problem because the fibration into projective spaces over a Zariski open set of $J^3(X)$ so constructed is a nontrivial Brauer–Severi variety over $\mathbb{C}(J^3(X))$. It does not admit a rational section.

Question 5.24 *Is the minimal class of the intermediate Jacobian $J^3(X)$ algebraic for a general cubic threefold X? Is the minimal class of a general principally polarized abelian variety of dimension 5 algebraic?*

Note that the question whether a smooth cubic of large dimension has universally trivial Chow group of zero-cycles is also completely open.

References

1. N. Addington, On two rationality conjectures for cubic fourfolds. Math. Res. Lett. **23**(1), 1–13 (2016)
2. N. Addington, R. Thomas, Hodge theory and derived categories of cubic fourfolds. Duke Math. J. **163**(10), 1885–1927 (2014)
3. M. Artin, D. Mumford, Some elementary examples of unirational varieties which are not rational. Proc. Lond. Math. Soc. **25**(3), 75–95 (1972)
4. M.F. Atiyah, F. Hirzebruch, Analytic cycles on complex manifolds. Topology **1**, 25–45 (1962)

5. A. Auel, J.-L. Colliot-Thélène, R. Parimala, Universal unramified cohomology of cubic fourfolds containing a plane, in *Brauer Groups and Obstruction Problems*. Progress in Mathematics, vol. 320 (Birkhäuser/Springer, Cham, 2017), pp. 29–55

6. L. Barbieri-Viale, Cicli di codimensione 2 su varietà unirazionali complesse, in *K-Theory (Strasbourg 1992)*. Astérisque, vol. 226 (Société mathématique de France, Paris, 1994), pp. 13–41

7. L. Barbieri-Viale, On the Deligne-Beilinson cohomology sheaves. Ann. K-Theory **1**(1), 3–17 (2016)

8. A. Beauville, Variétés de Prym et jacobiennes intermédiaires. Ann. Sci. Éc. Norm. Sup. **10**, 309–391 (1977)

9. A. Beauville, The Lüroth problem, in *Rationality Problems in Algebraic Geometry*. Springer Lecture Notes, vol. 2172 (Springer, Berlin, 2016), pp. 1–27

10. A. Beauville, R. Donagi, La variété des droites d'une hypersurface cubique de dimension 4. C. R. Acad. Sci. Paris Sér. I Math. **301**, 703–706 (1985)

11. S. Bloch, A. Ogus, Gersten's conjecture and the homology of schemes. Ann. Sci. Éc. Norm. Supér. IV. Sér. **7**, 181–201 (1974)

12. S. Bloch, V. Srinivas, Remarks on correspondences and algebraic cycles. Am. J. Math. **105**, 1235–1253 (1983)

13. L. Borisov, The class of the affine line is a zero divisor in the Grothendieck ring. J. Algebraic Geom. **27**, 203–209 (2018)

14. H. Clemens, P. Griffiths, The intermediate Jacobian of the cubic threefold. Ann. Math. Second Ser. **95**(2), 281–356 (1972)

15. J.-L. Colliot-Thélène, Quelques cas d'annulation du troisième groupe de cohomologie non ramifiée, in *Regulators*. Contemporary Mathematics, vol. 571 (American Mathematical Society, Providence, 2012), pp. 45–50

16. J.-L. Colliot-Thélène, M. Ojanguren, Variétés unirationnelles non rationnelles: au-delà de l'exemple d'Artin et Mumford. Invent. Math. **97**(1), 141–158 (1989)

17. J.-L. Colliot-Thélène, A. Pirutka, Hypersurfaces quartiques de dimension 3: non-rationalité stable. Ann. Sci. Éc. Norm. Supér. **49**(2), 371–397 (2016)

18. J.-L. Colliot-Thélène, C. Voisin, Cohomologie non ramifiée et conjecture de Hodge entière. Duke Math. J. **161**(5), 735–801 (2012)

19. A. Conte, J. Murre, The Hodge conjecture for fourfolds admitting a covering by rational curves. Math. Ann. **238**(1), 79–88 (1978)

20. S. Endrass, On the divisor class group of double solids. Manuscripta Math. **99**, 341–358 (1999)

21. W. Fulton, *Intersection Theory: Ergebnisse der Math. und ihrer Grenzgebiete 3 Folge, Band 2* (Springer, Berlin, 1984)

22. S. Galkin, E. Shinder, The Fano variety of lines and rationality problem for a cubic hypersurface (2014). arXiv:1405.5154

23. T. Graber, J. Harris, J. Starr, Families of rationally connected varieties. J. Am. Math. Soc. **16**(1), 57–67 (2003)

24. P. Griffiths, On the periods of certain rational integrals I, II. Ann. Math. **90**, 460–495 (1969); ibid. (2) 90 (1969) 496–541

25. B. Hassett, Special cubic fourfolds. Compos. Math. **120**(1), 1–23 (2000)

26. B. Hassett, A. Pirutka, Y. Tschinkel, Stable rationality of quadric surface bundles over surfaces. Acta Math. **220**(2), 341–365 (2018)

27. A. Höring, C. Voisin, Anticanonical divisors and curve classes on Fano manifolds. Pure Appl. Math. Quart. **7**(4), 1371–1393 (2011)

28. V.A. Iskovskikh, Y. Manin, Three-dimensional quartics and counterexamples to the Lüroth problem. Mat. Sb. (N.S.) **86**(128), 140–166 (1971)

29. B. Kahn, Torsion order of smooth projective surfaces. Comment. Math. Helv. **92**, 839–857 (2017)

30. M. Kerz, The Gersten conjecture for Milnor K-theory. Invent. Math. **175**(1), 1–33 (2009)

31. J. Kollár, Nonrational hypersurfaces. J. Am. Math. Soc. **8**(1995), 241–249 (1990)

32. J. Kollár, Lemma, in *Classification of Irregular Varieties*, ed. by E. Ballico, F. Catanese, C. Ciliberto. Lecture Notes in Mathematics, vol. 1515 (Springer, Berlin, 1992), p. 134

33. A. Kuznetsov, Derived categories of cubic fourfolds, in *Cohomological and Geometric Approaches to Rationality Problems*. Progress in Mathematics, vol. 282 (Birkhäuser, Boston, 2010)

34. A. Kuznetsov, Scheme of lines on a family of 2-dimensional quadrics: geometry and derived category. Math. Z. **276**(3–4), 655–672 (2014)

35. M. Larsen, V. Lunts, Motivic measures and stable birational geometry. Mosc. Math. J. **3**(1), 85–95 (2003)

36. D. Markushevich, A. Tikhomirov, The Abel–Jacobi map of a moduli component of vector bundles on the cubic threefold. J. Algebraic Geometry **10**, 37–62 (2001)

37. T. Matsusaka, On a characterization of a Jacobian variety. Memo. Coll. Sci. Univ. Kyoto. Ser. A. Math. **32**, 1–19 (1959)

38. T. Matsusaka, Algebraic deformations of polarized varieties. Nagoya Math. J. **31**, 185–245 (1968)

39. A. Merkur'ev, A. Suslin, K-cohomology of Severi-Brauer varieties and the norm residue homomorphism. Izv. Akad. Nauk SSSR Ser. Mat. **46**(5), 1011–1046, 1135–1136 (1982)

40. G. Mongardi, J. Ottem, Curve classes on irreducible holomorphic symplectic varieties (2018). arXiv:1806.09598

41. D. Mumford, Rational equivalence of 0-cycles on surfaces. J. Math. Kyoto Univ. **9**, 195–204 (1968)

42. J.P. Murre, Reduction of the proof of the non-rationality of a non-singular cubic threefold to a result of Mumford. Compos. Math. **27**, 63–82 (1973)

43. J.P. Murre, Applications of algebraic K-theory to the theory of algebraic cycles, in *Proceedings Conference on Algebraic Geometry, Sitjes 1983*. Lecture Notes in Mathematics, vol. 1124 (Springer, Berlin, 1985), pp. 216–261

44. A. Roitman, The torsion of the group of zero-cycles modulo rational equivalence. Ann. Math. **111**, 553–569 (1980)

45. S. Schreieder, Stably irrational hypersurfaces of small slopes (2018). arXiv:1801.05397

46. S. Schreieder, On the rationality problem for quadric bundles. Duke Math. J. **168**, 187–223 (2019)

47. J.-P. Serre, On the fundamental group of a unirational variety. J. Lond. Math. Soc. **34**, 481–484 (1959)

48. C. Soulé, C. Voisin, Torsion cohomology classes and algebraic cycles on complex projective manifolds. Adv. Math. **198**(1), 107–127 (2005)

49. T.A. Springer, Sur les formes quadratiques d'indice zéro. C. R. Acad. Sci. Paris **234**, 1517–1519 (1952)

50. Z. Tian, R. Zong, One-cycles on rationally connected varieties. Compos. Math. **150**(3), 396–408 (2014)

51. B. Totaro, Hypersurfaces that are not stably rational. J. Am. Math. Soc. **29**(3), 883–891 (2016)

52. V. Voevodsky, A nilpotence theorem for cycles algebraically equivalent to zero. Int. Math. Res. Not. **4**, 187–198 (1995)

53. V. Voevodsky, Motivic cohomology with $\mathbb{Z}/l$-coefficients. Ann. Math. **174**, 401–438 (2011)

54. C. Voisin, Remarks on zero-cycles of self-products of varieties, in *Moduli of Vector Bundles (Proceedings of the Taniguchi Congress on Vector Bundles)*, ed. by M.E. Decker (1994), pp. 265–285

55. C. Voisin, *Hodge Theory and Complex Algebraic Geometry II*. Cambridge Studies in Advanced Mathematics, vol. 77 (Cambridge University Press, Cambridge, 2003)

56. C. Voisin, On integral Hodge classes on uniruled and Calabi-Yau threefolds, in *Moduli Spaces and Arithmetic Geometry*. Advanced Studies in Pure Mathematics, vol. 45 (2006), pp. 43–73

57. C. Voisin, Some aspects of the Hodge conjecture. Jpn. J. Math. **2**(2), 261–296 (2007)

58. C. Voisin, Abel–Jacobi map, integral Hodge classes and decomposition of the diagonal. J. Algebraic Geom. **22**, 141–174 (2013)

59. C. Voisin, Remarks on curve classes on rationally connected varieties. Clay Math. Proc. **18**, 591–599 (2013)
60. C. Voisin, Unirational threefolds with no universal codimension 2 cycle. Invent. Math. **201**(1), 207–237 (2015)
61. C. Voisin, On the universal CH_0 group of cubic hypersurfaces. J. Eur. Math. Soc. **19**(6), 1619–1653 (2017)

Non rationalité stable sur les corps quelconques

Jean-Louis Colliot-Thélène

Abstract This is a survey on (lack of) stable rationality over arbitrary fields (including algebraically closed fields). Topics addressed include: Rationality and unirationality, R-equivalence on rational points, Chow groups of zero-cycles, Galois action on the Picard group, Brauer group, higher unramified cohomology, global differentials, specialisation method (via R-equivalence), geometrically rational surfaces, cubic hypersurfaces.

À la mémoire de Peter Swinnerton-Dyer

1 Introduction

Le présent rapport de synthèse est une version révisée des notes produites à l'occasion de l'École de printemps "Birational geometry of hypersurfaces", Palazzo Feltrinelli, Gargnano del Garda, 19–23 mars 2018.

Après les articles initiaux de C. Voisin [50] et de Colliot-Thélène et Pirutka [19], les divers articles qui ont établi la non rationalité stable de divers types de variétés classiques ont utilisé la spécialisation de Fulton du groupe de Chow des zéro-cycles. Je développe dans ce texte une remarque de [19] : on peut remplacer la spécialisation du groupe de Chow des zéro-cycles par la spécialisation de la R-équivalence.

Il y a quelques résultats nouveaux. On comparera les propositions 3.30 et 3.31 et le théorème 6.8 (ii) du présent texte avec les résultats de Totaro [49]. Signalons aussi les propositions 3.21 (c) et 7.2. La proposition 5.1 améliore un énoncé publié dans [12].

J.-L. Colliot-Thélène (✉)
CNRS et Université Paris Sud, Mathématiques, Orsay, France
e-mail: jlct@math.u-psud.fr

© Springer Nature Switzerland AG 2019

A. Hochenegger et al. (eds.), *Birational Geometry of Hypersurfaces*,
Lecture Notes of the Unione Matematica Italiana 26,
https://doi.org/10.1007/978-3-030-18638-8_2

Un schéma séparé de type fini sur un corps k est appelé une k-variété. Soit $\overline{k}$ une clôture algébrique de k. Étant donné un k-schéma X et F un corps contenant k, on note $X_F = X \times_k F$.

Une k-variété X est dite géométriquement intègre si le schéma $X_{\overline{k}}$ est intègre. Une k-variété X est dite géométriquement rationnelle si elle est géométriquement intègre et la $\overline{k}$-variété $X_{\overline{k}}$ est rationnelle, i.e. de corps des fonctions transcendant pur sur $\overline{k}$.

2 Entre rationalité et unirationalité

Lemme 2.1 *Soit k un corps. Soit X une k-variété géométriquement intègre de dimension d. Considérons les propriétés suivantes.*

(i) *La k-variété X est k-rationnelle, i.e. k-birationnelle à $\mathbf{P}_k^d$.*

(ii) *La k-variété X est stablement k-rationnelle, i.e. il existe un entier $n \geq 0$ tel que $X \times_k \mathbf{P}_k^n$ est k-birationnelle à $\mathbf{P}_k^{n+d}$.*

(iii) *La k-variété X est facteur direct d'une variété k-rationnelle, c'est-à-dire qu'il existe une k-variété Y géométriquement intègre telle que $X \times_k Y$ est k-birationnelle à un espace projectif $\mathbf{P}_k^m$.*

(iv) *La k-variété X est rétractilement k-rationnelle, c'est-à-dire qu'il existe un ouvert de Zariski non vide $U \subset X$, un ouvert de Zariski $V \subset \mathbf{P}_k^n$, et des k-morphismes $f : U \to V$ et $g : V \to U$ dont le composé $g \circ f$ est l'identité de U.*

(v) *La k-variété X est k-unirationnelle, c'est-à-dire qu'il existe $n \geq d$ et une k-application rationnelle dominante de $\mathbf{P}_k^n$ vers X.*

On a : (i) implique (ii) qui implique (iii), et (iv) implique (v). Si k est infini, (iii) implique (iv).

Démonstration. Expliquons ce dernier point, purement ensembliste. Par hypothèse, il existe un entier $n \geq 1$ et un ouvert non vide $U_0 \subset X \times_k Y$ qui est k-isomorphe à un ouvert U_1 de $\mathbf{P}_k^n$. Comme k est infini, il existe un k-point dans $U_1(k)$ et donc un point $(A, B) \in U_0(k) \subset X(k) \times Y(k)$. Soit $X_0 \subset X$ l'intersection de $X \times_k B \simeq X$ et de U_0 dans $X \times_k Y$. La composition $X_0 = X_0 \times B \subset U_0 \subset X \times Y \to X$, où la dernière flèche est la projection sur X, est l'inclusion naturelle $X_0 \subset X$. Soit U l'image réciproque de $X_0 \subset X$ dans U_0. On a alors la factorisation $X_0 \to U \to X_0$, et U est isomorphe à un ouvert non vide de $\mathbf{P}_k^n$.   □

L'hypothèse de (v) implique la même hypothèse avec $m = d$. C'est facile à établir pour k infini. Pour k quelconque, voir [41, Prop. 1.1].

On sait que (ii) n'implique pas (i), même sur $k = \mathbb{C}$ (Beauville, CT, Sansuc, Swinnerton-Dyer [4]). Sur un corps k non algébriquement clos convenable, on sait montrer que (iii) n'implique pas (ii). On ne sait pas ce qu'il en est sur $k = \mathbb{C}$. On ne sait pas si une variété rétractilement k-rationnelle est facteur direct d'une variété k-rationnelle, même si k est le corps des complexes. Pour p un nombre premier, et $PGL_p \subset GL_N$ un plongement de groupes, Saltman a montré que le quotient GL_N/PGL_p est rétractilement rationnel. On ne sait pas si cette variété est facteur direct d'une variété k-rationnelle. Pour $H \subset G$ des groupes réductifs connexes sur $\mathbb{C}$, on ne sait pas si G/H est rétractilement rationnel.

Remarque 2.2 Supposons le corps k algébriquement clos de caractéristique quelconque. Une k-variété projective et lisse X rétractilement rationnelle est rationnellement connexe par chaînes, comme est d'ailleurs toute k-variété unirationnelle. En caractéristique zéro, elle est donc séparablement rationnellement connexe, i.e. il existe un k-morphisme $f : \mathbf{P}^1 \to X$ tel que f^*T_X soit un fibré vectoriel ample. Pour ces notions, voir [34].

3 Invariants birationnels stables

3.1 R-équivalence

Définition 3.1 (Manin [37]) Soient k un corps et X une k-variété. On dit que deux k-points $A, B \in X(k)$ sont élémentairement R-liés s'il existe un ouvert $U \subset \mathbf{P}^1_k$ et un k-morphisme $h : U \to X$ tels que A, B soient dans $h(U(k))$. On dit que deux points $A, B \in X(k)$ sont R-équivalents s'il existe une chaîne $A = A_1, A_2, \ldots, A_n = B$ de k-points avec A_i et A_{i+1} élémentairement R-liés. On note $X(k)/\mathrm{R}$ le quotient de $X(k)$ par cette relation d'équivalence.

Si X est propre sur k, dans la définition ci-dessus, on peut prendre simplement $U = \mathbf{P}^1_k$.

Si $f : X \to Y$ est un k-morphisme, on a une application induite

$$X(k)/\mathrm{R} \to Y(k)/\mathrm{R}.$$

Si X est un ouvert d'un espace projectif $\mathbf{P}^n_k$, comme par deux k-points il passe une droite $\mathbf{P}^1_k$, deux k-points quelconques de X sont élémentairement R-liés, et $X(k)/\mathrm{R}$ a au plus un élément et a exactement un élément si k est infini.

Définition 3.2 Soient k un corps et X une k-variété intègre.

(i) On dit que X est R-triviale si, pour tout corps F contenant k, le quotient $X(F)/\mathrm{R}$ est d'ordre 1.

(ii) On dit que X est presque R-triviale s'il existe un ouvert de Zariski dense $U \subset X$ tel que, pour tout corps F contenant k, l'image de $U(F)$ dans $X(F)/\mathrm{R}$ est d'ordre 1.

Remarque 3.3

(a) Chacune de ces définitions implique que X possède un point k-rationnel.
(b) La condition (ii) n'implique pas la condition (i). Soit $X \subset \mathbf{P}_{\mathbb{R}}^2$ la cubique plane singulière d'équation homogène

$$y^2 t - x^2 (x - t) = 0.$$

Soit $U \subset X$ le complémentaire du point singulier réel isolé M donné par $(x, y, t) = (0, 0, 1)$. On a une désingularisation évidente $f : \mathbf{P}_{\mathbb{R}}^1 \to X$ qui est un isomorphisme au-dessus de U, et M n'est pas dans $f(\mathbf{P}^1(\mathbb{R}))$. On voit ainsi que $U(F)/\mathrm{R}$ a un élément pour tout corps F contenant $\mathbb{R}$, et donc que X est presque R-triviale, mais $X(\mathbb{R})/\mathrm{R}$ a deux éléments, donc X n'est pas R-triviale. Dans cet exemple, X et U sont des variétés $\mathbb{R}$-rationnelles.
(c) Dans la suite de ce texte on sera intéressé à la notion de presque R-trivialité dans une situation où U est lisse connexe mais où X n'est pas nécessairement lisse. On prendra garde qu'en l'absence de lissité de X la condition de presque R-trivialité n'est a priori pas très forte. Soit $Y \subset \mathbf{P}_k^n$ une k-variété quelconque et $X \subset \mathbf{P}_k^{n+1}$ le cône sur Y. Alors $X(k)/\mathrm{R} = 1$ car tout k-point de X est élémentairement R-lié au sommet $O \in X(k)$ du cône. Soit $U \subset X$ le complémentaire du sommet du cône. Si par exemple $k = \mathbb{C}$, $Y \subset \mathbf{P}_{\mathbb{C}}^2$ est une courbe elliptique E, alors $U(\mathbb{C})/\mathrm{R}$ est en bijection avec $E(\mathbb{C})/\mathrm{R} = E(\mathbb{C})$, mais l'application $U(\mathbb{C}) \to X(\mathbb{C})/\mathrm{R}$ a pour image un point. Les $\mathbb{C}$-variétés U et X ne sont pas rétractilement rationnelles.
(d) Si l'on suppose la k-variété X projective, lisse, géométriquement connexe et presque R-triviale, je ne sais pas si X est R-triviale.

Proposition 3.4 *Soient k un corps et X une k-variété intègre, de corps des fonctions $F = k(X)$ et de point générique η. Si X est presque R-triviale, alors il existe un k-point $m \in X(k)$ tel que, sur X_F, le point générique $\eta \in X(F)$ et le point $m_F \in X(F)$ soient R-équivalents.* $\square$

Définition 3.5 Soient k un corps et $f : X \to Y$ un k-morphisme. On dit que le morphisme f est R-trivial si, pour tout corps F contenant k, l'application induite $X_F(F)/\mathrm{R} \to Y_F(F)/\mathrm{R}$ est une bijection.

Un exemple est fourni par l'éclatement $X \to Y$ d'une sous-k-variété fermée lisse dans une k-variété lisse Y.

On a l'énoncé simple mais efficace suivant.

Proposition 3.6 *Soient k un corps et X une k-variété intègre. Si X est rétractilement k-rationnelle, alors il existe un ouvert non vide $U \subset X$ tel que, pour tout corps F contenant k, tout couple de points $A, B \in U(F)$ est élémentairement R-lié dans $U(F)$, et a fortiori dans $X(F)$. Si de plus k est infini, alors X est presque R-triviale.* $\square$

Démonstration. Pour établir le premier énoncé, il suffit de prendre un ouvert non vide U comme dans la définition de la k-rétractibilité rationnelle (Lemme 2.1). Pour établir le second énoncé, selon la définition 3.2, on veut $U(k) \neq \emptyset$. Pour $V \subset \mathbf{P}_k^n$ comme au (iv) du lemme 2.1, ceci résulte de $V(k) \neq \emptyset$, ce qui vaut dès que le corps k est infini. □

Théorème 3.7 *Soit k un corps de caractéristique zéro. Soient Y et X deux k-variétés projectives et lisses géométriquement intègres. S'il existe un ouvert $Y' \subset Y$, un k-morphisme dominant $Y' \to X$ et une k-section rationnelle de $Y' \to X$, alors il existe une application surjective $Y(k)/R \to X(k)/R$. En particulier, si X est rétractilement k-rationnelle, par exemple si X est stablement k-rationnelle, alors X est R-triviale.*

Démonstration. On commence par établir que, si $Y \to X$ est l'éclaté d'une sous-k-variété lisse Z fermée dans une k-variété projective et lisse X, alors l'application induite $Y(k)/R \to X(k)/R$ est une bijection.

Soit $Y \dashrightarrow X$ une k-application rationnelle dominante possédant une section rationnelle. D'après Hironaka, par éclatements successifs au-dessus de Y le long de sous-k-variétés fermées lisses, on peut obtenir un k-morphisme $W \to X$ qui couvre l'application rationnelle $Y \dashrightarrow X$. L'application induite $W(k)/R \to Y(k)/R$ est une bijection. Le k-morphisme $W \to X$ admet une section k-rationnelle. Appliquant le théorème de Hironaka à cette section, par éclatements successifs au-dessus de X le long de sous-k-variétés fermées lisses, on obtient une k-variété Z muni d'une application k-birationnelle $f : Z \to X$ et d'un k-morphisme $Z \to W$ tel que le composé $Z \to W \to X$ soit f. L'application composée induite $Z(k)/R \to W(k)/R \to X(k)/R$ est surjective, donc aussi $W(k)/R \to X(k)/R$, et l'application $W(k)/R \to Y(k)/R$ est une bijection. □

Le résultat ci-dessus est une variante de [21, Prop. 10]. Voir aussi [33, Cor. 8.6.3]. Kahn et Sujatha [33, Cor. 6.6.6, Thm. 7.3.1] obtiennent aussi des résultats au-dessus d'un corps k de caractéristique quelconque [33, Cor. 6.6.6, Thm. 7.3.1]. Voir aussi [33, §8.5, §8.6] pour une discussion du lien entre rationalité rétractile et R-trivialité.

Remarque 3.8 *C'est une question ouverte si une k-variété projective, lisse, connexe, qui est R-triviale est rétractilement k-rationnelle, et même si elle est facteur direct d'une k-variété k-rationnelle.*

Le cas particulier suivant est déjà très intéressant. Soit G un k-groupe algébrique (linéaire) réductif connexe. L'ensemble $G(k)/R$ est alors naturellement muni d'une structure de groupe. Si k est algébriquement clos, G est une variété rationnelle. Sur un corps k quelconque, un k-groupe algébrique réductif connexe G est k-unirationnel.

Pour un tel k-groupe G, les questions suivantes sont ouvertes. Sous des hypothèses particulières sur k ou sur G, elles ont fait l'objet de nombreux travaux [27].

(a) Le groupe $G(k)/R$ est-il commutatif ?
(b) Si G est R-trivial, G est-il rétractilement k-rationnel ?
(c) Si k est un corps de type fini sur le corps premier, le groupe $G(k)/R$ est-il fini ?

Remarque 3.9 Une k-variété X propre, lisse, connexe, presque R-triviale est géométriquement rationnellement connexe par chaînes (au sens de Kollár, Miyaoka, Mori [34]). Si k est de caractéristique zéro, elle est donc géométriquement séparablement rationnellement connexe : après extension du corps de base, il existe un morphisme $f : \mathbf{P}^1 \to X$ tel que f^*T_X soit un fibré vectoriel ample.

3.2   Groupe de Chow des zéro-cycles

Soit X une k-variété algébrique. On note $Z_0(X)$ le groupe abélien libre sur les points fermés de X. Pour P un tel point, de corps résiduel $k(P)$, on note $[k(P) : k]$ le degré de l'exension finie $k(P)/k$. Par linéarité, on obtient l'application degré

$$deg_k : Z_0(X) \to \mathbb{Z}$$

envoyant $\sum_i n_i P_i$ sur $\sum_i n_i [k(P_i) : k]$. Pour tout k-morphisme $f : Y \to X$ de k-variétés, on dispose d'une application induite $f_* : Z_0(X) \to Z_0(Y)$ qui est additive et envoie le point fermé $P \in X$ d'image le point fermé Q de Y sur $[k(P) : k(Q)]Q$. Cette application préserve le degré.

Si $f : C \to X$ est un k-morphisme propre de source une k-courbe normale intègre C, et si $g \in k(C)^*$ est une fonction rationnelle sur C, on dispose du zéro-cycle $f_*(div_C(g))$. On définit $CH_0(X)$ comme le quotient de $Z_0(X)$ par le sous-groupe engendré par tous les $f_*(div_C(g))$ pour tous les C, g, f comme ci-dessus.

Si la k-variété X est propre, le degré $deg_k : Z_0(X) \to \mathbb{Z}$, induit un homomorphisme $deg_k : CH_0(X) \to \mathbb{Z}$, car le degré du diviseur des zéros d'une fonction rationnelle sur une courbe propre est zéro. Plus généralement, pour $f : Y \to X$ un k-morphisme propre, l'application $f_* : Z_0(Y) \to Z_0(X)$ induit une application $f_* : CH_0(Y) \to CH_0(X)$. Pour X/k propre, on note

$$A_0(X) := \text{Ker}\,[deg_k : CH_0(X) \to \mathbb{Z}].$$

Définition 3.10 Soit X une k-variété propre. On dit que X est (universellement) CH_0-triviale si pour tout corps F contenant k, le degré

$$deg_F : CH_0(X_F) \to \mathbb{Z}$$

est un isomorphisme.

Proposition 3.11 *[Merkurjev [39, Thm. 2.11]] Soit X une k-variété propre, lisse, géométriquement intègre. Les propriétés suivantes sont équivalentes :*

(i) La k-variété X est CH_0-triviale.

(ii) X possède un zéro-cycle de degré 1 et, pour $F = k(X)$ le corps des fonctions de X, l'application $deg_F : CH_0(X_F) \to \mathbb{Z}$ est un isomorphisme.

(iii) La classe du point générique de X dans $CH_0(X_{k(X)})$ est dans l'image de l'application image réciproque $CH_0(X) \to CH_0(X_{k(X)})$. $\square$

Voir aussi [2, Lemma 1.3].

Définition 3.12 Soit $f : Y \to X$ un k-morphisme propre de k-variétés. On dit que f est un CH_0-isomorphisme (universel) si, pour tout corps F contenant k, l'application induite

$$f_{F*} : CH_0(Y_F) \to CH_0(X_F)$$

est un isomorphisme.

Lemme 3.13 ([10, p. 599] [26, Cor. 6.7]) *Soit X une variété quasi-projective régulière connexe sur un corps k. Étant donné un zéro-cycle z sur X et un ouvert de Zariski non vide $U \subset X$, il existe un zéro-cycle z' sur X dont le support est dans U et qui est rationnellement équivalent à z sur X.* $\square$

Lemme 3.14 *Soient k un corps et X une k-variété projective, lisse, géométrique- ment intègre. Si X est presque R-triviale, alors X est CH_0-triviale.*

Démonstration. Ceci résulte du lemme 3.13 (voir la démonstration de [19, Lemme 1.5]). $\square$

Proposition 3.15 ([19, Lemme 1.5]) *Soit k un corps. Soit X une k-variété projec- tive et lisse géométriquement intègre. Si X est rétractilement k-rationnelle, alors X est CH_0-triviale.*

Démonstration. Pour k un corps infini, ceci résulte immédiatement de la proposi- tion 3.6 et du lemme 3.14.

Le cas d'un corps fini $k = \mathbb{F}$ s'établit par l'argument de normes bien connu suivant. Le groupe de Galois absolu de $\mathbb{F}$ est $\hat{\mathbb{Z}} = \prod_{\ell} \mathbb{Z}_{\ell}$ où ℓ parcourt tous les nombres premiers et $\mathbb{Z}_{\ell}$ est le groupe des entiers ℓ-adiques. Fixons deux nombres premiers distincts p et q. Il existe des extensions infinies $\mathbb{F}_p$ et $\mathbb{F}_q$ de $\mathbb{F}$ telles que $\mathrm{Gal}(\mathbb{F}_p/\mathbb{F}) = \mathbb{Z}_p$ et $\mathrm{Gal}(\mathbb{F}_q/\mathbb{F}) = \mathbb{Z}_q$. Soit z un zéro-cycle sur X, de degré zéro. Le résultat sur un corps infini donne $z_{\mathbb{F}_p} = 0 \in CH_0(X_{\mathbb{F}_p})$ et $z_{\mathbb{F}_q} = 0 \in CH_0(X_{\mathbb{F}_q})$. Il existe donc une extension finie $\mathbb{F}_1/\mathbb{F}$ de degré une puissance de p et une extension finie $\mathbb{F}_2/\mathbb{F}$ de degré une puissance de q telles que $z_{\mathbb{F}_1} = 0 \in CH_0(X_{\mathbb{F}_1})$ et $z_{\mathbb{F}_2} = 0 \in CH_0(X_{\mathbb{F}_2})$. Pour un morphisme fini, on dispose d'une application norme sur les groupes de Chow, qui satisfait une propriété évidente par rapport à la restriction. Il existe donc des entiers r et s tels que $p^r.z = 0 \in CH_0(X)$ et $q^s.z = 0 \in CH_0(X)$. Par Bezout, on conclut $z = 0 \in CH_0(X)$. Pour conclure, il convient de rappeler que les estimations de Lang-Weil montrent que sur toute $\mathbb{F}$-variété géométriquement intègre, il existe un zéro-cycle de degré 1. Le degré $deg_{\mathbb{F}} : CH_0(X) \to \mathbb{Z}$ est donc un isomorphisme. $\square$

Remarque 3.16 On prendra garde qu'il existe des surfaces connexes projectives et lisses sur $\mathbb{C}$ qui sont CH_0-triviales mais sont de type général, et donc ne sont pas rationnellement connexes, et donc pas R-triviales (Voisin, [51, Cor. 2.2]; [2, Prop. 1.9]; voir déjà [5, p. 1252]). De telles surfaces satisfont $H^0(X, \Omega^i) = 0$ pour $i = 1, 2$ (Prop. 3.29 ci-dessous) mais ne satisfont pas $H^0(X, (\Omega^2)^{\otimes 2}) = 0$.

3.3   Action du groupe de Galois sur le groupe de Picard

Soit X une variété projective lisse, connexe, géométriquement rationnellement connexe sur un corps k. Si k est algébriquement clos de caractéristique zéro, tout revêtement fini galoisien étale connexe est trivial (Kollár, Miyaoka, Mori; Campana). Le groupe $\text{Pic}(X) = H^1_{\text{Zar}}(X, \mathbb{G}_m) = H^1_{\text{ét}}(X, \mathbb{G}_m)$ est donc un groupe abélien libre de type fini. Il n'y a pas là d'invariant qui détecterait la non rationalité. Si k n'est pas algébriquement clos, la situation change. On note k^s une clôture séparable de k, et $g = \text{Gal}(k^s/k)$ le groupe de Galois de k. Étant donné un module galoisien (i.e. un g-module continu discret) M, on note indifféremment $H^i(k, M)$ ou $H^i(g, M)$ (avec $i \in \mathbb{N}$) ses groupes de cohomologie.

Les invariants suivants ont tout d'abord été étudiés par Shafarevich, Manin [37], Iskovskikh, Voskresenskiĭ.

Théorème 3.17 *Soient k un corps, k^s une clôture séparable, et $g = \text{Gal}(k^s/k)$. On note $X^s = X \times_k k^s$. Soient X et Y deux k-variétés propres, lisses, géométriquement intègres.*

(a) Si X est k-birationnelle à Y, alors il existe des g-modules de permutation de type fini P_1 et P_2 et un isomorphisme de modules galoisiens

$$\text{Pic}(X^s) \oplus P_1 \simeq \text{Pic}(Y^s) \oplus P_2,$$

et l'on a $H^1(k, \text{Pic}(X^s)) \simeq H^1(k, \text{Pic}(Y^s))$.

(b) Supposons car.$(k) = 0$. Si X est CH_0-triviale, alors le module galoisien $\text{Pic}(X^s)$ est un facteur direct d'un g-module de permutation de type fini, et, pour tout corps F contenant k, on a $H^1(F, \text{Pic}(X \times_F F^s)) = 0$.

(c) Supposons car.$(k) = 0$. Si X est rétractilement k-rationnelle, alors le module galoisien $\text{Pic}(X^s)$ est un facteur direct d'un g-module de permutation de type fini, et, pour tout corps F contenant k, on a $H^1(F, \text{Pic}(X \times_F F^s)) = 0$.

Démonstration. Un g-module de permutation de type fini est un g-module qui est libre sur $\mathbb{Z}$ et admet une $\mathbb{Z}$-base globalement respectée par g. Pour l'élégante démonstration de (a) due à L. Moret-Bailly, voir [22, Prop. 2A1, p. 461]. Pour (b), voir l'appendice de [28]. La proposition 3.15 et (b) donnent (c).   □

Voici un exemple d'application.

Proposition 3.18 *Soit* k *un corps de caractéristique différente de 3. Soient* $a, b, c, d \in k^*$. *Si aucun des quotients* $ab/cd, ac/bd, ad/bc$ *n'est un cube dans* k^*, *alors la* k-*surface cubique lisse* $X \subset \mathbf{P}_k^3$ *d'équation*

$$ax^3 + by^3 + cz^3 + dt^3 = 0$$

n'est pas stablement k-*rationnelle, et si de plus* k *est de caractéristique zéro, elle n'est pas rétractilement* k-*rationnelle.*

Le cas $a = b = c = 1$ est traité dans le livre de Manin [37] et dans [22]. Le cas général est fait dans un article de CT-Kanevsky-Sansuc. On a une réciproque : si ab/cd est un cube, et X possède un k-point, alors X est k-birationnelle à $\mathbf{P}_k^2$.

Ceci est un cas particulier du théorème suivant de Swinnerton-Dyer, complétant un travail de B. Segre.

Théorème 3.19 ([48]) *Soient* k *un corps et* $X \subset \mathbf{P}_k^3$ *une surface cubique lisse. Les conditions suivantes sont équivalentes :*

(i) La k-*surface* X *est* k-*rationnelle.*
(ii) La k-*surface* X *contient un* k-*point et contient un* S_2, *ou un* S_3 *ou un* S_6.

Un S_n dans une telle surface est un ensemble de n droites contenues dans $X^s = X \times_k k^s \subset \mathbf{P}_{k^s}^3$ gauches deux à deux, globalement invariant sous l'action du groupe de Galois $\mathrm{Gal}(k^s/k)$. La démonstration de ce théorème utilise la théorie des systèmes linéaires à points bases, un ancêtre des méthodes de rigidité en géométrie complexe.

L'invariant $\mathrm{Pic}(X^s)$ est très intéressant pour les surfaces géométriquement rationnelles. Mais si X est une hypersurface lisse de degré $d \leq n$ dans $\mathbf{P}_k^n$ avec $n \geq 4$, alors $\mathbb{Z} = \mathrm{Pic}(\mathbf{P}_{k^s}^n) \xrightarrow{\sim} \mathrm{Pic}(X^s)$ et ceci ne donne aucune information sur l'éventuelle non k-rationalité de X.

3.4 Groupe de Brauer

Soit F un corps, F^s une clôture séparable, $g_F = \mathrm{Gal}(F^s/F)$. On note $\mathrm{Br}(F) = H^2(g_F, F_s^*)$ le groupe de Brauer de F. Soient k un corps et X une k-variété. On note $\mathrm{Br}(X) = H_{\text{ét}}^2(X, \mathbb{G}_m)$ le groupe de Brauer de X [30]. Si X est lisse et intègre, de corps des fonctions $k(X)$, on a une injection $\mathrm{Br}(X) \hookrightarrow \mathrm{Br}(k(X))$ (Auslander-Goldman, Grothendieck).

Le groupe $\mathrm{Br}(X)$ est un invariant k-birationnel des k-variétés propres et lisses, réduit à $\mathrm{Br}(k)$ pour $X = \mathbf{P}_k^n$ (Grothendieck [30] pour la torsion première à la caractéristique; Hoobler, Gabber, Česnavičius en général).

Rappelons l'énoncé bien connu [22, (1.5.0)] :

Proposition 3.20 *Soient k un corps, k^s une clôture séparable, et $g = \mathrm{Gal}(k_s/k)$. Pour toute k-variété projective, lisse, géométriquement connexe X on a une suite exacte*

$$0 \to \mathrm{Pic}(X) \to \mathrm{Pic}(X^s)^g \to \mathrm{Br}(k) \to \mathrm{Ker}[\mathrm{Br}(X) \to \mathrm{Br}(X^s)] \to H^1(k, \mathrm{Pic}(X^s))$$
$$\to H^3(k, k_s^*).$$

Si $X(k) \neq \emptyset$, on a un isomorphisme $\mathrm{Pic}(X) \overset{\simeq}{\to} \mathrm{Pic}(X^s)^g$ et on a la suite exacte

$$0 \to \mathrm{Br}(k) \to \mathrm{Ker}[\mathrm{Br}(X) \to \mathrm{Br}(X^s)] \to H^1(k, \mathrm{Pic}(X^s)) \to 0.$$

On voit donc que l'invariant "module galoisien $\mathrm{Pic}(X^s)$ à addition près de module de permutation" du théorème 3.17 raffine le sous-groupe "algébrique" $\mathrm{Ker}[\mathrm{Br}(X) \to \mathrm{Br}(X^s)]$ du groupe de Brauer de X.

La question du calcul de l'image de $\mathrm{Br}(X) \to \mathrm{Br}(X^s)$, qui est dans le groupe des invariants $\mathrm{Br}(X^s)^g$, et déjà de ce groupe des invariants, est délicate, c'est un problème "arithmétique", nous n'en parlerons pas ici.

Soit k un corps algébriquement clos de caractéristique zéro. Soit X une k-variété projective, lisse, connexe, de dimension d. Si X est rationnellement connexe, on a $H^1_{\text{ét}}(X, \mu_{\ell^n}) = \mathrm{Pic}(X)[\ell^n] = 0$ pour tout entier $n > 0$. On a donc $H^1_{\text{ét}}(X, \mathbb{Z}_\ell(1)) = 0$. Comme le groupe $\mathrm{Pic}(X)$ est sans torsion, il est égal au groupe de Néron-Severi $\mathrm{NS}(X)$, qui est donc lui-même sans torsion. La suite de Kummer en cohomologie étale (où $\mathbb{G}_m \to \mathbb{G}_m$ est donné par $x \mapsto x^{\ell^n}$) :

$$1 \to \mu_{\ell^n} \to \mathbb{G}_m \to \mathbb{G}_m \to 1$$

donne des suites exactes courtes compatibles (en n) :

$$0 \to \mathrm{Pic}(X)/\ell^n \to H^2_{\text{ét}}(X, \mu_{\ell^n}) \to \mathrm{Br}(X)[\ell^n] \to 0.$$

En passant à la limite projective, on obtient une suite exacte

$$0 \to \mathrm{NS}(X) \otimes \mathbb{Z}_\ell \to H^2_{\text{ét}}(X, \mathbb{Z}_\ell(1)) \to T_\ell(\mathrm{Br}(X)) \to 0.$$

Un module de Tate $T_\ell(A) = \mathrm{limproj}_n A[\ell^n]$ est toujours sans torsion. Ainsi, pour X comme ci-dessus, le groupe $H^2_{\text{ét}}(X, \mathbb{Z}_\ell(1))$ est sans torsion. Si $k = \mathbb{C}$, les théorèmes de comparaison entre cohomologie étale et cohomologie de Betti donnent alors $H^1_{Betti}(X, \mathbb{Z}) = 0$ et $H^2_{Betti}(X, \mathbb{Z})_{tors} = 0$.

Proposition 3.21 *Soit k un corps algébriquement clos de caractéristique zéro, et soit X une k-variété projective et lisse rationnellement connexe.*

(a) Si $k = \mathbb{C}$, alors $\mathrm{Br}(X) \xrightarrow{\sim} H^3_{Betti}(X(\mathbb{C}), \mathbb{Z})_{tors}$.

(b) En général, $\mathrm{Br}(X)$ est un groupe fini isomorphe à $\bigoplus_\ell H^3_{\text{ét}}(X, \mathbb{Z}_\ell(1))_{tors}$.

(c) Si F est un corps qui contient k, l'application naturelle $\mathrm{Br}(X) \to \mathrm{Br}(X_F)/\mathrm{Br}(F)$ est un isomorphisme.

Démonstration. Par passage à la limite inductive dans les suites exactes

$$0 \to \mathrm{Pic}(X)/\ell^n \to H^2_{\text{ét}}(X, \mu_{\ell^n}) \to \mathrm{Br}(X)[\ell^n] \to 0,$$

on obtient les suites exactes

$$0 \to \mathrm{Pic}(X) \otimes \mathbb{Q}_\ell/\mathbb{Z}_\ell \to H^2_{\text{ét}}(X, \mathbb{Q}_\ell/\mathbb{Z}_\ell(1)) \to \mathrm{Br}(X)\{\ell\} \to 0.$$

Le groupe $H^2_{\text{ét}}(X, \mathbb{Q}_\ell/\mathbb{Z}_\ell(1))$ est extension du groupe fini $H^3_{\text{ét}}(X, \mathbb{Z}_\ell(1))_{tors}$ par un quotient du groupe divisible $H^2_{\text{ét}}(X, \mathbb{Q}_\ell(1))$. Pour presque tout premier ℓ, on a $H^3_{\text{ét}}(X, \mathbb{Z}_\ell(1))_{tors} = 0$. Ainsi le groupe de Brauer de X est une extension du groupe fini $\bigoplus_\ell H^3_{\text{ét}}(X, \mathbb{Z}_\ell(1))_{tors}$ par un groupe divisible. Si X est rationnellement connexe, d'après [10, Prop. 11] il existe un entier $N > 0$ qui annule $A_0(X_F)$ pour tout corps F contenant k. Ceci implique que le groupe divisible est annulé par N, donc est nul. Ceci établit (a) et (b). Pour (c), considérons les inclusions $k \subset F \subset \overline{F}$. On a les applications naturelles

$$\mathrm{Br}(X) \to \mathrm{Br}(X_F) \to \mathrm{Br}(X_{\overline{F}}).$$

Fixons un k-point P de $X(k)$, et considérons les sous-groupes de ces divers groupes formés des éléments nuls en P. On a alors les applications

$$\mathrm{Br}^P(X) \to \mathrm{Br}^P(X_F) \to \mathrm{Br}^P(X_{\overline{F}}).$$

D'après (b), la composée est un isomorphisme.

Sur un corps algébriquement clos k de caractéristique zéro, le k-schéma de Picard d'une k-variété projective, lisse, connexe X est extension du k-groupe constant de type fini $\mathrm{NS}_{X/k}$ (groupe de Néron-Severi) par la k-variété abélienne $\mathrm{Pic}^0_{X/k}$ (variété de Picard). Pour X/k projective, lisse, rationnellement connexe, on a $\mathrm{Pic}^0_{X/k} = 0$ car l'espace tangent à l'origine est $H^1(X, \mathcal{O}_X) = 0$. Comme en outre on a $X(k) \neq \emptyset$, pour tout corps L contenant k, on a $\mathrm{Pic}(X_L) = \mathrm{Pic}_{X/k}(L)$. On a donc $\mathrm{Pic}(X) = \mathrm{Pic}(X_F) = \mathrm{Pic}(X_{\overline{F}})$, et la proposition 3.20 donne

$$\mathrm{Br}(F) = \mathrm{Ker}[\mathrm{Br}(X_F) \to \mathrm{Br}(X_{\overline{F}})]$$

et donc $\mathrm{Br}^P(X_F) \hookrightarrow \mathrm{Br}^P(X_{\overline{F}})$. On a donc $\mathrm{Br}^P(X) = \mathrm{Br}^P(X_F)$. $\qquad\square$

Remarque 3.22 Soit $X/\mathbb{C}$ une variété projective et lisse de dimension d. Si X est rétractilement rationnelle ou R-triviale, on a $\mathrm{Br}(X) = 0$. Comme on va voir au paragraphe 3.5, ceci vaut sous l'hypothèse plus générale que X est CH_0-triviale. Pour une telle variété, on a donc $H^1_{Betti}(X, \mathbb{Z}) = 0$, $H^2_{Betti}(X, \mathbb{Z})_{tors} = 0$ et $H^3_{Betti}(X, \mathbb{Z})_{tors} = 0$. Par diverses dualités, ceci implique $H^{2d-1}_{Betti}(X, \mathbb{Z}) = 0$. En utilisant la décomposition de la diagonale, un argument de correspondance et des désingularisations, C. Voisin établit ces résultats. Elle établit aussi $H^{2d-2}_{Betti}(X, \mathbb{Z})_{tors} = 0$.

3.5 *Cohomologie non ramifiée*

Références : [9, 17, 29, 44].

Soit A un anneau de valuation discrète, K son corps des fractions, κ_A son corps résiduel. Soit $n > 1$ un entier inversible dans κ_A. Pour tous entiers $j \in \mathbb{Z}$ et $i \geq 1$, on dispose d'une application résidu entre groupes de cohomologie galoisienne

$$\partial_A : H^i(K, \mu_n^{\otimes j}) \to H^{i-1}(\kappa_A, \mu_n^{\otimes j-1}).$$

Pour k un corps, X une k-variété lisse connexe de corps des fonctions $k(X)$ et $n > 0$ entier premier à la caractéristique de k, on définit

$$H^i_{nr}(X/k, \mu_n^{\otimes j}) := \bigcap_{x \in X^{(1)}} \mathrm{Ker}[\partial_x : H^i(k(X), \mu_n^{\otimes j}) \to H^{i-1}(k(x), \mu_n^{\otimes j-1})].$$

Ici x parcourt les points de codimension 1 de X et $k(x)$ est le corps résiduel de l'anneau de valuation discrète $\mathcal{O}_{X,x}$, anneau local de X au point x.

Soit $\mathcal{H}^i_X(\mu_n^{\otimes j})$ le faisceau Zariski sur X associé au préfaisceau $U \mapsto H^i_{\text{ét}}(U, \mu_n^{\otimes j})$.

La conjecture de Gersten pour la cohomologie étale (théorème de Bloch-Ogus–Gabber, voir [24]) implique :

- Le faisceau Zariski $\mathcal{H}^i_X(\mu_n^{\otimes j})$ est un sous-faisceau du faisceau constant défini par $H^i(k(X), \mu_n^{\otimes j})$.
- Pour X/k lisse connexe, on a $H^0(X, \mathcal{H}^i_X(\mu_n^{\otimes j})) = H^i_{nr}(X/k, \mu_n^{\otimes j})$.
- Si X est propre, lisse, connexe, ce groupe coïncide avec

$$H^i_{nr}(k(X)/k, \mu_n^{\otimes j}) := \bigcap_A \mathrm{Ker}[\partial_A : H^i(k(X), \mu_n^{\otimes j}) \to H^{i-1}(\kappa_A, \mu_n^{\otimes j-1})],$$

où A parcourt tous les anneaux de valuation discrète contenant k et de corps des fractions $k(X)$. Ceci peut aussi s'écrire :

$$H^i_{nr}(k(X)/k, \mu_n^{\otimes j}) := \bigcap_A H^i_{\text{ét}}(A, \mu_n^{\otimes j}) \subset H^i(k(X), \mu_n^{\otimes j}).$$

Sans restriction sur la caractéristique de k, on définit aussi

$$\text{Br}_{nr}(k(X)/k) := \bigcap_A \text{Br}(A) \subset \text{Br}(k(X))$$

où A parcourt tous les anneaux de valuation discrète contenant k et de corps des fractions $k(X)$. Pour X/k connexe, propre et lisse, un résultat de pureté assure $\text{Br}(X) = \text{Br}_{nr}(k(X)/k) \subset \text{Br}(k(X))$.

- Pour tout entier $m \geq 1$, on montre :

$$H^i(k, \mu_n^{\otimes j}) \xrightarrow{\simeq} H^i_{nr}(k(\mathbf{P}^m_k)/k, \mu_n^{\otimes j}).$$

Pour X/k connexe, propre et lisse, on a les propriétés suivantes :

$$H^1_{nr}(k(X)/k, \mathbb{Z}/n) = H^1_{\text{ét}}(X, \mathbb{Z}/n) \subset H^1(k(X), \mathbb{Z}/n)$$

$$H^2_{nr}(k(X)/k, \mu_n) = \text{Br}(X)[n] \subset \text{Br}(k(X))[n],$$

où $\text{Br}(X)[n]$ est le sous-groupe de n-torsion du groupe de Brauer de X.

Les groupes $H^i_{nr}(X/k, \mu_n^{\otimes j})$ sont fonctoriels contravariants pour les k-morphismes quelconques de k-variétés lisses, connexes. Ceci résulte de la formule

$$H^i_{nr}(X/k, \mu_n^{\otimes j}) = H^0(X, \mathcal{H}^i_X(\mu_n^{\otimes j})).$$

En particulier pour toute k-variété X propre, lisse, géométriquement connexe, pour tout corps F contenant k on dispose d'accouplements

$$X(F) \times H^i_{nr}(X_F/F, \mu_n^{\otimes j}) \to H^i(F, \mu_n^{\otimes j})$$

qui, par fonctorialité et utilisation de $H^i(F, \mu_n^{\otimes j}) \xrightarrow{\simeq} H^i_{nr}(F(\mathbf{P}^1_F)/F, \mu_n^{\otimes j})$, passent au quotient par la R-équivalence :

$$X_F(F)/\text{R} \times H^i_{nr}(X_F/F, \mu_n^{\otimes j}) \to H^i(F, \mu_n^{\otimes j}).$$

Pour toute F-variété propre, lisse, connexe X, on dispose d'accouplements

$$CH_0(X_F) \times H^i_{nr}(X_F/F, \mu_n^{\otimes j}) \to H^i(F, \mu_n^{\otimes j}).$$

Ceci vaut plus généralement dans le cadre des modules de cycles de Rost [39, Cor. 2.9]. Pour l'énoncé analogue pour l'accouplement avec le groupe de Brauer, y compris sa p-torsion en caractéristique p, voir [3].

La proposition 3.6 donne alors :

Proposition 3.23 *Soit X une k-variété propre, lisse, géométriquement connexe. Si X est presque R-triviale, pour tous i, j, $n > 1$ premier à la caractéristique de k et tout corps F contenant k, on a*

$$H^i(F, \mu_n^{\otimes j}) \overset{\simeq}{\to} H^i_{nr}(X_F/F, \mu_n^{\otimes j})$$

et $\mathrm{Br}(F) = \mathrm{Br}(X_F)$.

Démonstration. Il suffit de considérer le cas $F = k$. Pour le voir, il suffit de monter sur le corps $K = k(X)$ est d'utiliser le fait (Prop. 3.4) que le point générique est R-équivalent à un point de $X(k) \subset X(k(X))$. □

Corollaire 3.24 *Soit X une k-variété propre, lisse, géométriquement connexe. Si X est rétractilement k-rationnelle, alors, pour tous i, j, et tout corps F contenant k, on a*

$$H^i(F, \mu_n^{\otimes j}) \overset{\simeq}{\to} H^i_{nr}(X_F/F, \mu_n^{\otimes j}) \overset{\simeq}{\to} H^i_{nr}(F(X)/F, \mu_n^{\otimes j})$$

et $\mathrm{Br}(F) = \mathrm{Br}(X_F) = \mathrm{Br}_{nr}(F(X)/F)$.

Démonstration. Via la proposition 3.6, ceci résulte de l'énoncé précédent, sauf dans le cas d'un corps k fini. Dans ce cas on monte sur des extensions finies de k suffisamment grosses et on utilise un argument de norme. □

Proposition 3.25 *Si une k-variété propre, lisse et géométriquement connexe X est CH_0-triviale, alors, pour tous i, j, et tout corps F contenant k, on a*

$$H^i(F, \mu_n^{\otimes j}) \overset{\simeq}{\to} H^i_{nr}(X_F/F, \mu_n^{\otimes j}) \overset{\simeq}{\to} H^i_{nr}(F(X)/F, \mu_n^{\otimes j})$$

et $\mathrm{Br}(F) = \mathrm{Br}(X_F) = \mathrm{Br}_{nr}(F(X)/F)$.

Via la Proposition 3.15, cet énoncé implique le précédent, mais sa démonstration est un peu plus élaborée, car elle passe par l'accouplement avec le groupe de Chow.

On a des énoncés analogues aux précédents en remplaçant les groupes de cohomologie galoisienne $H^\bullet(F, \mu_n^{\otimes \bullet})$ des corps F par les modules de cycles de Rost des corps F, par exemples par les groupes $K_i^M(F)$ ($i \in \mathbb{N}$) de K-théorie de Milnor des corps. Voir à ce sujet l'article de Merkurjev [39], qui montre que la trivialité universelle de tous les invariants non ramifiés de tous les modules de cycles de Rost pour une k-variété propre, lisse, connexe donnée X est équivalente au fait que cette variété est CH_0-triviale [39, Thm. 2.11].

3.6   Calcul du groupe de Brauer non ramifié

Pour X une $\mathbb{C}$-variété propre, lisse, rationnellement connexe, la formule

$$\mathrm{Br}(X) \xrightarrow{\sim} H^3_{Betti}(X(\mathbb{C}), \mathbb{Z})_{tors}$$

donnée ci-dessus est théoriquement satisfaisante. Mais en pratique, quand on se donne une variété concrète, elle a tendance à être singulière. Il faudrait la désingulariser, ce qui en grande dimension est difficile, en outre il faut ensuite calculer sur un modèle projectif et lisse le groupe $H^3_{Betti}(X(\mathbb{C}), \mathbb{Z})_{tors}$. C'est ce qu'avaient fait Artin et Mumford [1] pour une variété de dimension 3 fibrée en coniques sur le plan projectif complexe.

Dans [17], avec M. Ojanguren, on a donné une autre façon d'établir $\mathrm{Br}_{nr}(\mathbb{C}(X)/\mathbb{C}) \neq 0$ pour des fibrations en coniques X sur le plan complexe.

Si X est une conique lisse C sur un corps k, sans k-point rationnel, de corps des fonctions $k(C)$, la suite exacte de la Proposition 3.20 se spécialise en une suite exacte

$$0 \to \mathbb{Z}/2 \to \mathrm{Br}(k) \to \mathrm{Br}(C) \to 0.$$

Si car.$(k) \neq 2$ et C est donnée par l'équation homogène $x^2 - ay^2 - bz^2 = 0$, le noyau de $\mathrm{Br}(k) \to \mathrm{Br}(C)$—qui est aussi le noyau de $\mathrm{Br}(k) \to \mathrm{Br}(k(C))$ car $\mathrm{Br}(C)$ s'injecte dans $\mathrm{Br}(k(C))$ puisque C est lisse—est engendré par la classe de l'algèbre de quaternions (a, b). Ce résultat remonte à Witt, et fut étendu aux variétés de Severi-Brauer par F. Châtelet.

Le point de vue "birationnel" adopté par Ojanguren et moi dans [17] est dans ses grandes lignes le suivant. On a une variété projective et lisse (non explicite) X sur $\mathbb{C}$ munie d'une fibration $p : X \to S = \mathbf{P}^2_{\mathbb{C}}$ dont la fibre générique est une conique $C/\mathbb{C}(S)$ sans point rationnel (i.e. la fibration n'a pas de section rationnelle). La fibration dégénère le long d'une union finie de courbes intègres $D_i \subset S$. On dispose de la classe $\alpha \in \mathrm{Br}(\mathbb{C}(S))$ de la conique générique, d'ordre 2, non nulle, qui engendre le noyau de l'application

$$\mathrm{Br}(\mathbb{C}(S)) \to \mathrm{Br}(\mathbb{C}(X)).$$

Comme $S = \mathbf{P}^2_{\mathbb{C}}$, on a $\mathrm{Br}(S) = 0$, et l'application résidu en tous les points de codimension 1 de S et le théorème de pureté pour le groupe de Brauer d'une variété lisse donnent une injection

$$\delta : \mathrm{Br}(\mathbb{C}(S)) \hookrightarrow \oplus_{x \in S^{(1)}} H^1(\mathbb{C}(x), \mathbb{Q}/\mathbb{Z}).$$

La classe α a un nombre fini de résidus non triviaux, correspondant aux points où la fibration dégénère. Sous des hypothèses sur la dégénérescence, on exhibe une autre classe $\beta \in \mathrm{Br}(\mathbb{C}(S))$ dont le résidu total $\delta(\beta)$ est non nul et formé d'un sous-ensemble propre des $\delta_x(\alpha)$. Par comparaison avec les résidus aux points de codimension 1 de X, qui implique une discussion précise de la situation aux points de codimension 2 de S, mais ne requiert pas la connaissance d'un modèle projectif et lisse explicite de X, ceci assure que β devient non ramifié dans $\mathrm{Br}(\mathbb{C}(X))$, et assure par ailleurs que β n'est pas dans $\mathbb{Z}/2 = \mathrm{Ker}[\mathrm{Br}(\mathbb{C}(S)) \to \mathrm{Br}(\mathbb{C}(X))]$. Ainsi $\mathrm{Br}_{nr}(\mathbb{C}(X)) \neq 0$, et la variété X n'est pas rétractilement rationnelle, ni même CH_0-triviale.

3.7 Calcul de la cohomologie non ramifiée de degré supérieur

Pour X une variété propre, lisse, rationnellement connexe sur $\mathbb{C}$, on ne dispose pas pour les invariants cohomologiques supérieurs $H^i_{nr}(\mathbb{C}(X)/\mathbb{C}, \mathbb{Q}/\mathbb{Z})$, $i \geq 3$, d'un analogue des différents énoncés de la proposition 3.21. De fait il est peu probable que ces invariants soient constants dans une famille projective et lisse de telles variétés (voir [23] pour une discussion).

Pour X comme ci-dessus, on a un certain nombre de résultats intéressants en degré $i = 3$, et quelques résultats en degré $i > 3$. Je renvoie ici le lecteur aux travaux [23], [50] et [11]. Dans [23], avec C. Voisin, on établit un lien entre $H^3_{nr}(\mathbb{C}(X)/\mathbb{C}, \mathbb{Q}/\mathbb{Z})$ et la conjecture de Hodge entière pour les cycles de codimension 2.

Le cas des hypersurfaces cubiques dans $\mathbf{P}^n_{\mathbb{C}}$, $n \geq 4$, a été particulièrement étudié, en particulier par C. Voisin [50], voir aussi [11]. Pour de telles hypersurfaces, on a $H^3_{nr}(\mathbb{C}(X)/\mathbb{C}, \mathbb{Q}/\mathbb{Z}) = 0$. Pour F un corps contenant $\mathbb{C}$, on sait que l'application

$$H^3(F, \mathbb{Q}/\mathbb{Z}) \to H^3_{nr}(F(X)/F, \mathbb{Q}/\mathbb{Z})$$

est un isomorphisme pour $n \geq 5$. Pour $n = 4$, la question est ouverte et importante (voir le corollaire 3.24 et la proposition 3.25).

Comme on a vu ci-dessus, le point de vue birationnel adopté dans [17] pour revisiter l'exemple d'Artin et Mumford repose sur le fait que sur un corps k de caractéristique différente de 2, et pour une conique C sur k d'équation homogène $x^2 - ay^2 - bz^2 = 0$, avec $a, b \in k^*$, le noyau de l'application $H^2(k, \mathbb{Z}/2) \to H^2(k(C), \mathbb{Z}/2)$ est d'ordre au plus 2, engendré par la classe de l'algèbre de quaternions (a, b), qui est aussi la classe du cup produit de la classe $a \in k^*/k^{*2} = H^1(k, \mathbb{Z}/2)$ et de la classe $b \in k^*/k^{*2} = H^1(k, \mathbb{Z}/2)$. Sur un corps k de caractéristique différente de 2, pour tout entier $n \geq 1$, et pour $a_1, \ldots, a_n \in k^*$, la n-forme de Pfister $<< a_1, \ldots, a_n >>$ est la forme quadratique en 2^n variables définie par $< 1, -a_1 > \otimes \cdots \otimes < 1, -a_n >$. De telles formes ont la propriété qu'elles sont hyperboliques dès qu'elles sont isotropes. On appelle voisine de Pfister d'une n-forme de Pfister ψ une sous-forme ϕ de ψ de rang strictement plus grand que 2^{n-1}. Les quadriques définies par une forme de Pfister et par une voisine de cette forme sont stablement k-birationnellement équivalentes. C'est ainsi le cas de la conique d'équation $x^2 - ay^2 - bz^2 = 0$ et de la quadrique de $\mathbf{P}_k^3$ d'équation $x^2 - ay^2 - bz^2 + abt^2 = 0$.

Une généralisation de la propriété remarquable des coniques décrites ci-dessus est le théorème suivant.

Théorème 3.26 *Soit k un corps de caractéristique différente de 2. Soit $<< a_1, \ldots, a_n >>$ une n-forme de Pfister. Soit $Q \subset \mathbf{P}^{2^n - 1}$ la quadrique lisse qu'elle définit. Le noyau de l'application naturelle de groupes de cohomologie galoisienne*

$$H^n(k, \mathbb{Z}/2) \to H^n(k(Q), \mathbb{Z}/2)$$

est engendré par le cup-produit $(a_1) \cup \cdots \cup (a_n)$, et il est non nul si et seulement si la forme de Pfister est anisotrope, i.e. si la quadrique Q n'a pas de k-point.

Ce théorème fut établi pour $n = 3$ en 1974 par Arason, avant les résultats spectaculaires de Merkurjev et Suslin en 1982. Il fut établi pour $n = 4$ par Jacob et Rost en 1989, et obtenu pour tout n par Orlov, Vishik et Voevodsky [42] en 2007 comme conséquence des travaux de Voevodsky sur la conjecture de Milnor.

Remarque 3.27 Le résultat d'Arason avait été précédé par un résultat analogue d'Arason et Pfister (voir [17, Thm. 1.7]) pour les groupes de Witt du corps des fonctions d'une quadrique (voisine) de Pfister, résultat fin mais nettement plus élémentaire qu'on peut aussi utiliser pour établir beaucoup des énoncés de non rationalité (voir l'appendice de [17] et le livre [41]).

Une fois le point de vue birationnel adopté dans [17], il est devenu clair comment étendre les résultats de non rationalité en dimension supérieure. Dans [17], avec Ojanguren, nous construisons des variétés a priori singulières Y munies d'une fibration sur $\mathbf{P}_{\mathbb{C}}^3$ dont la fibre générique est définie par une voisine d'une 3-forme de Pfister anisotrope $<< a_1, a_2, b_3 c_3 >>$ sur le corps $\mathbb{C}(\mathbf{P}^3)$, telle que la classe $\beta = (a_1, a_2, b_3) \in H^3(\mathbb{C}(\mathbf{P}^3), \mathbb{Z}/2)$ soit non nulle, car ramifiée sur $\mathbf{P}_{\mathbb{C}}^3$, différente de $\alpha = (a_1, a_2, b_3 c_3) \in H^3(\mathbb{C}(\mathbf{P}^3), \mathbb{Z}/2)$, car les ramifications sur $\mathbf{P}_{\mathbb{C}}^3$ diffèrent, et dont l'image $\beta_{\mathbb{C}(X)}$ est dans $H_{nr}^3(\mathbb{C}(X)/\mathbb{C}, \mathbb{Z}/2)$ car la ramification de β est "mangée" par celle de α. Comme on a

$$\beta \notin \{0, \alpha\} \subset H^3(\mathbb{C}(\mathbf{P}^3), \mathbb{Z}/2),$$

le théorème 3.26, dans le cas $n = 3$ (Arason) assure alors $\beta_{\mathbb{C}(X)} \neq 0$.

Pour accomplir le programme, il faut trouver les éléments $a_1, a_2, b_3, c_3 \in \mathbb{C}(\mathbf{P}^3)$. On les obtient dans [17] comme des produits d'un nombre assez grand de formes linéaires.

Dans [46], Schreieder a réussi à faire des constructions analogues sur $\mathbf{P}_{\mathbb{C}}^n$ pour tout n (les a_i, b_j, c_j faisant ici intervenir des formes homogènes de degré 2 sur $\mathbf{P}^n$). Le théorème 3.26 donne alors des variétés X munies d'une fibration sur $\mathbf{P}_{\mathbb{C}}^n$ dont la fibre générique est une (voisine d'une) n-quadrique de Pfister et qui satisfont $H_{nr}^n(\mathbb{C}(X), \mathbb{Z}/2) \neq 0$, et qui donc ne sont pas rétractilement rationnelles.

On trouve d'autres utilisations de ces idées dans des travaux d'E. Peyre et de A. Asok.

Remarque 3.28 Soit k un corps. Soit $Q \subset \mathbf{P}_k^n$, $n \geq 2$ une quadrique lisse. L'application $\mathrm{Br}(k) \to \mathrm{Br}(Q) = \mathrm{Br}_{nr}(k(Q)/k)$ est surjective. Pour $i \geq 3$, et car.$(k) \neq 2$, le conoyau de

$$H^i(k, \mathbb{Q}_2/\mathbb{Z}_2(i-1)) \to H^i_{nr}(k(Q)/k, \mathbb{Q}_2/\mathbb{Z}_2(i-1))$$

a été étudié par Kahn, Rost, Sujatha. Pour $i = 3$, ils ont montré que l'application est surjective, sauf peut-être si Q est définie par une forme d'Albert $< -a, -b, ab, c, d, -cd >$.

3.8 Différentielles

L'énoncé suivant est établi par Totaro dans [49].

Proposition 3.29 *Soit X une k-variété projective et lisse connexe sur un corps k. Si X est CH_0-triviale, alors $H^0(X, \Omega^i) = 0$ pour tout entier $i > 0$.*

La démonstration utilise des applications cycles à valeurs dans diverses théories cohomologiques, et des arguments de correspondances. $\square$

Proposition 3.30 *Soit k un corps. Soit Φ un foncteur contravariant de la catégorie des k-schémas vers la catégorie des ensembles. Supposons que, pour toute k-variété lisse intègre U, la flèche $\Phi(U) \to \Phi(\mathbf{P}_U^1)$ induite par la projection $\mathbf{P}_U^1 \to U$ soit une bijection. Soit X une k-variété propre, intègre, génériquement lisse. Si X est presque R-triviale, alors il existe un ouvert lisse non vide $U \subset X$ tel que*

$$Im(\Phi(X) \to \Phi(U)) = Im(\Phi(k) \to \Phi(U)),$$

la flèche $\Phi(k) \to \Phi(U)$ étant donnée par la projection $U \to \mathrm{Spec}(k)$.

Démonstration. Soit $U \subset X$ un ouvert lisse, et soit $g : \mathbf{P}^1 \times_k U \to X$ un k-morphisme. Soient f_1, f_2 deux sections de la projection $p : \mathbf{P}^1 \times_k U \to U$. Alors les applications $\Phi(X) \to \Phi(U)$ définies par $(g \circ f_1)^*$ et $(g \circ f_2)^*$ coïncident. En effet, pour tout $\alpha \in \Phi(X)$, on a $g^*(\alpha) = p^*(\beta)$, et donc $f_i^* \circ g^*(\alpha) = f_i^* \circ p^*(\beta) = \beta$ pour $i = 1, 2$.

Soit $F = k(X)$ le corps des fonctions de X et soit $\eta \in X$ le point générique. Comme X est presque R-triviale, il existe $n \in X(k)$ tel que, sur X_F, les points $\eta \in X_F(F)$ et $n_F \in X_F(F)$ sont R-équivalents. Comme X est propre sur k, ceci implique qu'il existe un ouvert lisse non vide $U \subset X$ et une famille finie de k-morphismes $f_i : \mathbf{P}^1 \times U \to X$, $i = 0, \ldots, s$, tels que $f_0(0, u) = u$, que $f_s(1, u) = n$, et que $f_i(1, u) = f_{i+1}(0, u)$ pour $0 \leq i < s$. L'énoncé résulte alors de ce qui précède. $\square$

Proposition 3.31 *Soient k un corps infini et X une k-variété propre et lisse, géométriquement connexe. Si X est presque R-triviale, alors $H^0(X, (\Omega^i)^{\otimes m}) = 0$ pour tout $i > 0$ et tout $m > 0$.*

Démonstration. Pour toute k-variété lisse intègre U, tout entier $i > 0$, tout entier $m > 0$, la flèche de restriction

$$H^0(U, (\Omega^i)^{\otimes m}) \to H^0(\mathbf{P}^1_U, (\Omega^i)^{\otimes m})$$

est un isomorphisme, comme on voit en utilisant la formule donnant le faisceau des différentielles sur un produit de k-variétés, et en utilisant le fait que sur la droite projective, on a $\Omega^1_{\mathbf{P}^1} = O_{\mathbf{P}^1}(-2)$, et donc, pour tout $m > 0$, toute section de $(\Omega^1_{\mathbf{P}^1})^{\otimes m}$ est nulle. On applique alors la proposition précédente au foncteur $U \mapsto H^0(U, (\Omega^i)^{\otimes m})$, et on utilise le fait que, pour la k-variété lisse X, l'application de restriction $H^0(X, (\Omega^i)^{\otimes m}) \to H^0(U, (\Omega^i)^{\otimes m})$ est injective. $\qquad\square$

3.9 Composantes connexes réelles

Théorème 3.32 *Soit $\mathbb{R}$ le corps des réels. Soit X une $\mathbb{R}$-variété projective, lisse, géométriquement connexe, de dimension d. Soit $s \geq 0$ le nombre de composantes connexes de $X(\mathbb{R})$.*

(a) L'entier s est un invariant birationnel stable.

(b) Si X est rétractilement $\mathbb{R}$-rationnelle, alors $s = 1$.

(c1) Pour $s \geq 1$, on a $CH_0(X)/2 = (\mathbb{Z}/2)^s$.

(c2) Si deux points de $X(\mathbb{R})$ sont rationnellement équivalents sur X, alors ils appartiennent à la même composante connexe de $X(\mathbb{R})$.

(d1) Si $s = 0$, pour tout entier $m \geq d + 1$, on a $H^m_{nr}(\mathbb{R}(X)/\mathbb{R}, \mathbb{Z}/2) = 0$.

(d2) Si $s \geq 1$, pour tout entier $m \geq d + 1$, on a $H^m_{nr}(\mathbb{R}(X)/\mathbb{R}, \mathbb{Z}/2) = (\mathbb{Z}/2)^s$.

(e) Si X est géométriquement rationnellement connexe, deux points de $X(\mathbb{R})$ sont R-équivalents si et seulement si ils sont dans la même composante connexe.

Démonstration. Pour (a), il suffit de voir que si $U \subset X$ est un ouvert de Zariski dont le complémentaire est de codimension au moins 2 dans X, alors $U(\mathbb{R}) \subset X(\mathbb{R})$ induit une bijection sur les composantes connexes. On utilise alors le fait qu'une k-application rationnelle d'une k-variété lisse dans une k-variété propre est définie en dehors d'un fermé de codimension au moins 2. Sous l'hypothèse de (b), il existe un ouvert de Zariski $U \subset X$ tel que l'image de $U(\mathbb{R})$ dans $X(\mathbb{R})$ soit formée de points directement R-liés sur X, donc dans la même composante connexe de $X(\mathbb{R})$. Comme pour la $\mathbb{R}$-variété lisse X tout point de $X(\mathbb{R})$ est limite de points de $U(\mathbb{R})$, ceci suffit à conclure que $X(\mathbb{R})$ est connexe. Pour (c), voir CT-Ischebeck [16]. Pour (d), voir CT-Parimala [18]. L'énoncé (e) fut établi par Kollár [35]. $\qquad\square$

En dimension $d = 1$, tous ces énoncés remontent à Witt. C'est B. Segre [47] qui le premier remarqua que les surfaces cubiques lisses X sur $\mathbb{R}$, qui sont toutes $\mathbb{R}$-unirationnelles, ne sont pas $\mathbb{R}$-rationnelles si $X(\mathbb{R})$ n'est pas connexe.

4   Surfaces géométriquement rationnelles

Théorème 4.1 (Enriques, Manin, Iskovskikh, Mori) *Soient k un corps et X une k-surface projective, lisse, géométriquement rationnelle. Alors X est k-birationnelle à une telle k-surface de l'un des deux types suivants :*

(i) Surface de del Pezzo de degré d, avec $1 \le d \le 9$.

(ii) Surface X munie d'une fibration relativement minimale $X \to D$, où D est une conique lisse, la fibre générique est une conique lisse, et toutes les fibres sont des coniques avec au plus un point singulier.

Rappelons que les surfaces de del Pezzo de degré 3 sont les surfaces cubiques lisses.

C'est une question ouverte depuis longtemps si une k-surface comme dans le théorème, dès qu'elle possède un k-point, est k-unirationnelle. C'est connu pour les surfaces cubiques. Une réponse affirmative impliquerait que les variétés complexes de dimension 3 fibrées en coniques sur le plan $\mathbf{P}^2_{\mathbb{C}}$ sont unirationnelles, ce qui est une question ouverte encore plus connue.

Une question générale (Sansuc et l'auteur) sur les surfaces du type ci-dessus est : dans quelle mesure le module galoisien $\mathrm{Pic}(X^s)$ (qui est un groupe abélien de type fini) et les objets qui lui sont attachés contrôlent-ils la géométrie et l'arithmétique de X ? En particulier, a-t-on la réciproque du théorème 3.17 (c) :

Question 1 (CT-Sansuc 1977) : *Si $X(k) \ne \emptyset$ et le module galoisien $\mathrm{Pic}(X^s)$ est un facteur direct d'un module de permutation, la k-surface X est-elle facteur direct birationnel d'un espace projectif $\mathbf{P}^n_k$?*

La K-théorie algébrique (idées de S. Bloch, théorème de Merkurjev-Suslin) a permis d'établir pour ces surfaces, sans analyse cas par cas, la réciproque du Théorème 3.17 (b).

Théorème 4.2 ([8]) *Soit X une k-surface projective, lisse, géométriquement rationnelle, possédant un zéro-cycle de degré 1. Si le module galoisien $\mathrm{Pic}(X^s)$ est un facteur direct d'un module de permutation, alors X est CH_0-triviale.*

Voici quelques rappels de CT-Sansuc [22]. Soit X une k-surface projective, lisse, géométriquement rationnelle. Soit $\mathrm{Pic}(X^s)$ le module galoisien défini par le groupe de Picard. C'est le groupe des caractères $\hat{S}$ d'un k-tore S. Pour tout k-tore T, on a une suite exacte de groupes abéliens

$$0 \to H^1(k, T) \to H^1_{\text{ét}}(X, T) \to \mathrm{Hom}_g(\hat{T}, \hat{S}) \to H^2(k, T) \to H^2_{\text{ét}}(X, T),$$

où la cohomologie est la cohomologie étale. Si X possède un k-point, la flèche $H^2(k, T) \to H^2_{\text{ét}}(X, T)$ a une rétraction, donc on a une suite exacte

$$0 \to H^1(k, T) \to H^1_{\text{ét}}(X, T) \to \text{Hom}_g(\hat{T}, \hat{S}) \to 0.$$

On appelle torseur universel sur X un torseur $\mathcal{T} \to X$ sous le k-tore S dont la classe dans $H^1_{\text{ét}}(X, S)$ a pour image l'identité dans $\text{Hom}_g(\hat{S}, \hat{S})$. Si X possède un k-point $P \in X(k)$, il existe un torseur universel, et on peut le fixer (à automorphisme de S-torseur près) en demandant que sa fibre en P soit triviale, ce qui équivaut au fait qu'il existe un k-point de $\mathcal{T}$ d'image P dans X.

Un torseur universel $\mathcal{T}$ sur une k-surface projective, lisse, géométriquement rationnelle est une k-variété géométriquement rationnelle (ouverte) de dimension $2 + \text{rang}(\text{Pic}(X^s))$.

La question 1 aurait une réponse affirmative s'il en était ainsi de la question suivante :

Question 2 (CT-Sansuc 1977). *Sur une k-surface projective et lisse géométriquement rationnelle X, les torseurs universels $\mathcal{T}$ avec un k-point sont-ils k-rationnels ?*

Ceci a été établi pour les surfaces fibrées en coniques au-dessus de $\mathbf{P}^1_k$ avec au plus 4 fibres géométriques non lisses. C'est d'ailleurs ce qui a mené aux exemples de variétés stablement k-rationnelles non k-rationnelles ([4], voir ci-dessous). La question est déjà ouverte pour les k-surfaces cubiques $X \subset \mathbf{P}^3_k$ d'équation

$$x^3 + y^3 + z^3 + at^3 = 0$$

avec $a \notin k^{*3}$.

En 1977, Sansuc et moi avons établi que si Y est une compactification lisse d'un k-torseur universel $\mathcal{T}$ alors $\text{Pic}(\overline{Y})$ est un g-module de permutation, et $\text{Br}(Y)/\text{Br}(k) = 0$. Pour tester l'éventuelle non rationalité des torseurs universels sur les surfaces géométriquement rationnelles, on peut essayer de calculer les invariants cohomologiques supérieurs $H^i_{nr}(k(\mathcal{T})/k, \mathbb{Q}/\mathbb{Z}(i-1))$. Dans sa thèse, Yang Cao a établi le théorème suivant, qui s'applique en particulier aux k-surfaces cubiques lisses, et donne $H^3(k, \mathbb{Q}/\mathbb{Z}(2) = H^3_{nr}(k(\mathcal{T})/k, \mathbb{Q}/\mathbb{Z}(2))$ pour $\mathcal{T}$ torseur universel avec un k-point au-dessus de $X \subset \mathbf{P}^3_k$ surface cubique d'équation

$$x^3 + y^3 + z^3 + at^3 = 0.$$

Théorème 4.3 *[6] Soit X une surface projective, lisse connexe, géométriquement rationnelle sur un corps k. Si X n'est pas k-birationnelle à une surface de del Pezzo k-minimale de degré 1, et si $\mathcal{T}$ est un torseur universel sur X avec un k-point, $H^3_{nr}(k(\mathcal{T})/k, \mathbb{Q}/\mathbb{Z}(2))/H^3(k, \mathbb{Q}/\mathbb{Z}(2))$ est un groupe de torsion 2-primaire.*

Soit k un corps de caractéristique différente de 2 possédant une extension finie $L = k[t]/P(t)$ de degré 3, de clôture galoisienne K/k de groupe $\mathfrak{S}_3$, et soit $k(\sqrt{a})$ l'extension discriminant. Dans [4], on a montré que la surface géométriquement rationnelle d'équation affine $y^2 - az^2 = P(x)$ est stablement k-rationnelle mais non k-rationnelle. Ceci fut utilisé dans [4] pour donner des exemples de variétés de dimension 3 sur $\mathbb{C}$ qui sont stablement rationnelles mais non rationnelles.

Hassett avait soulevé la question si de tels exemples de k-surfaces stablement k-rationnelles non k-rationnelles existent sur un corps k parfait dont la clôture algébrique est procyclique, par exemple sur un corps fini.

Le théorème suivant, qu'on confrontera avec la question 1 ci-dessus, n'admet pour l'instant qu'une démonstration extrêmement calculatoire, passant par l'analyse (résultat du travail de plusieurs auteurs sur une grande période de temps) de toutes les actions possibles du groupe de Galois absolu sur le groupe de Picard géométrique des surfaces de del Pezzo de degré 3, 2, 1, ce qui implique des groupes de Weyl de type E_6, E_7, E_8.

Théorème 4.4 ([14]) *Soient k un corps et X une k-surface projective, lisse, géométriquement rationnelle. Supposons que X possède un point k-rationnel et que X soit déployée par une extension cyclique de k. Si X n'est pas k-rationnelle, alors il existe une extension finie séparable k'/k telle que $\mathrm{Br}(X_{k'})/\mathrm{Br}(k') = H^1(k', \mathrm{Pic}(X^s)) \neq 0$, et alors X n'est pas stablement k-rationnelle, ni même rétractilement k-rationnelle.*

Soit X déployée par une extension cyclique de k. Si l'on suppose $\mathrm{Br}(X_{k'})/\mathrm{Br}(k') = H^1(k', \mathrm{Pic}(X^s)) = 0$ pour toute extension séparable k' de k, on peut montrer (Endo-Miyata) que $\mathrm{Pic}(X^s)$ est un facteur direct d'un module de permutation. Tout torseur universel $\mathcal{T}$ avec un k-point est alors k-birationnel à $X \times_k S$. Si de tels torseurs universels étaient automatiquement k-rationnels (question 2 ci-dessus), l'hypothèse $\mathrm{Br}(X_{k'})/\mathrm{Br}(k') = 0$ pour tout k'/k fini impliquerait que X est facteur direct d'une k-variété k-rationnelle.

5   Hypersurfaces cubiques

5.1   *Rationalité, unirationalité, CH_0-trivialité*

Soit $X \subset \mathbf{P}_k^n$ avec $n \geq 3$ une hypersurface cubique lisse avec $X(k) \neq \emptyset$. On sait que X est k-unirationnelle (B. Segre, Manin, Kollár). Si X contient une k-droite, alors X est k-unirationnelle de degré 2. Pour car.$(k) \neq 3$, il en est ainsi de l'hypersurface cubique de Fermat $X_n \subset \mathbf{P}_k^n$ définie par l'équation :

$$\sum_{i=0}^{n} x_i^3 = 0.$$

Je renvoie à [2] pour plus de rappels et des références à la littérature.

Pour tout $n = 2m + 1 \geq 3$ impair, il existe des hypersurfaces cubiques lisses $X \subset \mathbf{P}_k^n$ qui sont k-rationnelles. C'est le cas de celles qui contiennent une paire globalement k-rationnelle d'espaces linéaires Π_1, Π_2, de dimension m, chacun défini sur une extension au plus quadratique séparable de k, et sans point commun.

Il en est ainsi de l'hypersurface cubique de Fermat X_{2m+1}. Elle possède une paire globalement k-rationnelle de sous-espaces linéaires de dimension m gauches l'un à l'autre, à savoir

$$x_0 + jx_1 = x_2 + jx_3 = \cdots = x_{2m} + jx_{2m+1} = 0$$

et son conjugué (j est une racine primitive cubique de 1).

Pour simplifier, supposons dans la suite de ce paragraphe $k = \mathbb{C}$, et considérons des hypersurfaces cubiques lisses $X \subset \mathbf{P}_\mathbb{C}^n$, $n \geq 3$.

Toute hypersurface cubique $X \subset \mathbf{P}_\mathbb{C}^n$, $n \geq 3$ contient une droite, et est donc est unirationnelle de degré 2.

Si une hypersurface cubique est aussi unirationnelle de degré impair, alors elle est CH_0-triviale et tous les invariants de type cohomologie non ramifiée sont universellement triviaux. On ne sait pas si X est alors rétractilement rationnelle.

Un théorème fameux de Clemens et Griffiths dit qu'aucune X dans $\mathbf{P}_\mathbb{C}^4$ n'est rationnelle. Pour $n = 2m$ pair quelconque on ne connaît aucune X dans $\mathbf{P}_\mathbb{C}^{2m}$ qui soit rationnelle, ou même rétractilement rationnelle. Mais par ailleurs il n'en existe à ce jour aucune dont on sache qu'elle n'est pas rétractilement rationnelle.

C. Voisin a montré que sur une union dénombrable de fermés de codimension 3 de leur espace de modules, les hypersurfaces cubiques $X \subset \mathbf{P}_\mathbb{C}^4$ correspondantes sont CH_0-triviales.

On connaît des classes d'hypersurfaces cubiques de $\mathbf{P}_\mathbb{C}^5$ qui sont unirationnelles de degré impair (voir [31, Cor. 40]). Dans [12] on en donne dans $\mathbf{P}_\mathbb{C}^n$ pour tout n de la forme $6m - 1, 6m + 1, 6m + 3$. Elles sont "presque diagonales" mais plus générales que l'hypersurface de Fermat.

Hassett, et d'autres, ont décrit des sous-variétés de l'espace de modules des hypersurfaces cubiques de $\mathbf{P}_\mathbb{C}^5$ dont les hypersurfaces correspondantes sont rationnelles (outre celles contenant deux plans gauches congugués). Ces sous-variétés sont contenues dans une union dénombrable de diviseurs "spéciaux" de l'espace de modules.

C. Voisin [51] a montré que sur beaucoup de ces diviseurs spéciaux, les hypersurfaces cubiques correspondantes de $\mathbf{P}_\mathbb{C}^5$ sont CH_0-triviales.

Soit X une k-variété projective, lisse, connexe. Suivant [51], on définit la CH_0-dimension essentielle $\delta(X)$ de X comme la plus petite dimension d'une $\mathbb{C}$-variété Y projective, lisse, connexe munie d'un morphisme $Y \to X$ tel que pout tout corps F contenant $\mathbb{C}$, l'application induite $CH_0(Y_F) \to CH_0(X_F)$ soit surjective.

C. Voisin [51] a montré que pour les hypersurfaces cubiques lisses $X \subset \mathbf{P}_\mathbb{C}^n$ *très générales* avec $n = 5$ ou $n \geq 4$ pair, si $\delta(X) < \dim(X)$, alors $\delta(X) = 0$, i.e. X est CH_0-triviale.

5.2 Hypersurfaces cubiques presque diagonales

Dans [12] je donne en toute dimension des classes explicites d'hypersurfaces cubiques lisses complexes qui sont CH_0-triviales. Certains des résultats valent sur un corps non algébriquement clos, comme on va le voir. Voici une variation sur la proposition 3.5 de l'article [12].

Proposition 5.1 *Soient k un corps et X une k-variété projective et lisse telle que $H^1(X, O_X) = 0$, possédant un k-point. S'il existe une courbe Γ/k projective, lisse, connexe, avec un k-point, et un k-morphisme $\Gamma \to X$ tels que, pour tout corps F, l'application induite $CH_0(\Gamma_F) \to CH_0(X_F)$ soit surjective, alors, pour tout corps F, l'application $\deg_F : CH_0(X_F) \to \mathbb{Z}$ est un isomorphisme, en d'autres termes la k-variété X est CH_0-triviale.*

Démonstration. Soit J la jacobienne de Γ. Pour tout corps F, on a $A_0(\Gamma_F) = J(F)$. Notons $K = k(X)$ le corps des fonctions de X. L'hypothèse $H^1(X, O_X) = 0$ implique que la variété d'Albanese de X est triviale. Un point de $J(k(X))$ définit une k-application rationnelle de X dans J, donc un k-morphisme de X dans J car une application rationnelle d'une variété lisse dans une variété abélienne est partout définie (A. Weil). Mais comme la variété d'Albanese de X est triviale, tout tel morphisme est constant. On a donc $J(k) = J(k(X))$. Ainsi l'image de $A_0(\Gamma_K)$ dans $A_0(X_K)$ est dans l'image de l'application composée

$$J(k) = A_0(\Gamma) \to A_0(X) \to A_0(X_K).$$

Par hypothèse, l'application $A_0(\Gamma_K) \to A_0(X_K)$ est surjective. Ainsi la restriction $CH_0(X) \to CH_0(X_K)$ est surjective, et en particulier la classe du point générique η de X, qui définit un point de $X(K)$, a une classe dans $CH_0(X_K)$ qui est dans l'image de $CH_0(X)$. D'après la proposition 3.11 (Merkurjev), ceci assure que la k-variété X est CH_0-triviale.   □

Remarque 5.2 Soit $k = \mathbb{C}$. L'énoncé ci-dessus implique que si $CH_0(X) = \mathbb{Z}$, alors $\delta(X) \leq 1$ implique $\delta(X) = 0$. R. Mboro [38] a établi l'énoncé suivant. Supposons $CH_0(X) = \mathbb{Z}$, $H^2_{Betti}(X, \mathbb{Z})_{tors} = 0$ et $H^3_{Betti}(X, \mathbb{Z}) = 0$. Alors $\delta(X) \leq 2$ implique $\delta(X) = 0$.

Théorème 5.3 *Soit k un corps infini, de caractéristique différente de 3. Soient $f(x, y, z) \in k[x, y, z]$ et $g(u, v) \in k[u, v]$ des formes cubiques non singulières. Soit $X \subset \mathbf{P}^4_k$ l'hypersurface cubique lisse d'équation*

$$f(x, y, z) - g(u, v) = 0.$$

Faisons les hypothèses suivantes, qui sont satisfaites si k est un corps algébriquement clos :

(a) Il existe $a \in k^*$ tel que la surface cubique S de $\mathbf{P}^3_k$ donnée par l'équation $f(x, y, z) - at^3 = 0$ soit une surface k-rationnelle et que la courbe Γ de $\mathbf{P}^2_k$ d'équation $g(u, v) - at^3 = 0$ possède un k-point.

(b) L'hypersurface X contient une k-droite.

Alors l'hypersurface X est CH_0-triviale.

Démonstration. L'hypothèse (b) implique que X est k-unirationnelle de degré 2, ce qui implique $2A_0(X_F) = 0$ pour tout corps F contenant k. On a une application rationnelle dominante, de degré 3, de $S \times \Gamma$ vers X, qui envoie le produit des variétés affines $f(x, y, z) - a = 0$ et $g(u, v) - a = 0$ vers le point de coordonnées homogènes $(x, y, z, u, v) \in X \subset \mathbf{P}_k^4$. En utilisant le fait que S est k-rationnelle, on montre que pour tout corps F contenant k, il existe un k-morphisme $f : \Gamma \to X$ tel que pour tout corps F contenant k, on ait $3CH_0(X_F) \subset f_*(CH_0(\Gamma_F))$. Comme on a $2A_0(X_F) = 0$, on en déduit $CH_0(X_F) \subset f_*(CH_0(\Gamma_F))$. La proposition 5.1 donne alors que X est CH_0-triviale. $\square$

L'énoncé ci-dessus se généralise en dimension supérieure [12, Prop. 3.7]. Les arguments de [12, Prop. 3.7 (i)] et la proposition ci-dessus permettent d'établir que, sur tout corps k de caractéristique différente de 3, pour tout entier $n \geq 3$, impair ou non, l'hypersurface cubique de Fermat $X \subset \mathbf{P}_k^n$, $n \geq 3$, est CH_0-triviale. Pour tout $n = 2m \geq 4$, et $k = \mathbb{Q}$, c'est ainsi une question ouverte si cette hypersurface est rétractilement $\mathbb{Q}$-rationnelle, ou même stablement $\mathbb{Q}$-rationnelle.

Sur le corps $k = \mathbb{C}$, la méthode ci-dessus et des énoncés d'unirationalité plus ou moins classiques permettent d'établir l'énoncé général suivant.

Théorème 5.4 ([12, Thm. 3.8]) *Toute hypersurface cubique lisse $X \subset \mathbf{P}_{\mathbb{C}}^n$ de dimension au moins 2 dont l'équation est donnée par une forme $\sum_i \Phi_i$, où les Φ_i sont à variables séparées et chacune a au plus 3 variables, est CH_0-triviale.*

6 Spécialisation

6.1 Spécialisation de la **R**-équivalence et de l'équivalence rationnelle sur les zéro-cycles

L'énoncé suivant est "bien connu". Pour une démonstration détaillée pour $\mathcal{X}/A$ projectif, on consultera la note de D. Madore [36]. Voir aussi [33, Cor. 6.7.2].

Théorème 6.1 *Soit A un anneau de valuation discrète, K son corps des fractions, k son corps résiduel. Soit $\mathcal{X}$ un A-schéma propre, $X = \mathcal{X} \times_A K$ la fibre générique et $Y = \mathcal{X} \times_A k$ la fibre spéciale. L'application de réduction $X(K) = \mathcal{X}(A) \to Y(k)$ induit une application $X(K)/\mathrm{R} \to Y(k)/\mathrm{R}$.*

Démonstration. (Esquisse) Soit $\mathbf{P}_K^1 \to X$ un K-morphisme. Par éclatements successifs de points fermés sur $\mathbf{P}_A^1$, on obtient $Z \to \mathbf{P}_A^1$ et un A-morphisme $Z \to \mathcal{X}$ étendant l'application rationnelle. La fibre Z_k est géométriquement un arbre, dont les composantes sont des droites projectives. Comme on voit par récurrence sur le nombre d'éclatements, la réunion T des composantes de Z_k obtenues par éclatement de k-points forme elle-même un arbre formé de droites projectives $\mathbf{P}_k^1$, dont les intersections deux à deux sont égales à un unique k-point, et tout k-point de Z_k est contenu dans T. Les points 0 et ∞ de $\mathbf{P}^1(K) = Z(K)$ s'étendent en des sections

s_0 et s_∞ de $Z \to \mathrm{Spec}(A)$. Les spécialisations de ces sections au-dessus de $\mathrm{Spec}(k)$ sont des k-points de Z_k, qui sont dans le sous-arbre T. Les images de 0_K et ∞_K dans $Y(k)$ sont donc des points R-équivalents sur Y.   □

Soient A un anneau de valuation discrète et $\pi \in A$ une uniformisante. Soient $\mathcal{X}$ un A-schéma projectif et plat, $\mathcal{X}_K$ la fibre générique et $\mathcal{X}_k$ la fibre spéciale.

Étant donné un point fermé $P \in \mathcal{X}_K$, notons $\tilde{P}$ son adhérence dans $\mathcal{X}$. C'est un A-schéma fini. On a une immersion fermée $\tilde{P} \times_A \mathrm{Spec}(k) \hookrightarrow \mathcal{X}_k$. On associe à ce A-schéma fini une combinaison linéaire à coefficients entiers de points fermés de $\mathcal{X}_k$. Les coefficients sont définis par les longueurs évidentes. Le zéro-cycle obtenu sur $\mathcal{X}_k$ peut aussi être vu comme le zéro-cycle associé au k-schéma découpé par $\pi = 0$ sur $\tilde{P}$.

Ceci définit une application linéaire $Z_0(\mathcal{X}_K) \to Z_0(\mathcal{X}_k)$. On vérifie que ce processus est fonctoriel covariant en les morphismes (propres) de A-schémas projectifs et plats.

Le théorème suivant est un cas particulier d'un théorème de Fulton pour les groupes de Chow de cycles de dimension quelconque.

Théorème 6.2 (Fulton) *Soit A un anneau de valuation discrète, K son corps des fractions, k son corps résiduel, π une uniformisante. Soit $\mathcal{X}$ un A-schéma projectif et plat, $X = \mathcal{X} \times_A K$ la fibre générique et $Y = \mathcal{X} \times_A k$ la fibre spéciale. Il existe un unique homomorphisme de spécialisation*

$$CH_0(X) \to CH_0(Y)$$

qui associe à la classe d'un point fermé P de X d'adhérence $\tilde{P} \subset \mathcal{X}$ la classe du zéro-cycle associé au diviseur de Cartier découpé par $\pi = 0$ sur $\tilde{P}$.

C'est énoncé au début du §20.3 de [25], avec référence au §6.2 et au théorème 6.3. On part d'une suite exacte facile

$$CH_1(Y) \to CH_1(\mathcal{X}/A) \to CH_0(X) \to 0$$

établie au §1.8.

On utilise ensuite un homomorphisme de Gysin $i^! : CH_1(\mathcal{X}/A) \to CH_0(Y)$ introduit au §6.2. Il est démontré au Théorème 6.3 que le composé des applications $CH_1(Y) \to CH_1(\mathcal{X}/A) \to CH_0(Y)$ est nul, en utilisant le fait que Y est un diviseur de Cartier principal sur $\mathcal{X}$. Ceci induit un homomorphisme de spécialisation $CH_0(X) \to CH_0(Y)$.

Autant que je puisse voir, le §2, et la Proposition 2.6 de [25], qui utilisent un homomorphisme de Gysin $i^* : CH_1(\mathcal{X}/A) \to CH_0(Y)$, suffisent pour établir ces résultats. Ceci utilise un théorème fondamental, le Théorème 2.4 de [25]. La Définition 2.3 de [25] donne précisément la description de l'homomorphisme de spécialisation donnée dans l'énoncé ci-dessus.

Remarque 6.3 On peut facilement ramener la démonstration de l'énoncé ci-dessus au cas où $\mathcal{X}$ est une A-courbe plate, projective, connexe, régulière. Mais ce cas-là ne semble pas plus facile que le cas général, si la A-courbe n'est pas lisse. Or c'est tout le point : si la fibre spéciale $Y = \mathcal{X}_k$ est une union de diviseurs lisses Y_i/k (non principaux), on n'a pas en général de flèches $CH_0(X) \to CH_0(Y_i)$ qui par somme donneraient la flèche $CH_0(X) \to CH_0(Y)$. Par ailleurs, si Y/k n'est pas lisse, la flèche naturelle $\mathrm{Pic}(Y) \to CH_0(Y)$ n'est a priori ni injective ni surjective.

6.2 Non rationalité stable par spécialisation singulière

Les deux théorèmes suivants, qui généralisent un argument de C. Voisin [50], sont établis dans [19] en utilisant la spécialisation de Fulton des groupes de Chow (des zéro-cycles). Ils ont déjà été discutés dans divers textes, en particulier dans [43] et [44]. On développe ici la remarque 1.19 de [19] : on donne une démonstration qui utilise la spécialisation de la R-équivalence, plus simple à établir que celle du groupe de Chow des zéro-cycles. On comparera l'énoncé suivant avec [19, Thm. 1.12].

Théorème 6.4 *Soient A un anneau de valuation discrète, K son corps des fractions, et k son corps résiduel. Soient $\mathcal{X}$ un A-schéma projectif et plat, $X = \mathcal{X} \times_A K$ la fibre générique et $Y = \mathcal{X} \times_A k$ la fibre spéciale. Supposons X/K lisse et géométriquement intègre et Y/k géométriquement intègre. Supposons que $Y(k)$ est Zariski dense dans Y et qu'il existe une résolution des singularités projective $f : Z \to Y$ qui est un CH_0-isomorphisme. Sous l'une des hypothèses suivantes :*

(a) la K-variété X est R-triviale,
(b) la K-variété X est rétractilement K-rationnelle,

la k-variété Z est CH_0-triviale.

Démonstration. On procède au début comme dans [19, Thm. 1.12]. L'anneau local de $\mathcal{X}$ au point générique η de Y est un anneau de valuation discrète. On note B son hensélisé (ou son complété). Soit F son corps des fractions. La flèche $A \to B$ est un homomorphisme local, induisant $k \to k(Y)$ sur les corps résiduels et $K \to F$ sur les corps de fractions. On considère le B-schéma $\mathcal{X} \times_A B$. Sa fibre spéciale est $Y \times_k k(Y)$, qui admet la désingularisation $Z \times_k k(Y) \to Y \times_k k(Y)$. Le $k(Y)$-morphisme $Z \times_k k(Y) \to Y \times_k k(Y)$ est CH_0-trivial. Soit $U \subset Y_{lisse}$ un ouvert tel que $f^{-1}(U) \to U$ soit un isomorphisme. Soit $P \in U(k)$. Soit $M \in Z(k)$ son image réciproque sur $f^{-1}(U)$. Par Hensel, le point générique $\eta \in Y(k(Y))$ et le point $P_{k(Y)}$ se relèvent en des F-points de $X \times_K F$. Sous l'hypothèse (a), deux tels points sont R-équivalents sur $X \times_K F = \mathcal{X} \times_A F$. Pour K de caractéristique zéro, le théorème 3.7 (qui utilise le théorème de Hironaka) montre que l'hypothèse (b) implique l'hypothèse (a). Sans restriction sur la caractéristique, sous l'hypothèse (b), d'après la Proposition 3.6, il existe $W \subset X$ un ouvert Zariski de X tel que l'image de $W(F)$ dans $X_K(F)/R$ est réduite à un élément. Soit $T \subset X$ le complémentaire de W dans X. L'adhérence de T dans $\mathcal{X}$ ne contient pas Y. Comme $Y(k)$ est Zariski dense dans Y on peut choisir $P \in U(k) \subset Y(k)$ hors de cette adhérence. Le point $\eta \in Y(k(Y))$ et le point $P_{k(Y)}$ se relèvent alors en des

F-points de W, qui donc sont R-équivalents sur $X \times_K F$. Par spécialisation de la R-équivalence (Théorème 6.1), les points η et $P_{k(Y)}$ sont R-équivalents sur $Y_{k(Y)}$. Ils sont donc rationnellement équivalents sur $Y_{k(Y)}$. Soit ξ le point générique de Z d'image $\eta \in Y$. L'hypothèse que f est un CH_0-isomorphisme implique que $\xi_{k(Z)}$ est rationnellement équivalent à $M_{k(Z)}$ sur $Z_{k(Z)}$. D'après la proposition 3.11, ceci implique que la k-variété projective et lisse Z est CH_0-triviale. □

On en déduit une démonstration alternative de [19, Thm. 1.14] :

Théorème 6.5 *Soit A un anneau de valuation discrète, K son corps des fractions, k son corps résiduel supposé algébriquement clos. Soit $\overline{K}$ une clôture algébrique de K. Soit $\mathcal{X}$ un A-schéma projectif et plat, $X = \mathcal{X} \times_A K$ la fibre générique et $Y = \mathcal{X} \times_A k$ la fibre spéciale. Supposons X/K lisse et géométriquement intègre et Y/k géométriquement intègre. Supposons qu'il existe une résolution des singularités projective $f : Z \to Y$ qui est un CH_0-isomorphisme. Si la $\overline{K}$-variété $X \times_K \overline{K}$ est rétractilement rationnelle, alors la k-variété Z est CH_0-triviale.*

Démonstration. On commence par remplacer A et K par leurs complétés. Il existe une extension finie E/K sur laquelle $X \times_K E$ est rétractilement E-rationnelle. On remplace K par E et A par la clôture intégrale de A dans E, qui est un anneau de valuation discrète car A est complet. On applique alors le théorème 6.4, dont la démonstration utilise la spécialisation de la R-équivalence mais pas celle de l'équivalence rationnelle. □

Pour ce qui concerne l'hypothèse que la résolution des singularités est CH_0-triviale, rappelons le résultat facile suivant :

Proposition 6.6 *[19, Prop. 1.8] Soit $f : Z \to Y$ un morphisme propre de k-variétés. Pour établir que, sur tout corps F contenant k, l'homomorphisme $f_* : CH_0(Z_F) \to CH_0(Y_F)$ est un isomorphisme, il suffit de montrer que, pour tout point M du schéma Y, le $k(M)$-schéma fibre Z_M est CH_0-trivial.* □

Dans [44, §2.4], A. Pirutka développe une autre variante de la remarque 1.19 de [19].

Théorème 6.7 *Soit A un anneau de valuation discrète, de corps des fractions K et de corps résiduel k. Soit $\mathcal{X}$ un A-schéma projectif et plat, $X = \mathcal{X} \times_A K$ la fibre générique et $Y = \mathcal{X} \times_A k$ la fibre spéciale. Supposons X/K et Y/k géométriquement intègres, et $Y(k)$ Zariski dense dans Y. Supposons qu'il existe une résolution des singularités projective $f : Z \to Y$ qui soit R-triviale (au sens de la définition 3.5). Si X est rétractilement K-rationnelle, alors Z est presque R-triviale. En particulier, il existe un point $M \in Z(k)$ tel que le point générique de Z est, sur $Z_{k(Z)}$, R-équivalent à $M_{k(Z)}$.* □

Ceci nous permet d'établir un énoncé à comparer avec la méthode de Totaro [49], qui ne donne que $H^0(Z, \Omega^i) = 0$.

Théorème 6.8 *Soit A un anneau de valuation discrète, de corps des fractions K et de corps résiduel k algébriquement clos. Soit $\mathcal{X}$ un A-schéma projectif et plat,*

$X = \mathcal{X} \times_A K$ *la fibre générique et* $Y = \mathcal{X} \times_A k$ *la fibre spéciale. Supposons* X/K *et* Y/k *géométriquement intègres. Supposons qu'il existe une résolution des singularités projective* $f : Z \to Y$ *qui soit* R-*triviale (au sens de la définition 3.5). Supposons la* K-*variété* X *géométriquement rétractilement rationnelle. Alors :*

(i) *La* k-*variété* Z *est presque* R-*triviale. En particulier il existe un point* $M \in Z(k)$ *tel que le point générique de* Z *est, sur* $Z_{k(Z)}$, R-*équivalent à* $M_{k(Z)}$.

(ii) *Pour tous entiers* $i > 0$ *et* $m > 0$, *on a*

$$H^0(Z, (\Omega^i)^{\otimes m}) = 0.$$

Démonstration. Pour établir le point (i) à partir du théorème 6.7, on procède comme dans [19, Thm. 1.14], [44, Thm. 2.14] (où l'on s'est restreint à car$(k) = 0$) et dans les théorèmes 6.4 et 6.5 ci-dessus. La proposition 3.31 ci-dessus (appliquée à Z) donne alors le point (ii).   □

Pour établir dans des cas concrets qu'une résolution $f : Z \to Y$ est R-triviale, on peut utiliser l'énoncé suivant.

Proposition 6.9 *Soit* $f : Z \to Y$ *un* k-*morphisme propre. Si pour tout corps* F *contenant* k *et tout point* $M \in Y(F)$, *la* F-*variété fibre* Z_M *est* R-*triviale, alors le morphisme* f *est* R-*trivial.*   □

La démonstration de cet énoncé est facile, car les hypothèses impliquent que tout F-morphisme d'un ouvert de $\mathbf{P}_F^1$ vers Y_F se relève en un F-morphisme de cet ouvert vers Z_F. Mais, dans la pratique, établir que l'hypothèse sur les fibres Z_M vaut est l'une des principales difficultés.

6.3 Applications aux variétés algébriques complexes

Elles sont nombreuses. Certaines ont été décrites dans les rapports [43], [44].

Pour des familles projectives et lisses $\mathcal{X} \to S$ de variétés algébriques d'un "type donné", paramétrées par une variété algébrique complexe, on établit des théorèmes du type :

L'ensemble des points $s \in S(\mathbb{C})$ tels que la fibre $\mathcal{X}_s$ ne soit pas rétractilement rationnel est Zariski dense dans S.

On montre en fait que l'ensemble des points s où X_s est rétractilement rationnel est contenu dans une union dénombrable de fermés propres de S.

On s'intéresse bien sûr à des variétés projectives et lisses $X/\mathbb{C}$ qui sont "proches d'être rationnelles", en particulier qui sont rationnellement connexes (i.e. telles que $X(\mathbb{C})/R$ soit réduit à un point). C'est le cas des variétés de Fano.

On a étudié :

- les hypersurfaces lisses dans $\mathbf{P}_{\mathbb{C}}^n$ (de degré $d \leq n$)
- les revêtements cycliques ramifiés de $\mathbf{P}_{\mathbb{C}}^n$ (avec des conditions sur le degré du revêtement et le degré de l'hypersurface de ramification)

- des familles de quadriques de dimension relative d au moins 1 au-dessus de $\mathbf{P}_{\mathbb{C}}^n$
- des familles de surfaces de del Pezzo, et plus généralement de variétés de Fano, au-dessus de $\mathbf{P}_{\mathbb{C}}^n$

On procède par dégénérescence de ces variétés sur des variétés singulières Y/k, avec k éventuellement de caractéristique positive, pour lesquelles on trouve une résolution des singularités $Z \to Y$ qui soit un morphisme CH_0-trivial, et l'on montre que Z n'est pas CH_0-triviale, ou que Z n'est pas presque R-triviale en utilisant le groupe de Brauer ou la cohomologie non ramifiée ou bien, si le corps résiduel k est de caractéristique positive, l'invariant $H^0(Z, \Omega^i)$.

Il y a ici deux points qui demandent beaucoup de travail :

- Montrer que la résolution $Z \to Y$ est un morphisme CH_0-trivial (c'est une propriété indépendante de la résolution). En pratique, il faut faire la résolution explicite, et voir si les fibres sont CH_0-triviales, ce qui donne le résultat grâce à la Proposition 6.6.
- Montrer qu'un invariant (groupe de Brauer, cohomologie non ramifiée ...) n'est pas trivial sur Z.

La première méthode, avec H_{nr}^2, alias le groupe de Brauer, est celle qui a été utilisée par C. Voisin (doubles solides quartiques) puis dans [19] (quartiques lisses dans $\mathbf{P}^4$), puis par Beauville (doubles solides sextiques), et dans de nombreux articles subséquents de Hassett, Pirutka, Tschinkel, Kresch, Böhning, von Bothmer, Auel. C'est celle qui a permis le résultat spectaculaire de Hassett, Pirutka, Tschinkel [32] selon lequel la rationalité stable n'est pas forcément constante dans une famille lisse de dimension relative au moins 4.

La seconde méthode, avec les différentielles en caractéristique positive, a été initiée par B. Totaro [49]. Elle est inspirée d'un travail de Kollár de 1995, qui utilisait déjà un argument de spécialisation sur une variété singulière en caractéristique positive et $H^0(Z, \Omega^i)$. Totaro en a déduit des résultats très généraux sur la non rationalité stable des hypersurfaces très générales dans $\mathbf{P}^n$, de degré $d \leq n$ satisfaisant approximativement $d \geq 2n/3$. Elle a été poursuivie dans [20]. De nombreux autres résultats ont été ensuite obtenus par cette méthode par T. Okada, H. Ahmadinezhad, I. Krylov et par Chatzistamatiou et Levine pour d'autres types de variétés rationnellement connexes. *Peut-on envisager des applications intéressantes du Théorème 6.8 (ii) qu'on ne puisse obtenir à partir de la simple conclusion $H^0(Z, \Omega^i) = 0$?*

La première méthode, cette fois-ci avec les invariants cohomologiques supérieurs H_{nr}^i, vient d'être utilisée par Schreieder [45] pour des fibrations en quadriques de grande dimension au-dessus de l'espace projectif. À cette occasion, il a introduit *une variante importante de la méthode de spécialisation, qui évite dans certains cas de vérifier si la résolution de la fibre spéciale est CH_0-triviale (ou presque R-triviale).*

On trouvera ceci discuté dans le texte [15].

Par spécialisations successives à partir du cas des familles de quadriques de Pfister au-dessus d'un espace projectif, Schreieder [46] a fait progresser de façon spectaculaire le cas des hypersurfaces très générales de degré d dans $\mathbf{P}^n_{\mathbb{C}}$, obtenant leur non rationalité stable avec une condition du type $d \geq log(n)$.

7 Hypersurfaces cubiques non stablement rationnelles sur un corps non algébriquement clos

Soient k un corps et $X \subset \mathbf{P}^n_k$, $n \geq 3$, une hypersurface cubique lisse. On s'intéresse ici au cas où k n'est pas algébriquement clos.

Le défi ici est, pour un corps k de complexité arithmétique donné (corps fini, corps local, corps de nombres, corps de fonctions de d variables sur un de ces corps ou sur les complexes, corps de séries formelles itérées sur l'un de ces corps) de trouver des hypersurfaces cubiques lisses non rétractilement k-rationnelles $X \subset \mathbf{P}^n_k$ avec $X(k) \neq \emptyset$ et n aussi grand que possible.

7.1 Hypersurfaces cubiques réelles

Proposition 7.1 *Pour tout entier $n \geq 2$, il existe une hypersurface cubique lisse $X \subset \mathbf{P}^n_{\mathbb{R}}$ telle que le lieu des points réels $X(\mathbb{R})$ ait deux composantes connexes. En particulier, une telle hypersurface n'est pas rétractilement $\mathbb{R}$-rationnelle.*

Démonstration. Soit $n \geq 2$ et soient $x_0, \ldots, x_{n-2}, u, v$ des variables. Soit

$$\Phi(x_0, \ldots, x_{n-2}, u, v) = (\sum_i x_i^2)v - u(u - v)(u + v).$$

Soit $Y \subset \mathbf{P}^n_{\mathbb{R}}$ l'hypersurface cubique définie par l'équation

$$\Phi(x_0, \ldots, x_{n-2}, u, v) = 0.$$

Son lieu singulier est donné par $u = v = \sum_i x_i^2 = 0$, il n'a pas de point réel. On a donc $Y_{lisse}(\mathbb{R}) = Y(\mathbb{R})$. Les coordonnées (u, v) définissent une application continue $Y_{lisse}(\mathbb{R}) \to \mathbf{P}^1(\mathbb{R})$, dont l'image est la réunion des deux invervalles définis par $u(u - v)(u + v) \geq 0$. On vérifie ainsi que $Y(\mathbb{R})$ est une variété C^∞ avec deux composantes connexes. Soit $\Psi(x_0, \ldots, x_{n-2}, u, v) = \sum_i x_i^3 + u^3 + v^3$. Pour $\epsilon \in \mathbb{R}$ petit, l'hypersurface cubique définie par $\Phi + \epsilon\Psi = 0$ est lisse pour $\epsilon \neq 0$, pout tout $\epsilon \in \mathbb{R}$ petit, son lieu réel est une variété C^∞ lisse, et par le théorème d'Ehresmann, ce lieu est difféomorphe à $Y(\mathbb{R}) = Y_{lisse}(\mathbb{R})$. $\square$

Exercice Pour une hypersurface cubique $X \subset \mathbf{P}_{\mathbb{R}}^n$, $n \geq 2$, l'espace $X(\mathbb{R})$ a au plus deux composantes connexes.

7.2 Spécialisations à fibres réductibles

Dans le contexte de la spécialisation du groupe de Chow, Totaro [49] a utilisé des spécialisations à fibre réductible. On peut le faire aussi dans le cadre de la R-équivalence. L'énoncé suivant est inspiré par [49] et [7], mais est plus simple.

Proposition 7.2 *Soit A un anneau de valuation discrète, K son corps des fractions, k son corps résiduel. Soit $\mathcal{X}$ un A-schéma propre et plat. Supposons la fibre générique X/K lisse et géométriquement intègre. Soit Y la fibre spéciale. Supposons Y union de deux fermés $Y = V \cup W$, $T = V \cap W$, $T(k) = \emptyset$, $V_{lisse}(k) \neq \emptyset$ et $W_{lisse}(k) \neq \emptyset$. Alors la K-variété X n'est pas R-triviale et n'est donc pas rétractilement K-rationnelle.*

Démonstration. On peut supposer que A est hensélien. Soient $p \in V_{lisse}(k)$ et $q \in W_{lisse}(k)$. Par le lemme de Hensel, il existe des A-points P et Q de $\mathcal{X}(R) = X(K)$ qui se spécialisent l'un en p, l'autre en q. Par le théorème 6.1, l'application de spécialisation $X(K) = \mathcal{X}(A) \to Y(k)$ passe au quotient par la R-équivalence. Si X est R-triviale, il existe donc une chaîne de k-morphismes $\mathbf{P}_k^1 \to Y$ qui relie p et q. Il existe donc un k-morphisme $f : \mathbf{P}_k^1 \to Y$ tel que $f(0) \in V(k)$ et $f(\infty) \in W(k)$. La courbe $\mathbf{P}_k^1$ est alors couverte par les deux fermés non vides $v = f^{-1}(V)$ et $w = f^{-1}(W)$, qui contiennent chacun un k-point, et dont l'intersection n'a pas de k-point car $T(k) = \emptyset$. L'un des deux fermés, soit v, est égal à $\mathbf{P}_k^1$. Mais alors $w \subset v$, et tout k-point de w est dans v. Contradiction.   □

Exemples 7.3 Soient $n \geq 2$ et $f_0(x_1, \ldots, x_n) \in k[x_1, \ldots, x_n]$ une forme homogène de degré $d \geq 2$ sans zéro sur le corps k définissant une hypersurface lisse sur k. Soit $\alpha \in k^*$ une valeur (non nulle) de f sur k^n. Soit

$$f(x_0, \ldots, x_n) := \alpha x_0^d - f_0(x_1, \ldots, x_n) \in k[x_0, x_1, \ldots, x_n],$$

puis

$$g(x_0, \ldots, x_n) = x_0 . f(x_0, \ldots, x_n).$$

Soit $Y \subset \mathbf{P}_k^n$ l'hypersurface de degré $d + 1$ définie par $g = 0$. C'est l'union de V défini par $x_0 = 0$ et W défini par $f_0(x_1, \ldots, x_n) = 0$. L'intersection $T = V \cap W$ satisfait $T(k) = \emptyset$. On a $V(k) \neq \emptyset$ et $W(k) \neq \emptyset$.

Soit $g(x_0, \ldots, x_n) \in k[x_0, x_1, \ldots, x_n]$ une forme homogène de degré $d + 1$ définissant une hypersurface lisse dans $\mathbf{P}_k^n$. Notons $A = k[[t]]$ et $K = k((t))$. L'hypersurface $X \subset \mathbf{P}_K^n$ définie par $tg(x_0, \ldots, x_n) + f(x_0, \ldots, x_n) = 0$ n'est pas R-triviale, et n'est pas pas rétractilement K-rationnelle.

On peut aussi donner des exemples similaires avec A un anneau de valuation discrète complet d'inégale caractéristique.

En utilisant cette méthode dans le cas $d = 2$, on obtient des hypersurfaces cubiques lisses, avec un K-point, non rétractilement K-rationnelles dans $\mathbf{P}_K^N$ pour tout $N \leq 2^{r-1}$ sur $K = \mathbb{C}((u_1)) \ldots ((u_r))$ et dans $\mathbb{Q}_p((u_1)) \ldots ((u_{r-2}))$.

Sur $K = \mathbb{C}((u_1))((u_2))((u_3))$ on trouve donc des hypersurfaces cubiques dans $\mathbf{P}_K^4$. Ces bornes sont les mêmes que celles obtenues dans [7] et dans [13], qui établissent le résultat plus fort que les hypersurfaces cubiques concernées ne sont pas CH_0-triviales. La démonstration de ce dernier résultat utilise une variation due à Totaro de la technique de spécialisation de Voisin et CT-Pirutka pour les groupes de Chow de zéro-cycles.

7.3 Hypersurfaces cubiques diagonales et cohomologie non ramifiée

Ce paragraphe est extrait directement de l'article [13]. On utilise encore ici une technique de spécialisation, mais elle est différente de celles employées ci-dessus.

Théorème 7.4 *Soit k un corps de caractéristique différente de 3, possédant un élément a qui n'est pas un cube. Soient $0 \leq n \leq m$ des entiers. Soit F un corps avec*

$$k(\lambda_1, \ldots, \lambda_m) \subset F \subset F_m := k((\lambda_1)) \ldots ((\lambda_m)).$$

L'hypersurface cubique $X := X_{n,F}$ de $\mathbf{P}_F^{n+3}$ définie par l'équation

$$x^3 + y^3 + z^3 + aw^3 + \sum_{i=1}^n \lambda_i t_i^3 = 0$$

possède un point rationnel et n'est pas universellement CH_0-triviale, en particulier elle n'est pas rétractilement F-rationnelle.

Démonstration. Pour établir le résultat, on peut supposer que k contient une racine cubique primitive de l'unité, soit j, et que $F = F_m$. Le lemme 7.5 ci-dessous permet de supposer $n = m$. On fixe un isomorphisme $\mathbb{Z}/3 = \mu_3$ et on considère la cohomologie étale à coefficients $\mathbb{Z}/3$. On ignore les torsions à la Tate dans les notations. Etant donnés un corps L contenant k et des éléments $b_i, i = 1, \ldots, s$, de L^*, on note $(b_1, \ldots, b_s) \in H^s(L, \mathbb{Z}/3)$ le cup-produit, en cohomologie galoisienne, des classes $(b_i) \in L^*/L^{*3} = H^1(L, \mathbb{Z}/3)$.

On va démontrer par récurrence sur $n \neq 0$ l'assertion suivante, qui implique la proposition.

(A_n) Soient k, a, F_n et X_n/F_n comme ci-dessus. Le cup-produit

$$\alpha_n := ((x + jy)/(x + y), a, \lambda_1, \ldots, \lambda_n) \in H^{n+2}(F_n(X_n), \mathbb{Z}/3)$$

définit une classe de cohomologie non ramifiée (par rapport au corps de base F_n) qui ne provient pas d'une classe dans $H^{n+2}(F_n, \mathbb{Z}/3)$.

Le cas $n = 0$ est connu ([37, Chap. VI, §5] [22, §2.5.1]). Supposons l'assertion démontrée pour n.

La classe α_{n+1} sur la F_{n+1}-hypersurface $X_{n+1} \subset \mathbf{P}_{F_{n+1}}^{n+4}$ a ses résidus triviaux en dehors des diviseurs définis par $x + y = 0$ et $x + jy = 0$. Soit $\Delta \subset X_{n+1}$ le diviseur $x + y = 0$. Ce diviseur est défini par les équations

$$x + y = 0, z^3 + aw^3 + \sum_{i=1}^{n+1} \lambda_i t_i^3 = 0.$$

Le résidu de α_{n+1} au point générique de Δ est

$$\partial_\Delta(\alpha_{n+1}) = \pm(a, \lambda_1, \ldots, \lambda_{n+1}) \in H^{n+2}(F_{n+1}(\Delta), \mathbb{Z}/3).$$

Mais dans le corps des fonctions de Δ, on a

$$1 + a(w/z)^3 + \sum_{i=1}^{n+1} \lambda_i (t_i/z)^3 = 0$$

et cette égalité implique (cf. [40, Lemma 1.3]) :

$$(a, \lambda_1, \ldots, \lambda_{n+1}) = 0 \in H^{n+2}(F_{n+1}(\Delta), \mathbb{Z}/3).$$

Le même argument s'applique pour le diviseur défini par $x + jy = 0$. Ainsi α_{n+1} est une classe de cohomologie non ramifiée sur la F_{n+1}-hypersurface X_{n+1}.

Soit $\mathcal{X}_{n+1}$ le $F_n[[\lambda_{n+1}]]$-schéma défini par

$$x^3 + y^3 + z^3 + aw^3 + \sum_{i=1}^{n+1} \lambda_i t_i^3 = 0.$$

Le diviseur Z défini par $\lambda_{n+1} = 0$ sur $\mathcal{X}$ est le cône d'équation

$$x^3 + y^3 + z^3 + aw^3 + \sum_{i=1}^{n} \lambda_i t_i^3 = 0$$

dans $\mathbf{P}_{F_n}^{n+4}$, cône qui est birationnel au produit de $\mathbf{P}_{F_n}^1$ et de l'hypersurface cubique lisse $X_n \subset \mathbf{P}_{F_n}^{n+3}$ définie par

$$x^3 + y^3 + z^3 + aw^3 + \sum_{i=1}^{n} \lambda_i t_i^3 = 0.$$

Le corps des fonctions rationnelles de $\mathcal{X}_{n+1}$ est $F_{n+1}(X_{n+1})$.

On a

$$\partial_Z(\alpha_{n+1}) = \pm((x + jy)/(x + y), a, \lambda_1, \ldots, \lambda_n) \in H^{n+2}(F_n(Z), \mathbb{Z}/3).$$

Par l'hypothèse de récurrence

$$((x + jy)/(x + y), a, \lambda_1, \ldots, \lambda_n) \in H^{n+2}(F_n(X_n), \mathbb{Z}/3)$$

n'est pas dans l'image de $H^{n+2}(F_n, \mathbb{Z}/3)$. Ceci implique que

$$((x + jy)/(x + y), a, \lambda_1, \ldots, \lambda_n) \in H^{n+2}(F_n(Z)), \mathbb{Z}/3)$$

n'est pas dans l'image de $H^{n+2}(F_n, \mathbb{Z}/3)$. Du diagramme commutatif

$$\partial_Z : \quad H^{n+3}(F_{n+1}(X), \mathbb{Z}/3) \rightarrow H^{n+2}(F_n(Z), \mathbb{Z}/3)$$
$$\uparrow \qquad\qquad\qquad \uparrow$$
$$\partial_{\lambda_{n+1}=0} : \quad H^{n+3}(F_{n+1}, \mathbb{Z}/3) \quad \rightarrow \quad H^{n+2}(F_n, \mathbb{Z}/3)$$

on conclut que

$$\alpha_{n+1} := ((x + jy)/(x + y), a, \lambda_1, \ldots, \lambda_{n+1}) \in H^{n+3}(F_{n+1}(X), \mathbb{Z}/3)$$

n'est pas dans l'image de $H^{n+3}(F_{n+1}, \mathbb{Z}/3)$.

Ceci établit (A_n) pour tout entier n et implique (cf. [39]) que la F_n-variété X_n n'est pas universellement CH_0-triviale et n'est pas rétractilement F_n-rationnelle. □

Lemme 7.5 *Soit F un corps. Si une F-variété X projective, lisse, géométriquement connexe n'est pas universellement CH_0-triviale, alors la $F((t))$-variété $X \times_F F((t))$ n'est pas universellement CH_0-triviale, et donc n'est pas rétractilement $F((t))$-rationnelle.*

Démonstration. Sur tout corps L contenant F, on dispose de l'application de spécialisation $CH_0(X_{L((t))}) \rightarrow CH_0(X_L)$, et cette application est surjective et respecte le degré. □

Remarque 7.6 Il serait intéressant de comprendre la généralité de la construction faite dans le théorème 7.4. On utilise une classe de cohomologie non ramifiée non constante sur un modèle birationnel de la fibre spéciale d'une $k[[t]]$-schéma propre à fibres intègres, et on en tire une classe de cohomologie non ramifiée non constante de degré un de plus sur la fibre générique sur $k((t))$, essentiellement par cup-produit avec la classe d'une uniformisante de l'anneau $k[[t]]$.

On laisse au lecteur le soin d'établir l'analogue suivant du théorème 7.4.

Théorème 7.7 *Soient $p \neq 3$ un nombre premier et k un corps p-adique dont le corps résiduel contient les racines cubiques primitives de 1. Soit $a \in k^*$ une unité* *qui n'est pas un cube. Soit π une uniformisante de k. Soient $0 \leq n \leq m$ des entiers. Soit F un corps avec*

$$\mathbb{Q}(a)(\lambda_1, \ldots, \lambda_m) \subset F \subset k((\lambda_1)) \ldots ((\lambda_m)).$$

L'hypersurface cubique X_n de $\mathbf{P}_F^{n+4}$ définie par l'équation

$$x^3 + y^3 + z^3 + aw^3 + \pi t^3 + \sum_{i=1}^{n} \lambda_i t_i^3 = 0,$$

qui possède un point rationnel, n'est pas universellement CH_0-triviale et donc n'est pas rétractilement F-rationnelle.

Exemples En appliquant le théorème 7.4, on trouve $X_n \subset \mathbf{P}_F^{n+3}$ non rétractilement F-rationnelle avec

$$k(\lambda_1, \ldots, \lambda_n) \subset F \subset k((\lambda_1)) \ldots ((\lambda_n))$$

dans les situations suivantes.

(i) Le corps $k = \mathbb{F}$ est un corps fini de caractéristique différente de 3 contenant les racines cubiques de 1.
(ii) Le corps k, de caractéristique différente de 3, possède une valuation discrète, par exemple k est le corps des fonctions d'une variété complexe de dimension au moins 1, ou est un corps p-adique, ou est un corps de nombres.

On trouve ainsi des hypersurfaces cubiques lisses non rétractilement $\mathbb{C}(x_1, \ldots, x_m)$-rationnelles dans $\mathbf{P}_{\mathbb{C}(x_1, \ldots, x_m)}^n$, avec un point rationnel, pour tout entier n avec $3 \leq n \leq m + 2$.

En appliquant le théorème 7.7, sur un corps k p-adique ($p \neq 3$) contenant une racine cubique de 1, on trouve des hypersurfaces cubiques lisses non rétractilement $k(x_1, \ldots, x_m)$-rationnelles dans $\mathbf{P}_{k(x_1, \ldots, x_m)}^n$, avec un point rationnel, pour tout entier n avec $4 \leq n \leq m + 4$.

Remerciements. Je remercie Andreas Hochenegger, Manfred Lehn et Paolo Stellari de l'invitation à donner ce cours, et je remercie le rapporteur de ces notes pour sa lecture attentive.

References

1. M. Artin, D. Mumford, Some elementary examples of unirational varieties which are not rational. Proc. Lond. Math. Soc. (3) **25**, 75–95 (1972)
2. A. Auel, J.-L. Colliot-Thélène, R. Parimala, Universal unramified cohomology of cubic fourfolds containing a plane, in *Brauer Groups and Obstruction Problems: Moduli Spaces and Arithmetic* (Palo Alto, 2013). Progress in Mathematics, vol. 320 (Birkhäuser, Basel, 2017), pp. 29–56

3. A. Auel, A. Bigazzi, C. Böhning, H.-G. Graf von Bothmer, Universal triviality of the Chow group of 0-cycles and the Brauer group (2018), https://arxiv.org/abs/1806.02676

4. A. Beauville, J.-L. Colliot-Thélène, J.-J. Sansuc, Sir Peter Swinnerton-Dyer, Variétés stablement rationnelles non rationnelles. Ann. Math. (2) **121**(2), 283–318 (1985)

5. S. Bloch, V. Srinivas, Remarks on correspondences and algebraic cycles. Am. J. Math. **105**, 1235–1253 (1983)

6. Y. Cao, Troisième groupe de cohomologie non ramifiée des torseurs universels sur les surfaces rationnelles. Épijournal de Géométrie Algébrique **2** (2018). Article Nr. 12

7. A. Chatzistamatiou, M. Levine, Torsion orders of complete intersections. Algebra Number Theory **11**(8), 1779–1835 (2017)

8. J.-L. Colliot-Thélène, Hilbert's theorem 90 for K_2, with application to the Chow groups of rational surfaces. Invent. math. **71**, 1–20 (1983)

9. J.-L. Colliot-Thélène, Birational invariants, purity and the Gersten conjecture, in *K-theory and Algebraic Geometry: Connections with Quadratic Forms and Division Algebras* (Santa Barbara, CA, 1992). Proceedings of Symposia in Pure Mathematics, vol. 58.1 (American Mathematical Society, Providence, RI, 1995), pp. 1–64

10. J.-L. Colliot-Thélène, Un théorème de finitude pour le groupe de Chow des zéro-cycles d'un groupe algébrique linéaire sur un corps p-adique. Invent. math. **159**, 589–606 (2005)

11. J.-L. Colliot-Thélène, Descente galoisienne sur le second groupe de Chow: mise au point et applications. Documenta mathematica, extra volume: Alexander S. Merkurjev's Sixtieth Birthday (2015), pp. 195–220

12. J.-L. Colliot-Thélène, CH_0-trivialité universelle d'hypersurfaces cubiques presque diagonales. Algebr. Geom. **4**(5), 597–602 (2017)

13. J.-L. Colliot-Thélène, Non rationalité stable d'hypersurfaces cubiques sur des corps non algébriquement clos, in *Proceedings of the International Colloquium on K-Theory, TIFR, 2016*, vol. 19 (Tata Institute of Fundamental Research Publications, Mumbai, 2018)

14. J.-L. Colliot-Thélène, Surfaces stablement rationnelles sur un corps quasi-fini, à paraître dans Izvestija RAN Ser. Mat. Tom **83** (2019), volume dédié à V. A. Iskovskikh, https://arxiv.org/abs/1711.09595

15. J.-L. Colliot-Thélène, Introduction to work of Hassett-Pirutka-Tschinkel and Schreieder, in *Birational Geometry of Hypersurfaces*, ed. by A. Hochenegger et al. Lecture Notes of the Unione Matematica Italiana, vol. 26 (Springer, cham, 2019). https://doi.org/10.1007/978-3-030-18638-8_3

16. J.-L. Colliot-Thélène, F. Ischebeck, L'équivalence rationnelle sur les cycles de dimension zéro des variétés algébriques réelles. C. R. Acad. Sc. Paris, t. **292**, 723–725 (1981)

17. J.-L. Colliot-Thélène, M. Ojanguren, Variétés unirationnelles non rationnelles: au-delà de l'exemple d'Artin et Mumford. Invent. math. **97**, 141–158 (1989)

18. J.-L. Colliot-Thélène, R. Parimala, Real components of algebraic varieties and étale cohomology. Invent. math. **101**, 81–99 (1990)

19. J.-L. Colliot-Thélène, A. Pirutka, Hypersurfaces quartiques de dimension 3: non rationalité stable. Ann. Scient. Éc. Norm. Sup. 4e série, **49**(2), 371–397 (2016)

20. J.-L. Colliot-Thélène, A. Pirutka, Revêtements cycliques non stablement rationnels (en russe). Izvestija RAN Ser. Mat., Ser. Math. Tom **80**(4), 35–48 (2016)

21. J.-L. Colliot-Thélène, J.-J. Sansuc, La R-équivalence sur les tores. Ann. Sc. Éc. Norm. Sup. **10**, 175–229 (1977)

22. J.-L. Colliot-Thélène, J.-J. Sansuc, La descente sur les variétés rationnelles, II. Duke Math. J. **54**(2), 375–492 (1987)

23. J.-L. Colliot-Thélène, C. Voisin, Cohomologie non ramifiée et conjecture de Hodge entière. Duke Math. J. **161**, 735–801 (2012)

24. J.-L. Colliot-Thélène, B. Kahn, R. Hoobler, The Bloch-Ogus-Gabber theorem, in *Algebraic K-theory, Proceedings of the Great Lakes K-Theory Conference (Toronto 1996)*, ed. R. Jardine, V. Snaith. The Fields Institute for Research in Mathematical Sciences Communications Series, vol. 16 (American Mathematical Society, Providence, RI, 1997), pp. 31–94

25. W. Fulton, *Intersection Theory*. Ergebnisse der Math. und ihrer Grenzg. 3. Folge, Bd. 2 (Springer, Berlin, 1984/1998)
26. O. Gabber, Q. Liu, D. Lorenzini, The index of an algebraic variety. Invent. math. **192**(3), 567–626 (2013)
27. P. Gille, Le problème de Kneser-Tits, exposé Bourbaki no. 983, Astérisque **326**, 39–81 (2009)
28. S. Gille, Permutation modules and Chow motives of geometrically rational surfaces. J. Algebra **440**, 443–463 (2015)
29. P. Gille, T. Szamuely, *Central Simple Algebras and Galois Cohomology*, 2nd edn. Cambridge Studies in Advances Mathematics, vol. 165 (Cambridge University Press, Cambridge, 2017)
30. A. Grothendieck, Le groupe de Brauer I, II, III, in *Dix exposés sur la cohomologie des schémas* (Masson et North Holland, Paris, 1968)
31. B. Hassett, Cubic fourfolds, $K3$ surfaces, and rationality questions, in *Rationality Problems in Algebraic Geometry* (Levico Terme 2015). Lecture Notes in Mathematics, vol. 2172, Fond. CIME/CIME Found. Subser. (Springer, Cham, 2016), pp. 29–66
32. B. Hassett, A. Pirutka, Y. Tschinkel, Stable rationality of quadric surface bundles over surfaces. Acta Math. **220**(2), 341–365 (2018)
33. B. Kahn, R. Sujatha, Birational geometry and localisation of categories. With appendices by Jean-Louis Colliot-Thélène and by Ofer Gabber. Doc. Math. (2015), Extra vol.: Alexander S. Merkurjev's sixtieth birthday, pp. 277–334
34. J. Kollár, *Rational Curves on Algebraic Varieties*. Ergebnisse der Math. und ihrer Grenzg. 3. Folge, Bd. 32 (Springer, Berlin, 1996)
35. J. Kollár, Rationally connected varieties over local fields. Ann. Math. (2) **150**(1), 357–367 (1999)
36. D. Madore, Sur la spécialisation de la R-équivalence, prépublication (2005), http://perso.telecom-paristech.fr/~madore/specialz.pdf
37. Yu. I. Manin, *Formes cubiques : algèbre, géométrie, arithmétique (en russe)* (Nauka, Moscou, 1972)
38. R. Mboro, Remarks on approximate decompositions of the diagonal, https://arxiv.org/abs/1708.02422
39. A.S. Merkurjev, Unramified elements in cycle modules. J. Lond. Math. Soc. **78**, 51–64 (2008)
40. J. Milnor, Algebraic K-theory and quadratic forms. Invent. math. **9**, 318–344 (1970)
41. M. Ojanguren, *The Witt Group and the Problem of Lüroth*. With an introduction by Inta Bertuccioni. Dottorato di Ricerca in Matematica (ETS Editrice, Pisa, 1990)
42. D. Orlov, A. Vishik, V. Voevodsky, An exact sequence for $K_*^M/2$ with applications to quadratic forms. Ann. Math. (2) **165**(1), 1–13 (2007)
43. E. Peyre, Progrès en irrationalité [d'après C. Voisin, J.-L. Colliot-Thélène, B. Hassett, A. Kresch, A. Pirutka, B. Totaro, Y. Tschinkel et al.], dans Séminaire Bourbaki 69-ème année, exposé 1123, Astérisque (2016), pp. 1–26
44. A. Pirutka, Varieties that are not stably rational, zero-cycles and unramified cohomology, in *Algebraic Geometry: Salt Lake City 2015*. Proceedings of Symposia in Pure Mathematics, vol. 97.2 (American Mathematical Society, Providence, RI, 2018), pp. 459–483
45. S. Schreieder, On the rationality problem for quadric bundles. Duke Math. J. **168**, 187–223 (2019)
46. S. Schreieder, Stably irrational hypersurfaces of small slopes. J. Am. Math. Soc. (to appear), https://arxiv.org/abs/1801.05397
47. B. Segre, Sull'esistenza, sia nel campo razionale che nel campo reale, di involuzioni piane non birazionali. Rend. Acc. Naz. Lincei, Sc. fis. mat. e nat. **10**, 94–97 (1951)
48. H.P.F. Swinnerton-Dyer, The birationality of cubic surfaces over a given field. Michigan Math. J. **17**, 289–295 (1970)
49. B. Totaro, Hypersurfaces that are not stably rational. J. Am. Math. Soc. **29**, 883–891 (2016)
50. C. Voisin, Unirational threefolds with no universal codimension 2 cycle. Invent. math. **201**, 207–237 (2015)
51. C. Voisin, On the universal CH_0-group of cubic hypersurfaces. J. Eur. Math. Soc. **19**(6), 1619–1653 (2017)

Introduction to work
of Hassett-Pirutka-Tschinkel
and Schreieder

Jean-Louis Colliot-Thélène

Abstract Hassett, Pirutka and Tschinkel gave the first examples of families $X \to B$ of smooth, projective, connected, complex varieties having some rational fibres and some other fibres which are not even stably rational. This used the specialisation method of Voisin, as extended by Pirutka and myself. Under specific circumstances, a simplified version of the specialisation method was produced by Schreieder, leading to a simpler proof of the HPT example. I describe the method in its simplest form.

1 Introduction

Hassett, Pirutka and Tschinkel [13] gave the first examples of families $X \to B$ of smooth, projective, connected, complex varieties having some rational fibres and some other fibres which are not even stably rational. This used the specialisation method of Voisin, as extended by Pirutka and myself. Under specific circumstances, a simplified version of the specialisation method was produced by Schreieder [18, 19], leading to a simpler proof of the HPT example (no explicit resolution of singularities). In the following note I describe the method in its simplest form. For further developments, the reader is invited to read [1], which offers a different look at [13] as well as some generalizations, [4], and the papers [18–20] by Schreieder.

These notes were written on the occasion of the conference *Quadratic Forms in Chile 2018*, held at IMAFI, Universitad de Talca, 8–12 January 2018. They were further developed on the occasion of the School *Birational geometry of hypersurfaces*, Palazzo Feltrinelli, Gargnano del Garda, 19–23 March 2018. I thank Asher Auel for remarks on the typescript.

J.-L. Colliot-Thélène (✉)
CNRS, Université Paris-Sud, Université Paris-Saclay, Mathématiques, Orsay, France
e-mail: jlct@math.u-psud.fr

© Springer Nature Switzerland AG 2019

A. Hochenegger et al. (eds.), *Birational Geometry of Hypersurfaces*,
Lecture Notes of the Unione Matematica Italiana 26,
https://doi.org/10.1007/978-3-030-18638-8_3

2 Basics on the Brauer Group and on the Chow Group of Zero-Cycles

Grothendieck defined the Brauer group $\mathrm{Br}(X)$ of a scheme X as the second étale cohomology group $H^2_{\text{ét}}(X, \mathbb{G}_m)$ of X with values in the sheaf $\mathbb{G}_{m,X}$ on X. This is a contravariant functor with respect to arbitrary morphisms of schemes.

If $X = \mathrm{Spec}(K)$ is the spectrum of a field, then $\mathrm{Br}(X) = \mathrm{Br}(K)$, the more classical cohomological Brauer group $H^2(\mathrm{Gal}(K_s/K), K_s^*)$. Assume $2 \in K^*$. To the quaternion algebra (a, b) one associates a class $(a, b) \in \mathrm{Br}(K)[2]$. The quaternion algebra is isomorphic to a matrix algebra $M_2(K)$ if and only $(a, b) = 0 \in \mathrm{Br}(K)$, if and only if the diagonal quadratic form $< 1, -a, -b >$ has a nontrivial zero over K, if and only if the diagonal quadratic form $< 1, -a, -b, ab >$ has a nontrivial zero over K.

Proposition 2.1 *If X is an integral regular scheme and $K(X)$ is its field of rational functions, then the natural map $\mathrm{Br}(X) \to \mathrm{Br}(K(X))$ is injective.*

Proposition 2.2 *For R a discrete valuation ring with perfect residue field κ and field of fractions K, there is a natural exact sequence*

$$0 \to \mathrm{Br}(R) \to \mathrm{Br}(K) \to H^1(\kappa, \mathbb{Q}/\mathbb{Z}) \to 0.$$

The map $\partial_R : \mathrm{Br}(K) \to H^1(\kappa, \mathbb{Q}/\mathbb{Z})$ is the residue map.

Let R be a discrete valuation ring with perfect residue field κ and field of fractions K. Suppose $2 \in R^*$. Given $a, b \in K^*$ we may consider the element $(a, b) \in \mathrm{Br}(K)[2]$ associated to the quaternion algebra (a, b). Let $v : K^* \to \mathbb{Z}$ be the valuation map. The quotient $a^{v(b)}/b^{v(a)} \in K^*$ belongs to R^*. Let $cl((a^{v(b)}/b^{v(a)})$ denote its class in κ^*/κ^{*2}. One shows :

$$\partial_R((a, b)) = (-1)^{v(a).v(b)} cl((a^{v(b)}/b^{v(a)})) \in \kappa^*/\kappa^{*2} = H^1(\kappa, \mathbb{Z}/2) \subset H^1(\kappa, \mathbb{Q}/\mathbb{Z}).$$

Proposition 2.3 *Let $R \subset S$ be a local inclusion of discrete valuation rings, inducing an inclusion of fields $K \subset L$ and an inclusion of residue fields $\kappa \subset \lambda$. Assume $\mathrm{char}(\kappa) = 0$. Let e be the ramification index. Then there is a commutative diagram*

$$
\begin{array}{ccc}
\mathrm{Br}(K) & \to & H^1(\kappa, \mathbb{Q}/\mathbb{Z}) \\
\downarrow & & \downarrow \\
\mathrm{Br}(L) & \to & H^1(\lambda, \mathbb{Q}/\mathbb{Z})
\end{array}
$$

where $H^1(\kappa, \mathbb{Q}/\mathbb{Z}) \to H^1(\lambda, \mathbb{Q}/\mathbb{Z})$ is $e.\mathrm{Res}_{\kappa,\lambda}$.

Let K be a field and X an algebraic variety over K, i.e. a separated K-scheme of finite type. The group of zero-cycles $Z_0(X)$ is the free abelian group on closed points of X. Given any K-morphism $f : Y \to X$ of K-varieties, one defines the map

f_* : $Z_0(Y) \rightarrow Z_0(X)$ as the map sending a closed point $M \in Y$ with image the closed point $N = f(M) \in X$ to the zero cycle $[K(M) : K(N)]N \in Z_0(X)$.

Given a normal, connected curve C over K, and a rational function $g \in K(C)^*$, on associates to it its divisor $div_C(g) \in Z_0(C)$. Given a morphism $f : C \rightarrow X$, one may then consider the zero-cycle $f_*(div_C(g)) \in Z_0(X)$.

One then defines the Chow group $CH_0(X)$ of zero-cycles on X as the quotient of $Z_0(X)$ by the subgroup spanned by all $f_*(div_C(g))$, for $f : C \rightarrow X$ a *proper* K-morphism from a normal, integral K-curve to X and $g \in K(C)^*$.

If $\phi : X \rightarrow Y$ is a proper morphism of K-varieties, there is an induced map $\phi_* : CH_0(X) \rightarrow CH_0(Y)$. In particular, if X/K is proper, the structural map $X \rightarrow \mathrm{Spec}(K)$ induces a degree map $CH_0(X) \rightarrow \mathbb{Z}$.

If $\phi : U \rightarrow X$ is an open embedding of K-varieties, the natural restriction map $Z_0(X) \rightarrow Z_0(U)$ which forgets closed points outside of U induces a map $CH_0(X) \rightarrow CH_0(U)$.

Let X be a K-variety. There is a natural bilinear pairing

$$Z_0(X) \times \mathrm{Br}(X) \rightarrow \mathrm{Br}(K)$$

which sends a pair (P, α) with P a closed point of X and an element α in $\mathrm{Br}(X)$ to $Cores_{K(P)/K}(\alpha(P))$. If X/K is proper, this pairing induces a bilinear pairing

$$CH_0(X) \times \mathrm{Br}(X) \rightarrow \mathrm{Br}(K).$$

See [2, Prop. 3.1].

This pairing satisfies an obvious functoriality property with respect to (proper) K-morphisms of proper K-varieties.

3 Quadric Surfaces over a Field

The following proposition is classical. See [9, 21] and [3, Thm. 3.1].

Proposition 3.1 *Let K be a field, $char(K) \neq 2$, and let $X \subset \mathbb{P}^3_K$ be a smooth quadric surface. It is defined by a quadratic form q, which one may assume to be in diagonal form $q = < 1, -a, -b, abd >$, with $a, b, d \in K^*$. The class of d in K^*/K^{*2} is the discriminant, it does not depend on the choice of the quadratic form q defining the quadric X.*

The natural map $\mathrm{Br}(K) \rightarrow \mathrm{Br}(X)$ is surjective.

(a) If $d \notin K^{*2}$, the map $\mathrm{Br}(K) \rightarrow \mathrm{Br}(X)$ is an isomorphism.
(b) If $d \in K^{*2}$, the map $\mathrm{Br}(K) \rightarrow \mathrm{Br}(X)$ is surjective, and its kernel is of order at most 2, spanned by the class of the quaternion algebra (a, b), which is nontrivial if and only if $X(K) = \emptyset$.

4   A Special Quadric Surface over $\mathbb{P}^2_{\mathbb{C}}$

Reference: [13, 17].

Let $F(x, y, z) = x^2 + y^2 + z^2 - 2(xy + yz + zx)$.

Let $X \subset \mathbb{P}^3_{\mathbb{C}} \times \mathbb{P}^2_{\mathbb{C}}$ be the family of two-dimensional quadrics over $\mathbb{P}^2_{\mathbb{C}}$ given by the bihomogeneous equation

$$yzU^2 + zxV^2 + xyW^2 + F(x, y, z)T^2 = 0.$$

This family is smooth over the open set of $\mathbb{P}^2_{\mathbb{C}}$ whose complement is the octic curve defined by the determinant equation

$$\Delta = x^2.y^2.z^2.F(x, y, z) = 0.$$

Note that this is the union of the smooth conic $F = 0$ and (twice) three tangents to this conic. The family is flat over $\mathbb{P}^2_{\mathbb{C}}$ (all fibres are quadrics). The total space is not smooth.

Part (a) of the following proposition is a result of Hassett, Pirutka and Tschinkel [13, Prop. 11].

Part (b) is a special case of the general statement [19, Prop. 7], the proof of which builds upon results of Pirutka ([17, Thm. 3.17], [19, Thm. 4]), for which material is offered in Appendix 2 below.

As we shall see, the proof given for (a) in [13, Prop. 11] is easily modified to simultaneously give a proof of (b).

The proposition suffices for the special case described in this note; it dispenses us with the recourse to Appendix 2.

Proposition 4.1 *Let $\tilde{X} \to X$ be a projective desingularisation of X. Let α be the quaternion class $(x/z, y/z) \in \mathrm{Br}(\mathbb{C}(\mathbb{P}^2))$.*

(a) *The image β of α under the inverse map $p_2^* : \mathrm{Br}(\mathbb{C}(\mathbb{P}^2)) \to \mathrm{Br}(\mathbb{C}(X))$ is nonzero and lies in the subgroup $\mathrm{Br}(\tilde{X})$.*

(b) *For each codimension 1 subvariety Y of $\tilde{X}$ which does not lie over the generic point of $\mathbb{P}^2_{\mathbb{C}}$, the element $\beta \in \mathrm{Br}(\tilde{X})$ maps to $0 \in \mathrm{Br}(\mathbb{C}(Y))$.*

Proof The equation is symmetrical in (x, y, z). The class $\alpha = (x/z, y/z)$ is given by (x, y) in the open set $z \neq 0$, by $(x/z, 1/z) = (x, z)$ in the open set $y \neq 0$ and by $(1/z, y/z) = (y, z)$ in the open set $x \neq 0$. In view of the symmetry between (x, y, z) in the equation, we may restrict attention to the open set $\mathbb{A}^2_{\mathbb{C}}$ of $\mathbb{P}^2_{\mathbb{C}}$ defined by $z \neq 0$. From now on we use affine coordinates (x, y). In affine coordinates, the quaternion algebra (x, y) has nontrivial residues along $x = 0$ and $y = 0$.

Let $K = \mathbb{C}(\mathbb{P}^2)$. Let X_η/K be the (smooth) generic quadric. The discriminant of the quadratic form $q = \langle y, x, xy, F(x, y, 1) \rangle$ in K^* is not a square. Thus the map $\mathrm{Br}(K) \to \mathrm{Br}(X_\eta)$ is an isomorphism (see Sect. 3). Since the quaternion algebra

$\alpha = (x, y) \in \mathrm{Br}(\mathbb{C}(\mathbb{P}^2))$ has some nontrivial residues, it is nonzero in $\mathrm{Br}(\mathbb{C}(\mathbb{P}^2))$. Thus its image $\beta \in \mathrm{Br}(\mathbb{C}(X))$ is nonzero.   □

Let v be a discrete valuation of rank one on $L := K(X)$, let S be its valuation ring. Let κ_v denote the residue field. If $K \subset S$, then (x, y) is unramified. Suppose $S \cap K = R$ is a discrete valuation ring (of rank one). The image of the closed point of $\mathrm{Spec}(R)$ in $\mathbb{P}^2_{\mathbb{C}}$ is then either a point m of codimension 1 or a (complex) closed point m of $\mathbb{P}^2_{\mathbb{C}}$. By symmetry, for the argument we may assume that these points are in $\mathbb{A}^2_{\mathbb{C}}$.

Consider the first case. If the codimension 1 point m does not belong to $xy = 0$, then $\alpha = (x, y) \in \mathrm{Br}(K)$ is unramified at m on $\mathbb{A}^2_{\mathbb{C}}$ hence also in $\mathrm{Br}(L)$ at v. Moreover, the evaluation of β in $\mathrm{Br}(\kappa_v)$ is just the image under $\mathrm{Br}(\mathbb{C}(m)) \to \mathrm{Br}(\kappa_v)$ of the image of α in $\mathrm{Br}(\mathbb{C}(m))$, hence vanishes since $\mathrm{Br}(\mathbb{C}(m)) = 0$ (Tsen).

Suppose m is a generic point of a component of $xy = 0$. By symmetry, it is enough to examine the affine case where the point m of codimension 1 is the generic point of $x = 0$. In the function field L, we have an identity

$$yU^2 + xV^2 + xyW^2 + F(x, y, 1) = 0$$

with $yU^2 + xV^2 \neq 0$. In the completion of K at the generic point of $x = 0$, $F(x, y, 1)$ is a square, because $F(x, y, 1)$ modulo x is equal to $(y - 1)^2$, a nonzero square. Thus in the completion L_v we have an equality (with some other elements $U, V, W \in L_v$).

$$yU^2 + xV^2 + xyW^2 + 1 = 0.$$

This gives $(x, y)_{L_v} = 0 \in \mathrm{Br}(L_v)$. Hence $(x, y)_L$ is unramified at v, thus belongs to $\mathrm{Br}(R)$ and has image 0 in $\mathrm{Br}(\kappa_v)$.

Suppose we are in the second case, i.e. m is a closed point of $\mathbb{A}^2_{\mathbb{C}}$. There is a local map $O_{\mathbb{A}^2_{\mathbb{C}}, m} \to S$ which induces a map $\mathbb{C} \to \kappa_v$. If $x \neq 0$, then x becomes a nonzero square in the residue field $\mathbb{C}$ hence in κ_v, and the residue of $(x, y)_L$ at v is trivial. The analogous argument holds if $y \neq 0$. It remains to discuss the case $x = y = 0$. We have $F(0, 0, 1) = 1 \in \mathbb{C}^*$. Thus $F(x, y, 1)$ reduces to 1 in κ_v, hence is a square in the completion L_v. As above, in the completion L_v we have an equality

$$yU^2 + xV^2 + xyW^2 + 1 = 0,$$

which implies $(x, y)_{L_v} = 0 \in \mathrm{Br}(L_v)$. Hence $(x, y)_L$ is unramified at v, thus belongs to $\mathrm{Br}(S)$ and has image 0 in $\mathrm{Br}(\kappa_v)$.   □

As in the reinterpretation [5] of the Artin–Mumford examples, the intuitive idea behind the above result is that the quadric bundle $X \to \mathbb{P}^2_{\mathbb{C}}$ is ramified along $x.y.z.F(x, y, z) = 0$ on $\mathbb{P}^2_{\mathbb{C}}$ and that the ramification of the symbol $(x/z, y/z)$ on $\mathbb{P}^2_{\mathbb{C}}$, which is "included" in the ramification of the quadric bundle $X \to \mathbb{P}^2_{\mathbb{C}}$ disappears over smooth projective models of X : ramification eats up ramification (Abhyankar's lemma). Here one also uses the fact that the smooth conic defined by $F(x, y, z) = 0$ is tangent to each of the lines $x = 0$, $y = 0$, $z = 0$, and does not vanish at any of the intersection of these three lines.

5   The Specialisation Argument

The following theorem is an improvement by Schreieder [18, Prop. 26] of the specialisation method, as initiated by Voisin [22], in the format later proposed by Colliot-Thélène and Pirutka [6]. The assumptions in [18, Prop. 26] are more general than the ones given here. The generic fibre need not be smooth and one only requires that $f^{-1}(U) \to U$ be universally CH_0-trivial. There is a more general version which involves higher unramified cohomology with torsion coefficients. The proof is identical to the one given here with the Brauer group.

Schreieder's proof is cast in the geometric language of the decomposition of the diagonal. I provide a more "field-theoretic" proof. It is well known that both points of view are equivalent [3, 6]. I add a further, hopefully simplifying, twist by using specialization of R-equivalence on rational points instead of Fulton's specialisation theorem for the Chow group.

Theorem 5.1 *Let R be a discrete valuation ring, K its field of fractions, κ its residue field. Assume κ is algebraically closed and $char(\kappa) = 0$. Let $\overline{K}$ be an algebraic closure of K. Let $\mathcal{X}/R$ be an integral projective scheme over R, with generic fibre $X = \mathcal{X}_K/K$ smooth, geometrically integral, and with special fibre Z/κ geometrically integral. Assume there exists a nonempty open set $U \subset Z$ and a projective, birational desingularisation $f : \tilde{Z} \to Z$ such that $V := f^{-1}(U) \to U$ is an isomorphism, and such that the complement $\tilde{Z} \setminus V$ is a union $\cup_i Y_i$ of smooth irreducible divisors of $\tilde{Z}$. Assume that the $\overline{K}$-variety $X_{\overline{K}}$ is stably rational. If an element $\alpha \in \mathrm{Br}(\tilde{Z})$ vanishes on each Y_i, then $\alpha = 0 \in \mathrm{Br}(\tilde{Z})$.*

Proof To prove the result, one may assume $R = \kappa[[t]]$ (completion of the original R) and $K = \kappa((t))$. Assume $X_{\overline{K}}$ is stably rational. Then there exists a finite extension $K_1 = \kappa((t^{1/d}))$ of K over which X_{K_1} is K_1-stably rational. We replace $\mathcal{X}/R$ by $\mathcal{X} \times_R \kappa[[t^{1/d}]]$. This does not change the special fibre.

Changing notation once more, we now have $\mathcal{X}/R$ an integral projective scheme whose generic fibre X/K is stably rational over K and whose special fibre Z/κ is just as in the theorem. Fix $m \in V(\kappa)$, mapping to $n \in U(\kappa)$.

Let $L = \kappa(Z)$. We have the commutative diagram of exact sequences

$$
\begin{array}{ccccc}
\oplus_i CH_0(Y_{i,L}) & \to & CH_0(\tilde{Z}_L) & \to & CH_0(V_L) & \to & 0 \\
 & & \downarrow & & \downarrow \simeq & & \\
 & & CH_0(Z_L) & \to & CH_0(U_L) & \to & 0.
\end{array}
$$

where for each i, the closed embedding $\rho_i = Y_i \rightarrow \tilde{Z}$ induces

$$\rho_{i,*} : CH_0(Y_{i,L}) \rightarrow CH_0(\tilde{Z}_L),$$

the top exact sequence is the classical localisation sequence for the Chow group, the map $f_* : CH_0(\tilde{Z}_L) \rightarrow CH_0(Z_L)$ is induced by the proper map $f : \tilde{Z} \rightarrow Z$, the map $CH_0(V_L) \rightarrow CH_0(U_L)$ is the isomorphism induced by the isomorphism[1] $f : V \rightarrow U$, and the map $CH_0(Z_L) \rightarrow CH_0(U_L)$ is the obvious restriction map for the open set $U \subset Z$.

Let ξ be the generic point of $\tilde{Z}$ and η the generic point of Z.

Both η_L and n_L are smooth points of Y_L. There exists an extension $R \rightarrow S$ of complete dvr inducing $\kappa \rightarrow L$ on residue fields. Let F be the field of fractions of S. By Hensel's lemma, the points η_L and n_L lift to rational points of the generic fibre of $X \times_K F/F$ of $\mathcal{X}_S/S$. Since X/K is stably rational, all points of $X_F(F)$ are R-equivalent [7, Prop. 10] and [14, Cor. 6.6.6].

It is a well known fact [15, prop. 3.1] and [14, Comments after Thm. 6.6.2] that for a proper morphism $\mathcal{X}_S \rightarrow S$ over a discrete valuation ring S there is an induced map on R-equivalence classes $X(F)/R \rightarrow Z(L)/R$. This implies $\eta_L - n_L = 0 \in CH_0(Z_L)$.[2]

From the above diagram we conclude that

$$\xi_L = m_L + \sum_i \rho_{i*}(z_i) \in CH_0(\tilde{Z}_L)$$

with $z_i \in CH_0(Y_{i,L})$.

For the proper variety $\tilde{Z}_L$, there is a natural bilinear pairing

$$CH_0(\tilde{Z}_L) \times \mathrm{Br}(\tilde{Z}) \rightarrow \mathrm{Br}(L).$$

For the smooth, proper, integral variety $\tilde{Z}$, on the generic point $\xi \in \tilde{Z}_L(L)$, this pairing induces the embedding $\mathrm{Br}(\tilde{Z}) \hookrightarrow \mathrm{Br}(\kappa(Z))$. Suppose $\alpha \in \mathrm{Br}(\tilde{Z})$ vanishes in each $\mathrm{Br}(Y_i)$ (which follows from the vanishing in $\mathrm{Br}(\kappa(Y_i))$ because Y_i is smooth). The evaluation of α on m_L is just the image of $\alpha(m) \in \mathrm{Br}(\kappa) = 0$. The above equality implies $\alpha(\xi) = 0 \in \mathrm{Br}(L)$, hence $\alpha = 0 \in \mathrm{Br}(\tilde{Z})$. $\qquad\square$

---

[1]Instead of assuming that $f^{-1}(U) \rightarrow U$ is an isomorphism, it would be enough, as in [18], to assume that this morphism is a universal CH_0-isomorphism.

[2]Alternatively, one could argue as follows. Since X is stably rational over K, over any field F containing K, the degree map $CH_0(X_F) \rightarrow \mathbb{Z}$ is an isomorphism (for a simple proof, see [6, Lemme 1.5]). One could then invoke Fulton's specialisation theorem for the Chow group of a proper scheme over a dvr [11, §2, Prop. 2.6], to get $\eta_L - n_L = 0 \in CH_0(Z_L)$. Fulton's specialisation theorem is a nontrivial theorem. The argument via R-equivalence (cf. [6, Remarque 1.19]) looks simpler.

6 Stable Rationality is not Constant in Smooth Projective Families

We now complete the simplified proof of the theorem of Hassett, Pirutka and Tschinkel [13].

Theorem 6.1 *There exist a smooth projective family of complex fourfolds* f : $X \to T$ *parametrized by an open set* T *of the affine line* $\mathbb{A}^1_{\mathbb{C}}$ *and points* $m, n \in T(\mathbb{C})$ *such that the fibre* X_n *is rational and the fibre* X_m *is not stably rational.*

Proof One considers the universal family of quadric bundles over $\mathbb{P}^2_{\mathbb{C}}$ given in $\mathbb{P}^3_{\mathbb{C}} \times \mathbb{P}^2_{\mathbb{C}}$ by a bihomogenous form of bidegree $(2, 2)$. This is given by a symmetric $(4, 4)$ square matrix with entries $a_{i,j}(x, y, z)$ homogeneous quadratic forms in three variables (x, y, z). If its determinant is nonzero, it is a homogeneous polynomial of degree 8.

We thus have a parameter space B given by a projective space of dimension 59 (the corresponding vector space being given by the coefficients of ten quadratic forms in three variables). We have the map $X \to B$ whose fibres X_m are the various quadric bundles $X_m \to \mathbb{P}^2_{\mathbb{C}}$, for $X_m \subset \mathbb{P}^3_{\mathbb{C}} \times \mathbb{P}^2_{\mathbb{C}}$ given by the vanishing of a nonzero complex bihomogeneous form of bidegree $(2, 2)$.

Using Bertini's theorem, one shows that there exists a nonempty open set $B_0 \subset B$ such that the fibres of $X \to B$ over points of $m \in B_0$ are flat quadric bundles $X_m \to \mathbb{P}^2_{\mathbb{C}}$ which are smooth as $\mathbb{C}$-varieties.

Using Bertini's theorem, one also shows that there exist points $m \in B_0$ with the property that the corresponding quadric bundle has $a_{1,1} = 0$, which implies that the fibration $X_m \to \mathbb{P}^2_{\mathbb{C}}$ has a rational section (given by the point $(1, 0, 0, 0)$), hence that the generic fibre of $X_m \to \mathbb{P}^2_{\mathbb{C}}$ is rational over $\mathbb{C}(\mathbb{P}^2)$, hence that the $\mathbb{C}$-variety X_m is rational over $\mathbb{C}$. [Warning : this Bertini argument uses the fact that we consider families of quadric surfaces over $\mathbb{P}^2_{\mathbb{C}}$. It does not work for families of conics over $\mathbb{P}^2_{\mathbb{C}}$.]

These Bertini arguments are briefly described in [18, Lemma 20 and Thm. 47] and are tacitly used in [19, p. 3].

By Proposition 4.1, the special example in Sect. 4 defines a point $P_0 \in B(\mathbb{C})$ whose fibre is $Z = X_{P_0}$ and which admits a projective birational desingularisation $f : \tilde{Z} \to Z$ satisfying :

(a) there exists a nonempty open set $U \subset Z$, such that the induced map $V := f^{-1}(U) \to U$ is an isomorphism;
(b) the complement $\tilde{Z} \setminus V$ is a union $\cup_i Y_i$ of smooth irreducible divisors of $\tilde{Z}$;
(c) there is a nontrivial element $\alpha \in \mathrm{Br}(\tilde{Z})$ which vanishes on each Y_i.

Theorem 5.1 then implies that the generic fibre of $X \to B$ is not geometrically stably rational. There are various ways to conclude from this that there are many points $m \in B_0(\mathbb{C})$ such that the fibre X_m is not stably rational.

Take one such point $m \in B_0(\mathbb{C})$ and a point $n \in B_0(\mathbb{C})$ such that X_n is rational. Over an open set of the line joining m and n we get a projective family of smooth varieties with one fibre rational and with one fibre not stably rational. □

The proof by Hassett, Pirutka and Tschinkel [13] uses an explicit desingularisation of the variety Z in Sect. 4, with a description of the exceptional divisors appearing in the process. Schreieder's improvement of the specialisation method enables one to bypass this explicit desingularisation.

Note that [13] and [19] contain many more results on families of quadrics surfaces over $\mathbb{P}^2$ than Theorem 6.1.

Appendix 1: Conics over a Discrete Valuation Ring

Let R be a dvr with residue field k of characteristic not 2. Let K be the fraction field. A smooth conic over K admits a regular model $\mathcal{X}$ given in $\mathbb{P}^2_R$ either by an equation

$$x^2 - ay^2 - bz^2 = 0$$

with $a, b \in R^*$ (case (I)) or a regular model $\mathcal{X}$ given by an equation

$$x^2 - ay^2 - \pi z^2 = 0$$

with $a \in R^*$ and π a uniformizing parameter (case (II)). Moreover, in the second case one may assume that a is not a square in the residue field κ.

Proposition 1 *Let R be a dvr with residue field k of characteristic not 2. Let K be the fraction field. Let $W \to \mathrm{Spec}(R)$ be a proper flat morphism with W regular and connected. Assume that the generic fibre is a smooth conic over K. Then:*

(a) The natural map $\mathrm{Br}(R) \to \mathrm{Br}(W)$ is onto.

(b) For $Y \subset W$ an integral divisor contained in the special fibre of $W \to \mathrm{Spec}(R)$, and $\beta \in \mathrm{Br}(W)$, the image of β under restriction $\mathrm{Br}(W) \to \mathrm{Br}(Y)$ belongs to the image of $\mathrm{Br}(\kappa) \to \mathrm{Br}(Y)$.

Proof By purity for the Brauer group of a two-dimensional regular scheme, to prove (a), one may assume that $W = \mathcal{X}$ as above. Let $X = \mathcal{X} \times_R K$. It is well known that the map $\mathrm{Br}(K) \to \mathrm{Br}(X)$ is onto, with kernel spanned by the quaternion symbol $(a, b)_K$ in case (I) and by $(a, \pi)_K$ in case (II).

Let $\beta \in \mathrm{Br}(\mathcal{X}) \subset \mathrm{Br}(X)$. Let $\alpha \in \mathrm{Br}(K)$ be some element with image β_K. We have the exact sequence

$$0 \to \mathrm{Br}(R)\{2\} \to \mathrm{Br}(K)\{2\} \to H^1(\kappa, \mathbb{Q}_2/\mathbb{Z}_2)$$

Comparison of residues on $\mathrm{Spec}(R)$ and on $\mathcal{X}$ shows that the residue $\delta_R(\alpha)$ is either 0 or is equal to the nontrivial class in $H^1(k(\sqrt{\overline{a}})/k, \mathbb{Z}/2)$, and this last case may happen only in case (II). In the first case, we have $\alpha \in \mathrm{Br}(R)$, hence $\beta - \alpha_\mathcal{X} = 0$ in $\mathrm{Br}(X)$ hence also in $\mathrm{Br}(\mathcal{X})$ since $\mathcal{X}$ is regular. In the second case, we have

$$\delta_R(\alpha) = \delta_R((a, \pi))$$

hence $\alpha = (a, \pi) + \gamma$ with $\gamma \in \mathrm{Br}(R)$. We then get

$$\beta = (a, \pi)_{K(X)} + \gamma_{K(X)} \in \mathrm{Br}(K(X)).$$

But $(a, \pi)_{K(X)} = 0$. Thus $\beta - \gamma_\mathcal{X} \in \mathrm{Br}(\mathcal{X}) \subset \mathrm{Br}(K(X))$ vanishes, hence $\beta = \gamma_\mathcal{X} \in \mathrm{Br}(\mathcal{X})$. The map $\mathrm{Br}(R) \to \mathrm{Br}(\mathcal{X})$ is thus surjective. This gives (a) for $\mathcal{X}$ hence for W, and (b) immediately follows.   □

Exercise Artin-Mumford type examples are specific singular conic bundles X in the total space of a rank 3 projective bundle over $\mathbb{P}^2_\mathbb{C}$ whose unramified Brauer group is non trivial. Using Proposition 1 and Theorem 5.1, deform such examples into conic bundles of the same type with smooth ramification locus and whose total space is not stably rational. As in Sect. 4, there is no need to compute an explicit resolution of singularities of X.

Appendix 2: Quadric Surfaces over a Discrete Valuation Ring

The following section was written up to give details on some tools and results used in [19, Thm. 4]. As demonstrated above, this section turns out not to be necessary to vindicate the HPT example. But it is useful for more general examples.

References: [21], [8, §3], [9, Thm. 2.3.1], [17, Thm. 3.17].

Let R be a discrete valuation ring, K its fraction field, π a uniformizer, $\kappa = R/(\pi)$ the residue field. Assume $char(\kappa) \neq 2$.

Let $X \subset \mathbb{P}^3_K$ be a smooth quadric, defined by a nondegenerate four-dimensional quadratic form q. Up to scaling and changing of variables, there are four possibilities.

(I) $q = <1, -a, -b, abd>$ with $a, b, d \in R^*$.
(II) $q = <1, -a, -b, \pi>$ with $a, b \in R^*$ and π a uniformizing parameter of R.
(III) $q = <1, -a, \pi, -\pi.b>$ with $a, b \in R^*$ and π a uniformizing parameter of R. The class of $a.b \in R^*$ represents the discriminant of the quadratic form. Its image $\overline{a}.\overline{b} \in \kappa^*$ is a square if and only if the discriminant of q is a square in the completion of K for the valuation defined by R.

Let $\mathcal{X} \subset \mathbb{P}^3_R$ be the subscheme cut out by q. Let Y/κ be the special fibre.
In case (I), $\mathcal{X}/R$ is smooth.
In case (II), X is regular, the special fibre Y is a cone over a smooth conic.
In case (III), the special fibre is given by the equation $x^2 - \overline{a}y^2 = 0$ in $\mathbb{P}^3_\kappa$. If $\overline{a}$ is a square, this is the union of two planes intersecting along the line $x = y = 0$. If $\overline{a}$ is not a square, this is an integral scheme which over $\kappa(\sqrt{a})$ breaks up as the

union of two planes. In both cases, the scheme $\mathcal{X}$ is singular at the two points given by $x = y = 0$, $z^2 - \overline{d}t^2 = 0$. See [21, §2].

Proposition 1 *Let us assume char*$(\kappa) = 0$.

In case (III), let $W \to \mathcal{X}$ be a projective, birational desingularisation of $\mathcal{X}$.

In case (I), the map $\mathrm{Br}(R) \to \mathrm{Br}(\mathcal{X})$ *is onto. If $d \in R$ is not a square, it is an isomorphism. If d is a square, the kernel is spanned by the class* $(a, b) \in \mathrm{Br}(R)$.

In case (II), the map $\mathrm{Br}(R) \to \mathrm{Br}(\mathcal{X})$ *is an isomorphism.*

In case (III), assume $\overline{a}.\overline{b}$ is not a square in κ. Then $\mathrm{Br}(R) \to \mathrm{Br}(W)$ *is onto.*

In case (III), if either $\overline{a}$ or $\overline{b}$ is a square, or if $\overline{a}.\overline{b}$ is not a square, then $\mathrm{Br}(R) \to \mathrm{Br}(W)$ *is onto. An element of* $\mathrm{Br}(K)$ *whose image in* $\mathrm{Br}(X)$ *lies in* $\mathrm{Br}(W)$ *belongs to* $\mathrm{Br}(R)$.

In case (III), assume $\overline{a}.\overline{b}$ is a square in κ. Then the image of $(a, \pi) \in \mathrm{Br}(K)$ in $\mathrm{Br}(X)$ *belongs to* $\mathrm{Br}(W)$. *It spans the quotient of* $\mathrm{Br}(W)$ *by the image of* $\mathrm{Br}(R)$. *If moreover $\overline{a}$ is not a square in κ, then it does not belong to the image of* $\mathrm{Br}(R)$.

Proof Let x be a codimension 1 regular point on $\mathcal{X}$ or on W, lying above the closed point of $\mathrm{Spec}(R)$. Let e_x denote its multiplicity in the fibre. We have a commutative diagram

$$
\begin{array}{ccc}
\mathrm{Br}(K) & \to & H^1(\kappa, \mathbb{Q}/\mathbb{Z}) \\
\downarrow & & \downarrow \\
\mathrm{Br}(X) & \to & H^1(\kappa(x), \mathbb{Q}/\mathbb{Z})
\end{array}
$$

The kernel of $\mathrm{Br}(K) \to H^1(\kappa, \mathbb{Q}/\mathbb{Z})$ is $\mathrm{Br}(R)$.

In case (I) and (III), the special fibre Y is geometrically integral over κ, the multiplicity is 1, the map $H^1(\kappa, \mathbb{Q}/\mathbb{Z}) \to H^1(\kappa(x), \mathbb{Q}/\mathbb{Z})$ is thus injective. This is enough to prove the claim.

Let us consider case (III). The map $\mathrm{Br}(K) \to \mathrm{Br}(X)$ is onto. Let $\alpha \in \mathrm{Br}(K)$. Let $\rho \in H^1(\kappa, \mathbb{Q}/\mathbb{Z})$ be its residue. On the (singular) normal model given by $q = <1, -a, \pi, -\pi.b>$ over R, if $\overline{a} \in \kappa$ is a square, the fibre Y contains geometrically integral components of multiplicity 1 given by the components of $x^2 - \overline{a}y^2 = 0$. By the above diagram, $\rho = 0 \in H^1(\kappa, \mathbb{Q}/\mathbb{Z})$. We can also use the model given by $q = <1, -b, \pi, -\pi.a>$. If $\overline{b} \in \kappa$ is a square, we conclude that $\rho = 0 \in H^1(\kappa, \mathbb{Q}/\mathbb{Z})$. Let us assume that $\rho \neq 0 \in H^1(\kappa, \mathbb{Q}/\mathbb{Z})$. Thus $\overline{a}$ and $\overline{b}$ are nonsquares. On the first model, the kernel of $H^1(\kappa, \mathbb{Q}/\mathbb{Z}) \to H^1(\kappa(Y), \mathbb{Q}/\mathbb{Z})$ coincides with the kernel of $H^1(\kappa, \mathbb{Q}/\mathbb{Z}) \to H^1(\kappa(\sqrt{\overline{a}}, \mathbb{Q}/\mathbb{Z})$, which is the $\mathbb{Z}/2$-module spanned by the class of $\overline{a}$ in $\kappa^*/\kappa^{*2} = H^1(\kappa, \mathbb{Z}/2)$. On the second model, the kernel of $H^1(\kappa, \mathbb{Q}/\mathbb{Z}) \to H^1(\kappa(Y), \mathbb{Q}/\mathbb{Z})$ is the $\mathbb{Z}/2$-module spanned by the class of $\overline{b}$ in $\kappa^*/\kappa^{*2} = H^1(\kappa, \mathbb{Z}/2)$. We thus conclude that $\overline{a}.\overline{b}$ is a square in κ, and that the residue of α coincides with $\overline{a}$, i.e. is equal to the residue of $(a, \pi) \in \mathrm{Br}(K)$ (or to the residue of (b, π)).

It remains to show that if $\overline{a}.\overline{b}$ is a square in κ, then (a, π) has trivial residues on W and more generally with respect to any rank one discrete valuation v on the function field $K(X)$ of X. One may restrict attention to those v which induce the

R-valuation on K. Let $S \subset K(X)$ be the valuation ring of v and let λ be its residue field. There is an inclusion $\kappa \subset \lambda$. In $K(X)$ we have an equality

$$(x^2 - ay^2) = \pi.(z^2 - b),$$

where both sides are nonzero. Thus in $\mathrm{Br}(K(X))$, we have the equality

$$(a, \pi) = (a, x^2 - ay^2) + (a, z^2 - b) = (a, z^2 - b),$$

where the last equality comes from the classical $(a, x^2 - ay^2) = 0$. To compute residues, we may go over to completions. In the completion of R, $a.b$ is a square. It is thus a square in the completion of $K(X)$ at v. But then in this completion $(a, z^2 - b) = (b, z^2 - b) = 0$ Hence the residue of (a, π) at v is zero. $\qquad\square$

Proposition 2 *Assume* $char(\kappa) = 0$. *Let* $\mathcal{X} \subset \mathbb{P}_R^3$ *be as above, and let* $W \to \mathcal{X}$ *be a proper birational map with* W *regular. Let* $\beta \in \mathrm{Br}(W)$ *and let* $Y \subset W$ *be an integral divisor contained in the special fibre of* $W \to \mathrm{Spec}(R)$. *Then the image of* β *in* $\mathrm{Br}(\kappa(Y))$ *belongs to the image of* $\mathrm{Br}(\kappa) \to \mathrm{Br}(\kappa(Y))$.

Proof In case (I) and (II), and in case (III) when $\overline{a}.\overline{b}$ is a square in κ, this is clear since then the map $\mathrm{Br}(R) \to \mathrm{Br}(W)$ is onto.

Suppose we are in case (III). To prove the result, we may make a base change from R to its henselisation. Then ab is square in R. The group $\mathrm{Br}(W)$ is spanned by the image of $\mathrm{Br}(R)$ and the image of the class (a, π). The equation of the quadric may now be written

$$X^2 - aY^2 + \pi Z^2 - a\pi T^2 = 0.$$

This implies that $(a, -\pi)$ vanishes in the Brauer group of the function field $\kappa(W)$ of W. Since W is regular, the map $\mathrm{Br}(W) \to \mathrm{Br}(\kappa(W))$ is injective. Since $(a, -\pi)_{\kappa(W)}$ belongs to $\mathrm{Br}(W)$ and spans $\mathrm{Br}(W)$ modulo the image of $\mathrm{Br}(R)$, this completes the proof. $\qquad\square$

One may rephrase the above results in a simpler fashion.

Proposition 3 *Assume* $char(\kappa) = 0$. *Let* $\mathcal{X} \subset \mathbb{P}_R^3$ *be as above, and let* $W \to \mathcal{X}$ *be a proper birational map with* W *regular.*

(i) *If* R *is henselian, then the map* $\mathrm{Br}(R) \to \mathrm{Br}(W)$ *is onto.*

(ii) *For any element* $\beta \in \mathrm{Br}(W)$ *and* $Y \subset W$ *an integral divisor contained in the special fibre of* $W \to \mathrm{Spec}(R)$, *the image of* β *under restriction* $\mathrm{Br}(W) \to \mathrm{Br}(Y)$ *belongs to the image of* $\mathrm{Br}(\kappa) \to \mathrm{Br}(Y)$.

Upon use of Merkurjev's geometric lemmas [16, §1], and use of Tsen's theorem, one then gets [19, Prop. 7] of Schreieder.

Appendix 3: A Remark on the Vanishing of Unramified Elements on Components of the Special Fibre

The following proposition, found in June 2017, gives some partial explanation for the vanishing on components of the special fibre which occurs in [18, Prop. 6, Prop. 7] and [19, Prop. 7] or in Proposition 4.1 above. Unfortunately the proof requires that the component be of multiplicity one in the fibre. Since this was written, in the case of quadric bundles, Schreieder [20, §9.2] has managed to use arguments as in [8, §3] to get information on what happens with the other components.

Proposition 1 *Let $A \hookrightarrow B$ be a local homomorphism of discrete valuation rings and let $K \subset L$ be the inclusion of their fraction fields. Let $\kappa \subset \lambda$ be the induced inclusion on their residue fields.*

Let ℓ be a prime invertible in A.

Let $i \geq 2$ be an integer and let $\alpha \in H^i(K, \mu_\ell^{\otimes i})$.

Assume:

(i) B is unramified over A.

(ii) The image of α in $H^i(L, \mu_\ell^{\otimes i})$ is unramified, and in particular is the image of a (well defined) element $\beta \in H^i(B, \mu_\ell^{\otimes i})$.

Then $\beta(\lambda) \in H^i(\lambda, \mu_\ell^{\otimes i})$ is in the image of $H^i(\kappa, \mu_\ell^{\otimes i}) \to H^i(\lambda, \mu_\ell^{\otimes i})$.

Proof We may assume that A and B are henselian. Then the residue map $\partial_A : H^i(K, \mu_\ell^{\otimes i}) \to H^{i-1}(\kappa, \mu_\ell^{\otimes(i-1)})$ is part of a split exact sequence [10, Appendix B] and [12, Cor. 6.8.8]

$$0 \to H^i(A, \mu_\ell^{\otimes i}) \to H^i(K, \mu_\ell^{\otimes i}) \to H^{i-1}(\kappa, \mu_\ell^{\otimes(i-1)}) \to 0,$$

and all reduction maps $H^j(A, \mu_\ell^{\otimes i}) \to H^j(\kappa, \mu_\ell^{\otimes i})$, denoted $\rho \mapsto \rho(\kappa)$, are isomorphisms. We have the analogous split exact sequence

$$0 \to H^i(B, \mu_\ell^{\otimes i}) \to H^i(L, \mu_\ell^{\otimes i}) \to H^{i-1}(\lambda, \mu_\ell^{\otimes(i-1)}) \to 0,$$

Let $\pi \in A$ be a uniformizer. Given $\alpha \in H^i(K, \mu_\ell^{\otimes i})$, the residue $\partial_A(\alpha) \in H^{i-1}(\kappa, \mu_\ell^{\otimes(i-1)})$ is the image of some unique $\gamma \in H^{i-1}(A, \mu_\ell^{\otimes(i-1)})$. Let us denote by (π) the class of π in $K^*/K^{*\ell} = H^1(K, \mu_\ell)$. Then the difference $\alpha - (\pi) \cup \gamma \in H^i(K, \mu_\ell^{\otimes i})$ has trivial residue. Thus there exists $\zeta \in H^i(A, \mu_\ell^{\otimes i})$ such that

$$\alpha = \zeta + (\pi) \cup \gamma \in H^i(K, \mu_\ell^{\otimes i}).$$

By hypothesis, the restriction β of α to L is unramified. Thus $(\pi) \cup \gamma \in H^i(L, \mu_\ell^{\otimes i})$ is unramified. Since B is unramified over A, the uniformizer π is also a uniformizer of B. Thus

$$0 = \partial_B((\pi) \cup \gamma) = \gamma_\lambda \in H^{i-1}(\lambda, \mu_\ell^{\otimes(i-1)})$$

hence $\gamma = 0 \in H^{i-1}(B, \mu_\ell^{\otimes(i-1)})$, from which follows $\beta = \zeta \in H^i(B, \mu_\ell^{\otimes i})$ and $\beta(\lambda) \in H^i(\lambda, \mu_\ell^{\otimes i})$ is the image of $\zeta(\kappa) \in H^i(\kappa, \mu_\ell^{\otimes i})$.   □

References

1. A. Auel, C. Boehning, H.-G. Graf von Bothmer, A. Pirutka, Conic bundles over threefolds with nontrivial unramified Brauer group (2016), https://arXiv:1610.04995
2. A. Auel, A. Bigazzi, C. Böhning, H.-G. Graf von Bothmer, Universal triviality of the Chow group of 0-cycles and the Brauer group (2018), https://arxiv.org/abs/1806.02676
3. A. Auel, J.-L. Colliot-Thélène, R. Parimala, Universal unramified cohomology of cubic fourfolds containing a plane, in *Brauer Groups and Obstruction Problems: Moduli Spaces and Arithmetic*. Progress in Mathematics, vol. 320 (Birkhäuser, Basel, 2017), pp. 29–56
4. A. Auel, C. Boehning, A. Pirutka, Stable rationality of quadric and cubic surface bundle fourfolds. Eur. J. Math. **4**(3), 732–760 (2018)
5. J.-L. Colliot-Thélène, M. Ojanguren, Variétés unirationnelles non rationnelles : au-delà de l'exemple d'Artin et Mumford. Invent. Math. **97**, 141–158 (1989)
6. J.-L. Colliot-Thélène, A. Pirutka, Hypersurfaces quartiques de dimension 3 : non rationalité stable. Ann. Sc. Éc. Norm. Sup. **49**, 371–397 (2016)
7. J.-L. Colliot-Thélène, J.-J. Sansuc, La R-équivalence sur les tores. Ann. Sci. École Norm. Sup. **10**, 175–229 (1977)
8. J.-L. Colliot-Thélène, A.N. Skorobogatov, Groupe de Chow des zéro-cycles sur les fibrés en quadriques. K-Theory **7**, 477–500 (1993)
9. J.-L. Colliot-Thélène, P. Swinnerton-Dyer, Hasse principle and weak approximation for pencils of Severi-Brauer and similar varieties. J. für die reine und angew. Math. **453**, 49–112 (1994)
10. J.-L. Colliot-Thélène, R.T. Hoobler, B. Kahn, The Bloch-Ogus–Gabber theorem, in *Proceedings of the Great Lakes K-Theory Conference (Toronto 1996)*, ed. by R. Jardine, V. Snaith. The Fields Institute for Research in Mathematical Sciences Communications Series, vol. 16 (A.M.S., Providence, 1997), pp. 31–94
11. B. Fulton, *Intersection Theory*. Ergebnisse der Math. und ihr. Grenzg. vol. 2 (Springer, Berlin, 1998)
12. P. Gille, T. Szamuely, *Central Simple Algebras and Galois Cohomology*. Cambridge Studies in Advanced Mathematics, vol. 165, 2nd edn. (Cambridge University Press, Cambridge, 2017)
13. B. Hassett, A. Pirutka, Yu. Tschinkel, Stable rationality of quadric surface bundles over surfaces. Acta Math. **220**(2), 341–365 (2018)
14. B. Kahn, R. Sujatha, Birational geometry and localisation of categories. With appendices by Jean-Louis Colliot-Thélène and by Ofer Gabber. Doc. Math. **2015**, 277–334 (2015), Extra vol.: Alexander S. Merkurjev's sixtieth birthday
15. D.A. Madore, Sur la spécialisation de la R-équivalence, https://perso.telecom-paristech.fr/~madore/specialz.pdf
16. A.S. Merkurjev, Unramified elements in cycle modules. J. Lond. Math. Soc. **78**, 51–64 (2008)
17. A. Pirutka, Varieties that are not stably rational, zero-cycles and unramified cohomology, in *Algebraic Geometry: Salt Lake City 2015*, pp. 459–483. *Proceedings of Symposia in Pure Mathematics*, vol. 97(2) (American Mathematical Society, Providence, 2018)

18. S. Schreieder, On the rationality problem for quadric bundles. Duke Math. J. **168**, 187–223 (2019)
19. S. Schreieder, Quadric surface bundles over surfaces and stable rationality. Algebra Number Theory **12**, 479–490 (2018)
20. S. Schreieder, Stably irrational hypersurfaces of small slopes. J. Am. Math. Soc. (to appear). https://arxiv.org/abs/1801.05397
21. A. Skorobogatov, Arithmetic on certain quadric bundles of relative dimension 2. I. J. für die reine und angew. Math. **407**, 57–74 (1990)
22. C. Voisin, Unirational threefolds with no universal codimension 2 cycle. Invent. math. **201**, 207–237 (2015)

Part II
Hypersurfaces

The Rigidity Theorem
of Fano–Segre–Iskovskikh–Manin–
Pukhlikov–Corti–Cheltsov–de Fernex–
Ein–Mustaţă–Zhuang

János Kollár

Abstract We prove that n-dimensional smooth hypersurfaces of degree $n + 1$ are superrigid. Starting with the work of Fano in 1915, the proof of this Theorem took 100 years and a dozen researchers to construct. Here I give complete proofs, aiming to use only basic knowledge of algebraic geometry and some Kodaira type vanishing theorems.

The classification theory of algebraic varieties—developed by Enriques for surfaces and extended by Iitaka and then Mori to higher dimensions—says that every variety can be built from three basic types:

- (General type) the canonical class K_X is ample,
- (Calabi-Yau) K_X is trivial and
- (Fano) $-K_X$ is ample.

Moreover, in the Fano case the truly basic ones are those that have *class number* equal to 1. That is, every divisor D on X is linearly equivalent to a (possibly rational) multiple of $-K_X$.

If two varieties X_1, X_2 on the basic type list are birationally equivalent then they have the same type. In the general type case they are even isomorphic and in the Calabi-Yau case the possible birational maps are reasonably well understood, especially for threefolds, see [28, 29].

By contrast, Fano varieties are sometimes birationally equivalent in quite unexpected ways and the Noether–Fano method aims to understand what happens.

Definition 1 (Weak Rigidity and Superrigidity) I call a Fano variety X with class number 1 *weakly rigid* if it is not birational to any other Fano variety Y with class number 1, and *weakly superrigid* if every birational map $\Phi : X \dashrightarrow Y$ to another Fano variety Y with class number 1 is an isomorphism.

J. Kollár (✉)
Princeton University, Princeton, NJ, USA
e-mail: kollar@math.princeton.edu

© Springer Nature Switzerland AG 2019   129
A. Hochenegger et al. (eds.), *Birational Geometry of Hypersurfaces*,
Lecture Notes of the Unione Matematica Italiana 26,
https://doi.org/10.1007/978-3-030-18638-8_4

The adjective "weakly" is not standard; it allows us to define these notions without first discussing terminal singularities and Mori fiber spaces. The definitions of *rigid* and *superrigid* are similar, but allow Y to have terminal singularities and to be a Mori fiber space; see [3, 49].

There are many Fano varieties, especially in dimensions 2 and 3, that are rigid but not superrigid. Superrigidity is the more basic notion, though, in dimension 3, the theory of rigid Fano varieties is very rich.

The aim of these notes is to explain the proof of the following theorem. From now on we work over a field of characteristic 0. It is not important, but we may as well assume that it is algebraically closed.

Main Theorem 2 *Every smooth hypersurface $X \subset \mathbb{P}^{n+1}$ of dimension $n \geq 3$ and of degree $n + 1$ is weakly superrigid.*

The proofs in the theory are designed to prove superrigidity, and the optimal version of Theorem 2 says that a smooth Fano hypersurface $X \subset \mathbb{P}^{n+1}$ of dimension ≥ 3 is superrigid if and only if $\deg X = n + 1$; see [3, 52]. The proof of this version needs only some new definitions and minor changes in Step 10.1.

A smooth hypersurface $X \subset \mathbb{P}^{n+1}$ of dimension $n \geq 3$ has class number 1 by Lefschetz's theorem (see [39] or [17, p. 156]). If $n = 2$ then X is a cubic surface, hence it has class number 7. However, if the base field is not algebraically closed, it frequently happens that X has class number 1, in which case it is weakly rigid but usually not weakly superrigid by Segre [59]; see [37, Chap. 2] for a modern treatment.

3 (The History of Theorem 2) The first similar result is Max Noether's description of all birational maps $\mathbb{P}^2 \dashrightarrow \mathbb{P}^2$ [46]. Noether's method formed the basis of all further developments. It was used by Segre to study birational maps of cubic surfaces over arbitrary fields [59], and later generalized by Manin and Iskovskikh to a birational theory of all del Pezzo surfaces and two-dimensional conic bundles [21, 42].

Theorem 2 was first stated by Fano for threefolds [13, 14]. His arguments contain many of the key ideas, but they also have gaps. I call this approach the *Noether–Fano method*. The first complete proof for threefolds, along the lines indicated by Fano, is in Iskovskikh-Manin [23]. Iskovskikh and his school used this method to prove similar results for many other threefolds, see [20, 22, 24, 57]. This approach was gradually extended to higher dimensions by Pukhlikov [48, 50, 51] and Cheltsov [2]. These results were complete up to dimension 8, but needed some additional general position assumptions in higher dimensions. A detailed survey of this direction is in [52].

The theory of Fano varieties may be the oldest topic of higher dimensional birational geometry, but for a long time it grew almost independently of Mori's Minimal Model Program. The Fano–Iskovskikh classification of Fano threefolds using extremal rays and flops was first treated by Mori [44] and later improved by Takeuchi [64].

The Noether–Fano method and the Minimal Model Program were brought together by Corti [5]. Corti's technique has been very successful in many cases, especially for threefolds; see [7] for a detailed study and [37, Chap. 5] for an introduction. However, usually one needs some special tricks to make the last steps work, and a good higher dimensional version proved elusive for a long time.

New methods involving multiplier ideals were introduced by de Fernex et al. [11]; these led to a more streamlined proof that worked up to dimension 12. The proof of Theorem 2 was finally completed by de Fernex [9].

The recent paper of Zhuang [67] makes the final step of the Corti approach much easier in higher dimensions. The papers [10, 41, 62, 63, 67] contain more general results and applications.

The name Fano–Segre–Iskovskikh–Manin–Pukhlikov–Corti–Cheltsov–de Fernex–Ein–Mustaţă–Zhuang theorem was chosen to give credit to all those with a substantial contribution to the proof, though this under emphasizes the major contributions of Fano, Iskovskikh and Pukhlikov.

The methods apply to many other Fano varieties for which $-K_X$ is a generator of the class group; see [52, 67] for several examples. One of the big challenges is to understand what happens if $-K_X$ is a multiple of the generator, see [53].

Open Problems About Hypersurfaces

The following questions are stated in the strongest forms that are consistent with the known examples. I have no reasons to believe that the answer to either of them is positive, and there may well be rather simple counter examples. As far as I know, there has been very little work on low degree hypersurfaces beyond cubics in dimension 4.

Question 4 Is every smooth hypersurface of degree ≥ 4 non-rational?

Non-rationality of a smooth hypersurface $X \subset \mathbb{P}^{n+1}$ is obvious if $\deg X \geq n+2$. For very general hypersurfaces of degree $\geq \frac{2}{3}n + 3$, non-rationality was proved in [31], a major improvement by Schreieder [58] shows this for $\deg X \geq \log_2 n + 2$.

Question 5 Is every smooth hypersurface of degree ≥ 5 weakly superrigid?

Here ≥ 5 is necessary since there are some smooth quartics with nontrivial birational maps.

Example 6 Let $X \subset \mathbb{P}^{2n+1}$ be a quartic hypersurface that contains 2 disjoint linear subspaces L_1, L_2 of dimension n. For every $p \in \mathbb{P}^{2n+1} \setminus (L_1 \cup L_2)$ there is a unique line ℓ_p through p that meets both L_1, L_2. This line meets X in 4 points, two of these are on L_1, L_2. If $p \in X$ then this leaves a unique 4th intersection point, call it $\Phi(p)$. Clearly Φ is an involution which is not defined at p if either $p \in L_1 \cup L_2$ or if $\ell_p \subset X$.

1   Rigidity and Superrigidity, an Overview

In the following outlines and in subsequent Sections I aim to put the pieces together and write down a simple proof of the superrigidity of smooth hypersurfaces $V_{n+1}^n \subset \mathbb{P}^{n+1}$, where most steps are either easy or are direct applications of some general principle of the Minimal Model Program.

The key notion we need is canonical and log canonical pairs involving linear systems.

Definition 7 (Log Resolution)   Assume that we have a variety X, a (not necessarily complete) linear system $|M|$ on X and a divisor D on X. A *log resolution* of these data is a proper birational morphism $\pi : X' \to X$ such that

(1) X' is smooth,
(2) $\pi^*|M| = |M'| + B$ where $|M'|$ is base-point free and B is the fixed part of $\pi^*|M|$, and
(3) $B + \pi_*^{-1}(D) + \mathrm{Ex}(\pi)$ is a simple normal crossing divisor.

(Here $\mathrm{Ex}(\pi)$ denotes the exceptional set of π, $\pi_*^{-1}(D)$ denotes the birational transform of D and *simple normal crossing* means that the irreducible components are smooth and they intersect transversally. The adjective "log" loosely refers to condition (3).)

The existence of log resolutions was proved by Hironaka; see [34, Chap. 3] for a recent treatment.

Definition 8 (Canonical, Log Canonical, Etc.)   Let X be a smooth variety and $|M|$ a linear system on X. Let $\pi : X' \to X$ be a log resolution of $|M|$ as in Definition 7. Write $\pi^*|M| = |M'| + B$ where $|M'|$ is base-point free and B is the fixed part of $\pi^*|M|$, and $K_{X'} \sim \pi^* K_X + E$ where E is effective and π-exceptional. For any nonnegative rational number c we can thus formally write

$$K_{X'} + c|M'| \sim_{\mathbb{Q}} \pi^*\big(K_X + c|M|\big) + (E - cB), \tag{1}$$

where $A_1 \sim_{\mathbb{Q}} A_2$ means that $N \cdot A_1$ is linearly equivalent to $N \cdot A_2$ for some $N > 0$.

A pair $\big(X, c|M|\big)$ is called *canonical* (resp. *log canonical*) if every divisor appears in $E - cB$ with coefficient ≥ 0 (resp. ≥ -1). This is independent of the log resolution [36, 2.32].

Note that if $r > 0$ is an integer then $\big(X, c|M|\big)$ is canonical (resp. log canonical) iff $\big(X, \frac{c}{r}|rM|\big)$ is. (Keep in mind that $|rM|$ is the linear system spanned by sums of the form $M_1 + \cdots + M_r$ where $M_i \in |M|$.) Thus we can always restrict to dealing with pairs $\big(X, c|M|\big)$ where $c < 1$; this is frequently convenient.

If $|M|$ is base point free then $B = 0$, thus $\big(X, c|M|\big)$ is canonical for any c. In all other cases, $\big(X, c|M|\big)$ is canonical (resp. log canonical) for small values of c but not for large values. (The transitional value of c is called the canonical (resp. log canonical) *threshold*.) Roughly speaking, small threshold corresponds to very singular base locus.

Write $E - cB = \sum_i a_i E_i$. Then E_i is called a *non-canonical divisor* (resp. a *non-log-canonical divisor*) of $(X, c|M|)$ iff $a_i < 0$ (resp. $a_i < -1$). The corresponding image $\pi(E_i) \subset X$ is a *non-canonical center* (resp. *non-log-canonical center*) of $(X, c|M|)$. These centers are always contained in the base locus of $|M|$. It is not very important for us, but, as we run through all log resolutions and all divisors on them, we might get infinitely many non-(log)-canonical centers, however their union is the closed subset $\bigcup_{a_i < 0} \pi(E_i)$ (resp. $\bigcup_{a_i < -1} \pi(E_i)$); see [36, 2.31].

Using Remark 9, these formulas also define the above notions for pairs (X, Δ) where Δ is an effective divisor and pairs (X, I^c) where I is an ideal sheaf.

Remark 9 (Divisors, Linear Systems and Ideal Sheaves) Much of the Minimal Model Program literature works with pairs (X, Δ) where Δ is a divisor (with rational or real coefficients), see [35, 36]. For rigidity questions, the natural object seems to be a pair $(X, c|M|)$ where $|M|$ is a linear system and c is a rational or real coefficient. It is easy to see that if $c \in [0, 1)$ (which will always be the case for us) and $D \in |M|$ is a general divisor then the definitions and theorems for $(X, c|M|)$ and (X, cD) are equivalent.

As we noted in Definition 8, working with $(X, |M|)$ is equivalent to working with $(X, \frac{1}{2}|2M|)$, but one version may give a clearer picture than the other. As an illustration, consider the linear system $|\lambda x + \mu(x - y^r) = 0|$ in the plane. A general member of it is a smooth curve and the role of the y^r term is not immediately visible. By contrast the linear system $|2M|$ is $|\lambda x^2 + \mu x(x - y^r) + \nu(x - y^r)^2 = 0|$, its general member is (after a local analytic coordinate change) of the form $(x^2 - y^{2r} = 0)$. Now we see both the original smoothness (since x^2 is there) and the order of tangency between two members of $|M|$ (shown by y^{2r}). While computationally this is not important, conceptually it seems clearer that information about intersections of two divisors in $|M|$ is now visible on individual divisors in $|2M|$.

Let X be an affine variety and $I \subset \mathcal{O}_X$ an ideal sheaf. Many authors, for example [12, 38], work with pairs (X, I^c) where c is viewed as a formal exponent. If I is generated by global sections $g_1, \ldots, g_m$, we can consider the linear system $|M| := |\sum \lambda_i g_i = 0|$. Again we find that the definitions and theorems for $(X, c|M|)$ and (X, I^c) are equivalent.

Here I follow the language of linear systems, since this seems best suited to our current aims. I will also always assume that c is rational. This is always the case in our applications and makes some statements simpler. However, it does not cause any essential difference at the end.

We discuss the canonical and log canonical property of linear systems in detail in Sect. 4. For now we mainly need to know that canonical means mild singularities and log canonical means somewhat worse singularities. In some sense the main question of the theory was how to describe these properties in terms of other, better understood, measures of singularities.

10 (Main Steps of the Proof) The proof can be organized into six fairly independent steps. Roughly speaking, Steps 1 and 2 are essentially in the works of Fano, at least for threefolds. Steps 3 and 4 are substantial reinterpretations of the classical ideas while Steps 5 and 6 give a new way of finishing the proof.

Notation For the rest of this section I write Y for a smooth, projective variety, X for a smooth, projective Fano variety with class number 1 and V (or V_{n+1} or V_{n+1}^n) for a smooth hypersurface $V_{n+1}^n \subset \mathbb{P}^{n+1}$ of degree $n + 1$ and of dimension $n \geq 3$. The base field has characteristic 0.

Step 10.1 (Noether-Fano Criterion, Sect. 2) *A smooth Fano variety X of class number 1 is (weakly) superrigid if for every movable linear system $|M| \subset |-mK_X|$ the pair $\left(X, \frac{1}{m}|M|\right)$ is canonical.*

Comments 10.1.1 Movable means that there are no fixed components, some authors use *mobile* instead. If $\dim X = 2$ then $\left(X, \frac{1}{m}|M|\right)$ is not canonical iff $\mathrm{mult}_x |M| > m$ for some point $x \in X$ by Lemma 28; this equivalence made Noether's and Segre's proofs work well. If $\dim X = 3$ then Fano tried to prove that if $\left(X, \frac{1}{m}|M|\right)$ is not canonical then either $\mathrm{mult}_C |M| > m$ for some curve $C \subset X$ or $\mathrm{mult}_x |M| > 2m$ for some point $x \in X$. Fano understood that the latter condition for points is not right, one needs instead only a consequence of it: The local intersection number at x is $(M \cdot M \cdot H)_x > 4m^2$, where H is a hyperplane through x. In higher dimensions it does not seem possible to define canonical in terms of just multiplicities and intersection numbers, this is one reason why the above form of Step 10.1 was established only in [5]. We prove Step 10.1 in Theorem 14. Although historically the notion of "canonical" was first defined starting from varieties of general type (see [36, 54]), the Noether-Fano criterion leads to the exact same notion.

If X is not (weakly) superrigid then there is a movable linear system $|M| \subset |-mK_X|$ such that $\left(X, \frac{1}{m}|M|\right)$ is not canonical, thus it has some non-canonical divisors and centers as in Definition 8. (The "worst" non-canonical centers are called *maximal centers* by the Iskovskikh school.) From now on we focus entirely on understanding movable linear systems and their possible non-canonical centers on X. There are two persistent problems that we encounter.

- We can usually bound the *multiplicities* of $|M|$, but there is a gap—growing with the dimension—between multiplicity and the canonical property.
- We are better at understanding when a pair is *log* canonical, instead of canonical.

While we try to make statements about arbitrary Fano varieties, at some point we need to use special properties of the V_{n+1}^n. The following bounds, going back to Fano and Segre, were put into final form by Pukhlikov [51, Prop. 5] and later generalized by Cheltsov [4, Lem. 13] and Suzuki [63, 2.1] to complete intersections.

Step 10.2 (Multiplicity Bounds, Fano, Segre, Pukhlikov, Sect. 3) *Let $Y \subset \mathbb{P}^{n+1}$ be a smooth hypersurface and $|H|$ the hyperplane class on Y. Let $D \in |mH|$ be a divisor, $|M| \subset |mH|$ a movable linear system and $Z \subset Y$ an irreducible subvariety.*

(a) *If $\dim Z \geq 1$ then $\mathrm{mult}_Z D \leq m$.*
(b) *If $\dim Z \geq 2$ then $\mathrm{mult}_Z(M \cdot M) \leq m^2$,*

where $M \cdot M$ denotes the intersection of two general members of $|M|$.

Comments 10.2.1 Note that Step 10.1 works with $|M| \subset |-mK_X|$ and Step 10.2 with $|M| \subset |mH|$. The two match up iff $-K_X \sim H$; the latter holds for $X = V_{n+1}^n$, the case that we are considering. In general, the method works best for those Fano varieties where every divisor is an integral multiple of $-K_X$ (up to linear equivalence).

Next we need to understand the relationship between the multiplicity bounds in Step 10.2 and the canonical property. This is rather easy for Step 10.2.a. Combining it with Step 10.3.a we get that $\left(Y, \frac{1}{m}D\right)$ is canonical, except at a finite point set $P \subset Y$. We already mentioned this in Comments 10.1.1; see Lemma 28 or [37, 6.18] for proofs.

Relating Step 10.2.b to the canonical property was less obvious; it was done by Corti [6, 3.1] (see also [37, Sec. 6.6]), then very much generalized by de Fernex et al. [12] and sharpened by Liu [40].

Step 10.3 (Non-(log)-canonical points and multiplicity, Corti, Sects. 4 and 5) *Let $|M|$ be a movable linear system on a smooth variety Y.*

(a) *If $\left(Y, \frac{1}{m}|M|\right)$ is not canonical at $\mathbf{p} \in Y$ then* $\mathrm{mult}_{\mathbf{p}} |M| > m$.
(b) *If $\left(Y, \frac{1}{m}|M|\right)$ is not log canonical at $\mathbf{p} \in Y$ then* $\mathrm{mult}_{\mathbf{p}}(M \cdot M) > 4m^2$.

Comments 10.3.1 Both of these bounds are sharp as shown by the examples

$$\left(\mathbb{A}^2, \tfrac{1}{m}|\lambda x^{m+1} + \mu y^{m+1} = 0|\right) \quad \text{and} \quad \left(\mathbb{A}^2, \tfrac{1}{m}|\lambda x^{2m+1} + \mu y^{2m+1} = 0|\right),$$

which have a non-canonical (resp. non-log-canonical) center at the origin. Surprisingly, part (a) can not be improved for non-log-canonical centers, as shown by

$$\left(\mathbb{A}^2, \tfrac{1}{m}|\lambda x^{m+1} + \mu y^{(m+1)^2} = 0|\right).$$

(This can be computed by hand or see Theorem 29.)

Using a—by now standard—method called *inversion of adjunction,* which we discuss in Sect. 5, both parts follow from claims about linear systems on algebraic surfaces:

Claim 10.3.2 Let $|M|$ be a movable linear system on a smooth surface S.

(a) If $\left(S, c|M|\right)$ is not canonical at $s \in S$ then $\mathrm{mult}_s |M| > 1/c$.
(b) If $\left(S, c|M|\right)$ is not log canonical at $s \in S$ then $\mathrm{mult}_s(M \cdot M) > 4/c^2$.

It would be very nice to continue the claims Step 10.3.a, b to stronger and stronger inequalities for higher codimension non-log-canonical centers. This was done in [12]. This is very useful if by chance the base locus of $|M|$ has codimension > 2. However, in many cases the base locus of $|M|$ has codimension 2 and it is not easy to apply the estimates of [12] directly.

Fano always aimed to reduce questions about Fano threefolds (for him these meant $X \subset \mathbb{P}^n$ such that $-K_X \sim H$) to their hyperplane sections. These are K3 surfaces, whose geometry was quite well understood. In higher dimensions, the hyperplane sections are Calabi-Yau varieties, whose geometry is much less known. Thus the modern focus is on the change of the singularities as we restrict a linear systems to a hyperplane section.

Step 10.4 (Cutting by Hyperplanes, Sect. 5) *Let $|M|$ be a movable linear system on a smooth variety Y. Fix a point $p \in Y$ and let W be a general member of a very ample linear system $|H|$ that passes through p. Then*

(a) *If $(Y, c|M|)$ is (log) canonical outside a closed subset $Z \subset Y$ then $(W, c|M|_W)$ is (log) canonical outside $\{p\} \cup (Z \cap W)$.*

(b) *If $(Y, c|M|)$ is not log canonical at p then $(W, c|M|_W)$ is also not log canonical at p.*

(c) *If p is a non-canonical center of $(Y, c|M|)$ then p is a non-log-canonical point of $(W, c|M|_W)$.*

Warning Note that in (c) the point p needs to be a non-canonical *center* on Y (see Definition 8) and then it is a non-*log*-canonical point on W.

Comments 10.4.1 The multiplicity versions of these go back to Bertini and Fano, but the above form of (c) may have been first made explicit in [6]. By now these are special cases of the theory of *adjunction* for log canonical pairs, we discuss this in Sect. 5.

Note that cutting by a hyperplane has a very curious effect on the singularities.

If $(X, c|M|)$ is a canonical (resp. log canonical) pair then its restriction to a general member of a base point free linear system is still canonical (resp. log canonical); this is an easy Bertini-type theorem, see Proposition 35.1. Applying this to $X := Y \setminus Z$ gives (a). Part (b) is quite a bit harder to prove but it fits the general pattern that singularities do not get better by cutting with a hyperplane.

The surprising part is (c) which says that the singularity is made *worse* by restriction to a general hypersurface through a non-canonical center. This is in marked contrast with multiplicity, which is preserved by such restrictions. We discuss this in Sect. 5.

The first application of these ideas is the following.

Rigidity of Quartic Threefolds, Corti's Variant. 10.4.2 Let $X \subset \mathbb{P}^4$ be a smooth quartic threefold and assume that we have a linear system $|M| \subset |mH|$ such that $(X, \frac{1}{m}|M|)$ is not canonical. One dimensional non-canonical centers are excluded by Steps 10.2.a and 10.3.a. If $x \in X$ is a zero-dimensional non-canonical center then let W be a general hyperplane section passing through x. Then x is a non-log-canonical center of $(W, \frac{1}{m}|M|_W)$ by Step 10.4.c, hence the local intersection number $(M \cdot M \cdot W)_x$ is $> 4m^2$ by Step 10.3.b. Therefore $(M \cdot M \cdot W) > 4m^2$. On the other hand, $M \sim mH$ and hence $(M \cdot M \cdot W) = 4m^2$, a contradiction. Thus smooth quartic threefolds are weakly superrigid.   □

More generally, the method described so far works well if $-K_X$ generates the class group and $(-K_X)^n \leq 4$. Among hypersurfaces in $\mathbb{P}^{n+1}$, this holds only for the quartic threefolds. However there are smooth hypersurfaces in weighted projective spaces with these properties. For example, fix $r > 1$ and let X be a smooth hypersurface of degree $4r + 2$ and dimension $2r$ in the weighted projective space $\mathbb{P}(1^{2r}, 2, 2r+1)$ (the notation means that we have $2r$ coordinates of weight 1, see [37, 3.48] for an introduction). Then $-K_X \sim H$ and $(-K_X)^{2r} = 1$. With small changes the method proves that they are superrigid; see [37, 5.22] for details. $\qquad\square$

In the above proof we have used Step 10.3.b on $W \in |H|$. The new idea of [9] is to use it directly on Y.

Step 10.5 (Doubling the Linear System, de Fernex, Sects. 4 and 5) *Instead of working only with* $(Y, \frac{1}{m}|M|)$, *we should focus on the interaction between* $(Y, \frac{1}{m}|M|)$ *and* $(Y, \frac{1}{m}|2M|)$.

In order to contrast the two cases, let $Y \subset \mathbb{P}^{n+1}$ be a smooth hypersurface and $|M| \subset |mH|$ a movable linear system. Combining Step 10.2.a with Step 10.3.a and Step 10.2.b with Step 10.3.b gives the following.

(a) $(Y, \frac{1}{m}|M|)$ *is canonical outside a finite set of points* $P \subset Y$ *and*

(b) $(Y, \frac{1}{m}|2M|)$ *is log canonical outside a finite set of curves* $C \subset Y$.

One should think of these as saying that $|M|$ and $|2M|$ are very singular at p but less singular almost everywhere else. A key insight of [9] is that (b) is much stronger than (a). In order to understand this, let us see how one can use the information provided by Step 10.5.a, b.

Comments 10.5.1 Let Y be a smooth, projective variety, H an ample divisor and $\Delta \sim_{\mathbb{Q}} H$ a $\mathbb{Q}$-divisors with an isolated non-log-canonical center at a point $p \in Y$. The observation that this leads to a global section of $\mathcal{O}_Y(K_Y + H)$ that does not vanish at p has been an important ingredient of the Kawamata–Reid–Shokurov approach to the cone theorem (cf. [36, Chap. 3]) and is central in the works around Fujita's conjecture (cf. [33, Secs. 5–6] or [38, Sec. 10.4]). In all these applications the aim is to get at least 1 section that does not vanish at a given point. Although it was known that the process can be used to get several sections, this has not been the focus in the past.

At first sight, Step 10.5.a is better suited to use this method. If $(Y, \frac{1}{m}|M|)$ is not canonical at some $p \in P$, then, by Step 10.4.c, after restricting to a general hyperplane section $p \in W \subset Y$, we get $(W, \frac{1}{m}|M|_W)$ that is not log canonical at p but is canonical outside $P \cap W$. This leads to a section of $\mathcal{O}_W(K_W + H)$ that does not vanish at p. However, in our cases $\mathcal{O}_W(K_W + H)$ is very ample, so there is no contradiction.

The problem seems to be that while we have been thinking of canonical as "much better" than log canonical, from the numerical point of view the difference seems small. We saw an instance of this in Step 10.3.a, where both the non-canonical and non-log-canonical cases yield the same inequality; see Step 10.3.1.

In Step 10.5.b we "gain" since $|M|$ is replaced by $|2M|$ but also "lose" since canonical is replaced by log canonical and the finite set of points P is replaced by a finite set of curves C. However, when we switch to a hyperplane section $W \subset Y$, we focus on the non-log-canonical property anyhow, and $C \cap W$ becomes a finite set of points. So the "losses" do not matter at the end but the "gain" stays with us. Thus we get that

(a) $\left(W, \frac{1}{m}|M|_W\right)$ is not log canonical at some $p \in P \cap W$, but even
(b) $\left(W, \frac{1}{m}|2M|_W\right)$ is log canonical outside $P \cap W$.

A fundamental claim of [9] is that this in itself leads to a contradiction. While the argument at the end of [9] is worded differently, unraveling the proofs of [9, Lems. 3–4] gives a quadratic lower bound for $h^0\left(W, \mathcal{O}_W(K_W + 2H)\right)$, almost enough to get a contradiction without further work. Building on [9], a key observation of [67] is that a suitable modification of this method leads to an exponential lower bound and a quick numerical contradiction.

Remark 10.5.2 Once the technical details are settled, we see that there is lot of room in Step 10.5.b. Namely, if we know only that, for some fixed $\epsilon > 0$ and d, $\left(Y, \frac{1}{m}|(1 + \epsilon)M|\right)$ is log canonical outside a subset of dimension $\leq d$, that is still enough to prove Theorem 2 for n sufficiently large (depending on ϵ and d).

Step 10.6 (Zhuang, Sect. 6) *Let Y be a smooth projective variety of dimension d and L an ample divisor on Y. Further let $|M| \subset |mL|$ be a movable linear system and $P \subset Y$ a finite (nonempty) subset of Y. Assume that*

(a) $\left(Y, \frac{1}{m}|M|\right)$ *is not log canonical at some $p \in P$, but*
(b) $\left(Y, \frac{1}{m}|2M|\right)$ *is log canonical outside P.*

Then

$$h^0\left(Y, \mathcal{O}_Y(K_Y + 2L)\right) \geq \tfrac{1}{2}3^d.$$

Comments 10.6.1 One should think of this as saying that if $|M|$ is much more singular at a finite set of points than elsewhere then the linear system $|K_Y + 2L|$ is very large. I stated the case where we compare the singularities of $\frac{1}{m}|M|$ and $\frac{1}{m}|2M|$, the complete version in [67] also applies if we work with $c|M|$ and $(c + \epsilon)|M|$ for some $\epsilon > 0$.

It is quite remarkable that there is also a rather easy converse.

Let $|L|$ be any linear system on Y and $y \in Y$ a point. If $\dim |2L| \geq \binom{3d}{d}$ then there is a linear subsystem $|N| \subset |2L|$ that has multiplicity $> 2d$ at y. In particular, $\left(Y, \frac{1}{2}|N|\right)$ is not log canonical at y. As $d \to \infty$, $\binom{3d}{d}$ grows like 6.75^d.

Thus if $\dim |2L| \geq 6.75^d$ then we can find a linear system $|N|$ that satisfies Step 10.6.a, and usually also Step 10.6.b. Informally we can restate Step 10.6 as

Principle 10.6.2 There are no accidental isolated singularities.

11 (Proof of Theorem 2 Using Steps 10.1–6) Let $V \subset \mathbb{P}^{n+1}$ be a smooth hypersurface of degree $n + 1 \geq 4$. If V is not weakly superrigid, then, by Step 10.1 we get $\left(V, \frac{1}{m}|M|\right)$ that is not canonical. Thus Steps 10.3–5 give a $W = W_{n+1}^{n-1} \subset \mathbb{P}^n$ and $|M|_W$ such that

(a) $\left(W, \frac{1}{m}|M|_W\right)$ is not log canonical at finitely many points $P \subset W$, but

(b) $\left(W, \frac{1}{m}|2M|_W\right)$ is log canonical outside P.

By Step 10.6 this implies that

$$h^0\left(W, \mathcal{O}_W(K_W + 2H)\right) \geq \tfrac{1}{2}3^{n-1}.$$

On the other hand $h^0\left(W, \mathcal{O}_W(K_W + 2H)\right) = h^0\left(W, \mathcal{O}_W(2H)\right) = h^0\left(\mathbb{P}^n, \mathcal{O}_{\mathbb{P}^n}(2)\right)$
$= \binom{n+2}{2}$, so

$$\binom{n+2}{2} \geq \tfrac{1}{2}3^{n-1}.$$

The left hand side is quadratic in n, the right hand side is exponential, so for $n \gg 1$ this can not hold. (In fact, we have a lot of room, leading to many other cases where the method applies in large dimensions; see [67].)

By direct computation, we get a contradiction for $n \geq 5$, hence we get the superrigidity of $V^n_{n+1} \subset \mathbb{P}^{n+1}$ for $n \geq 5$.

One can improve the lower bound in Step 10.6 to $\tfrac{1}{2}3^d + \tfrac{3}{2}$, and then for $n = 4$ we get an equality $\binom{6}{2} = 15 = \tfrac{1}{2}3^3 + \tfrac{3}{2}$. So there is no contradiction, but it is quite likely that a small change can make the proof work. However, the $n = 3$ case does not seem to follow, but this was already treated in Step 10.4.2.   □

12 (Attribution of the Steps) In rereading many of the contributions to the proof I was really struck by how gradual the progress was and how difficult it is to attribute various ideas to a particular author or paper.

Fano's papers are quite hard to read, and some people who spent years on trying to learn from them came away with feeling that Fano got most parts of the proof wrong. Others who looked at Fano's works feel that he had all the essential points right. In particular, the attribution of Step 2 has been controversial.

I think of Corti's work [6] as a major conceptual step forward, but some authors felt that it did not add anything new, at least initially. The idea of doubling the linear system is in retrospect already in [6], but the new viewpoint of de Fernex [9] turned out to be very powerful and, as we discussed in Step 10.5.1, the latter contains many of the ingredients of Step 6. I had a hard time formulating Steps 3–6 in a way that shows the differences between them meaningfully while highlighting the new ideas of the main contributor. Nonetheless, at least in hindsight, each of the Steps represents a major new idea, though this was not always immediately understood.

No doubt several people will feel that my presentation is flawed in many ways. Luckily the reader can consult the excellent survey [3] and books [7, 52] for different viewpoints.

13 (What Is Missing?) My aim was to write down a proof of Theorem 2 that is short and focuses on the key ideas. My preference is for steps that follow from general results and techniques of the MMP. Thus several important developments have been left out.

After proving rigidity for quartic threefolds, the Russian school went on to study other Fano threefolds. They found that they are frequently rigid but not superrigid and the main question is how to find generators for $\mathrm{Bir}(X)$. The contributions of Iskovskikh, Sarkisov, Pukhlikov and Cheltsov are especially significant. These results and their higher dimensional extensions are surveyed in [3, 52].

The first major applications of the Corti method were also in dimension 3, see [7] for a survey and [18] for a higher dimensional extension.

In our proof we need to understand zero-dimensional log canonical centers, but the theory of arbitrary log canonical centers has been quite important in higher dimensional geometry. The first structure theorems were proved by Ambro [1]; see [35, Chaps. 4–5] and [15] for later treatments and generalizations.

2   The Noether-Fano Method

We start the proof of Theorem 2 by establishing Step 10.1.

Theorem 14 (Noether-Fano Inequality) *Let* $\Phi : X \dashrightarrow X'$ *be a birational map between smooth Fano varieties of class number 1. Then*

(1) *either* Φ *is an isomorphism,*
(2) *or there is a movable linear system* $|M| \subset |-mK_X|$ *for some* $m > 0$ *on* X *such that* $\left(X, \frac{1}{m}|M|\right)$ *is not canonical.*

Proof Let Z be the normalization of the closure of the graph of Φ with projections $p : Z \to X$ and $q : Z \to X'$. Pick any base-point-free linear system $|M'| \subset |-m'K_{X'}|$ and let $|M| := \Phi_*^{-1}|M'|$ denote its birational transform on X. Set $|M_Z| = q^*|M'|$. Since the class number of X is 1, $|M| \sim_{\mathbb{Q}} -mK_X$ for some $m > 0$. (If m is not an integer, we replace $|M'|$ by a suitable multiple. Thus we may as well assume that $|M| \subset |-mK_X|$.) We define a q-exceptional divisor E_q and p-exceptional divisors E_p, F_p by the formulas

$$K_Z = q^*K_{X'} + E_q, \quad |M_Z| = q^*|M'| \quad \text{and} \tag{2}$$
$$K_Z = p^*K_X + E_p, \quad |M_Z| = p^*|M| - F_p.$$

Since X', X are smooth, E_q, E_p are effective (cf. [60, III.6.1]) and F_p is effective since $p_*|M_Z| = |M|$.

For any rational number c we can rearrange (2) to get

$$K_Z + c|M_Z| \sim_{\mathbb{Q}} q^*\left(K_{X'} + c|M'|\right) + E_q \quad \text{and} \tag{3}$$
$$K_Z + c|M_Z| \sim_{\mathbb{Q}} p^*\left(K_X + c|M|\right) + E_p - cF_p.$$

First we set $c = \frac{1}{m'}$. Then $K_{X'} + \frac{1}{m'}|M'| \sim_{\mathbb{Q}} 0$, hence

$$K_Z + \frac{1}{m'}|M_Z| \sim_{\mathbb{Q}} q^*\left(K_{X'} + \frac{1}{m'}|M'|\right) + E_q \sim_{\mathbb{Q}} E_q \geq 0.$$

Pushing this forward to X we get that

$$K_X + \frac{1}{m'}|M| = p_*\left(K_Z + \frac{1}{m'}|M_Z|\right) \sim_{\mathbb{Q}} p_*(E_q) \geq 0.$$

Since

$$p_*(E_q) \sim_{\mathbb{Q}} K_X + \frac{1}{m'}|M| \sim_{\mathbb{Q}} K_X - \frac{1}{m'}mK_X = \frac{m-m'}{m'}(-K_X), \tag{4}$$

we see that $m \geq m'$.

Next set $c = \frac{1}{m}$. Then we get that

$$K_Z + \frac{1}{m}|M_Z| \sim_{\mathbb{Q}} p^*\left(K_X + \frac{1}{m}|M|\right) + E_p - \frac{1}{m}F_p \sim_{\mathbb{Q}} E_p - \frac{1}{m}F_p.$$

Pushing this forward to X yields

$$K_{X'} + \frac{1}{m}|M'| = q_*\left(K_Z + \frac{1}{m}|M_Z|\right) \sim_{\mathbb{Q}} q_*(E_p - \frac{1}{m}F_p).$$

As in (4) we obtain that

$$\frac{m'-m}{m}(-K_{X'}) \sim_{\mathbb{Q}} K_{X'} + \frac{1}{m}|M'| \sim_{\mathbb{Q}} q_*(E_p - \frac{1}{m}F_p). \tag{5}$$

Basic Alternative 14.1

- If $E_p - \frac{1}{m}F_p$ is not effective, then we declare the linear system $|M|$ to be "very singular." In our terminology, $(X, \frac{1}{m}|M|)$ is not canonical. This is case (2).
- If $E_p - \frac{1}{m}F_p$ is effective, then we declare the linear system $|M|$ to be "mildly singular." In our terminology, $(X, \frac{1}{m}|M|)$ is canonical. We need to prove that in this case Φ is an isomorphism.

Thus assume from now on that $E_p - \frac{1}{m}F_p$ is effective. Then (5) implies that $m' \geq m$. Combining it with (4) gives that $m' = m$ and then (4) shows that $p_*(E_q) = 0$. That is, Supp E_q is p-exceptional. Since X' is smooth, the support of E_q is the whole q-exceptional divisor $\mathrm{Ex}(q)$. Thus every q-exceptional divisor is also p-exceptional.

To see the converse, let $D \subset Z$ be an irreducible divisor that is not q-exceptional. Then $q_*(D) \sim_{\mathbb{Q}} r|M'|$ for some $r > 0$. Thus

$$r|M_Z| \sim_{\mathbb{Q}} q^*(r|M'|) \sim_{\mathbb{Q}} D + (q\text{-exceptional divisor}).$$

Pushing forward to X now gives that $r|M| \sim_{\mathbb{Q}} p_*(D)$, since every q-exceptional divisor is also p-exceptional. Here $p_*(D) \neq 0$ since $r > 0$, so D is not p-exceptional. This shows that $\mathrm{Ex}(p) = \mathrm{Ex}(q)$.

Finally set $Z := p\big(\mathrm{Ex}(p)\big) \subset X$, $Z' := q\big(\mathrm{Ex}(q)\big) \subset X'$ and apply the following result of Matsusaka and Mumford [43] to conclude that Φ is an isomorphism   □

Lemma 15 *Let* $\Psi : Y \dashrightarrow Y'$ *be a birational map between smooth projective varieties. Let* $Z \subset Y$ *and* $Z' \subset Y'$ *be closed sets of codimension* ≥ 2 *such that* Ψ *restricts to an isomorphism* $Y \setminus Z \cong Y' \setminus Z'$. *Let* H *be an ample divisor on* Y *such that* $H' := \Psi_* H$ *is also ample. Then* Ψ *is an isomorphism.*

Proof We may assume that H' and H are both very ample. Then

$$|H'| = \big|H'_{Y' \setminus Z'}\big| = \Psi_* \big|H_{Y \setminus Z}\big| = \Psi_* |H|.$$

Thus $\Psi_* |H|$ is base point free, hence Ψ^{-1} is everywhere defined. The same argument, with the roles of Y, Y' reversed, shows that Ψ is also everywhere defined. So Ψ is an isomorphism.   □

Remark 16 The proof of Theorem 14 also works if X has canonical singularities, X' has terminal singularities and they both have class number 1.

3 Subvarieties of Hypersurfaces

Our aim is to prove that a subvariety of a smooth hypersurface can not be unexpectedly singular along a large dimensional subset. The claim and the method go back to Fano and Segre; the first complete statement and proof is in [51, Prop. 5].

Theorem 17 *Let* $X \subset \mathbb{P}^{n+1}$ *be a smooth hypersurface,* $Z \subset X$ *an irreducible subvariety and* $W \subset X$ *a pure dimensional subscheme such that* $\dim Z + \dim W \geq \dim X$. *Assume that either* $\dim Z < \dim W$ *or* $\dim Z = \dim W = \frac{1}{2} \dim X$ *and* W *is a complete intersection in* X. *Then*

$$\mathrm{mult}_Z W \leq \frac{\deg W}{\deg X}. \tag{6}$$

We define the multiplicity $\mathrm{mult}_Z W$ in Paragraph 20. See [4, Lem. 13] and [63, 2.1] for generalizations of the theorem to complete intersections.

18 (Proof of Step 10.2) For part (a) set $W := D \in |mH|$. By Bézout's theorem, $\deg W = m \deg X$ so $\mathrm{mult}_Z D \leq m$. For part (b) set $W = M \cdot M$. Then $\deg W = m^2 \deg X$ so $\mathrm{mult}_Z(M \cdot M) \leq m^2$. Note that $M \cdot M$ is a complete intersection in X, so the Theorem applies even if $n = 4$ and $\dim Z = 2$.   □

Remark 19 The simplest special case of the theorem is when W is an intersection of X with a hyperplane. Then $\deg W = \deg X$ hence we claim that W has only finitely many singular points. Equivalently, a given hyperplane can be tangent to a smooth hypersurface only at finitely many points. I encourage the reader to prove this; there are very easy proofs but also messy ones. Note that this is truly a projective statement. For example, $(z - y^2 x = 0)$ is a smooth surface in $\mathbb{A}^3$ and the plane $z = 0$ is tangent to it everywhere along the x-axis.

Consider next the case when W is an intersection of X with a hypersurface of degree d. Then (6) says that W has multiplicity $\leq d$ at all but finitely many of its points $\mathbf{p}$. The easy geometric way to prove this would be to find a line ℓ in X that passes through $\mathbf{p}$ but not contained in W. This sounds like a reasonable plan if $\deg X \leq n$, since in these cases there is a line through every point of X, see [32, V.4.3], which also shows that if $\deg X \geq 2n$ then a general X does not contain any lines.

In Proposition 22, as replacements of lines, we construct certain auxiliary subvarieties Z^* that have surprisingly many intersections with W.

The extra assumption in case $\dim Z = \dim W = \frac{1}{2}\dim X$ is necessary. Indeed, there are smooth hypersurfaces $X \subset \mathbb{P}^{2n+1}$ that contain a linear space L of dimension n. Setting $Z = W = L$ we get that $\mathrm{mult}_Z W = 1$ but $\frac{\deg W}{\deg X} = \frac{1}{\deg X}$.

20 (Multiplicity) The simplest measure of a singularity is its *multiplicity*. Let $X = (h = 0) \subset \mathbb{A}^n$ be an affine hypersurface and $\mathbf{p} = (p_1, \ldots, p_n)$ a point on X. We can write the equation as

$$h = \sum a_{i_1,\ldots,i_n} (x_1 - p_1)^{i_1} \cdots (x_n - p_n)^{i_n}.$$

The multiplicity of X at $\mathbf{p}$, denoted by $\mathrm{mult}_\mathbf{p} X$, is defined as

$$\mathrm{mult}_\mathbf{p} X := \min\{i_1 + \cdots + i_n : a_{i_1,\ldots,i_n} \neq 0\}. \tag{7}$$

The definition of multiplicity for other varieties is, unfortunately, more complicated. Let $Y \subset \mathbb{A}^n$ be a variety of dimension m and $\mathbf{p} = (p_1, \ldots, p_n)$ a point on Y. The following give the correct definition of the multiplicity $\mathrm{mult}_\mathbf{p} Y$, see [45, Chap. 5] for details.

(1) Let $\pi : \mathbb{A}^n \to \mathbb{A}^{m+1}$ be a general projection. Then $\pi(Y)$ is a hypersurface and $\mathrm{mult}_\mathbf{p} Y = \mathrm{mult}_{\pi(\mathbf{p})} \pi(Y)$.
(2) If we are over $\mathbb{C}$, we can fix a small Euclidean ball $B(\epsilon)$ around $\mathbf{p}$, a general linear subspace L of dimension $n - m$ through $\mathbf{p}$ and count the number of those intersection points of X with a general small translate of L that are contained in $B(\epsilon)$.
(3) The multiplicity also equals the limit

$$\lim_{r \to \infty} \tfrac{m!}{r^m} \dim_k k[x_1, \ldots, x_n]/\big(I_Y, (x_1 - p_1, \ldots, x_n - p_n)^r\big).$$

The first two are old-style definitions that capture the essence but are not easy to work with rigorously, the third is easy to use algebraically but it is not even obvious that the limit exists; see [56]. The most complete modern treatment is given in [16].

Finally we set

$$\text{mult}_Z Y := \min\{\text{mult}_\mathbf{p} Y : \mathbf{p} \in Z\}, \tag{8}$$

and note that the minimum is achieved on a dense open subset.

We will also need the following.

Theorem 20.1 Let $X \subset \mathbb{P}^{n+1}$ be a smooth hypersurface and $Z, W \subset X$ irreducible subvarieties such that $Z \cap W$ is finite and $\dim Z + \dim W = \dim X$. Assume furthermore that neither of them has dimension $\frac{n}{2}$. Then

$$\sum_\mathbf{p} \text{mult}_\mathbf{p} Z \cdot \text{mult}_\mathbf{p} W \leq \frac{\deg Z \cdot \deg W}{\deg X}.$$

Comments on the Proof There are several theorems rolled into one here.

Intersection theory says that if X is any smooth projective variety and $Z, W \subset X$ irreducible subvarieties such that $\dim Z + \dim W = \dim X$, then they have a natural *intersection number*, denoted by $(Z \cdot W)$. Intersection theory can be developed completely algebraically, but working over $\mathbb{C}$ there is a shortcut. Both Z, W have a homology class $[Z] \in H_{2\dim Z}(X(\mathbb{C}), \mathbb{Z})$ and $[W] \in H_{2\dim W}(X(\mathbb{C}), \mathbb{Z})$ and then

$$(Z \cdot W) = [Z] \cap [W] \in H_0(X(\mathbb{C}), \mathbb{Z}) \cong \mathbb{Z}. \tag{9}$$

Furthermore, if $Z \cap W$ is finite then their intersection number $(Z \cdot W)$ is the sum of local terms, denoted by $(Z \cdot W)_\mathbf{p}$, computed at each $\mathbf{p} \in Z \cap W$. Next we need that

$$(Z \cdot W)_\mathbf{p} \geq \text{mult}_\mathbf{p} Z \cdot \text{mult}_\mathbf{p} W. \tag{10}$$

This very useful inequality does not seem to be included in introductory books. It is easy to derive it from [45, Cor. A.14], see also [56, p. 95] or [16, Cor. 12.4].

Assume next that $X \subset \mathbb{P}^{n+1}$ is a smooth hypersurface of degree d and W is obtained as the intersection of X by $n - r$ hypersurfaces of degrees $m_{r+1}, \ldots, m_n$. If $\dim W = r$ then, by Bézout's theorem,

$$\deg W = d \cdot m_{r+1} \cdots m_n \quad \text{and} \quad (Z \cdot W) = \deg Z \cdot m_{r+1} \cdots m_n.$$

Thus we obtain that

$$(Z \cdot W) = \frac{\deg Z \cdot \deg W}{\deg X}. \tag{11}$$

It is not at all true that every W can be obtained this way, but, by the Lefschetz hyperplane theorem (see [39] or [17, p. 156]), the homology class of W is a rational multiple of a power of the hyperplane class, provided $\dim W \neq \frac{n}{2}$. Thus the above computation applies to every W as in Theorem 17.

21 (Proof of Theorem 17) If $\dim Z + \dim W > \dim X$ and the claim holds for all subvarieties $Z' \subset Z$ of codimension 1 then it also holds for Z. Thus we may assume from now on that $\dim Z + \dim W = \dim X$.

Both the multiplicity and the degree is linear in irreducible components, so write $W = \sum m_i W_i$ where the W_i are irreducible subvarieties of X.

In Proposition 22 we construct a subvariety $Z^* \subset X$ such that $\dim Z^* = \dim Z$, $\deg Z^* = (d-1)^r \deg Z$, $Z \cap Z^*$ consists of at least $(d-1)^r \deg Z$ distinct points and $W \cap Z^*$ is finite.

There is nothing to prove if $Z \not\subset W_i$. Otherwise, at each point of $Z \cap Z^*$ the intersection multiplicity of Z^* and W_i is at least $\mathrm{mult}_Z W_i$ by (10). Therefore

$$\big((d-1)^r \deg Z\big) \cdot \mathrm{mult}_Z W_i \le (W_i \cdot Z^*). \tag{12}$$

Next we use that $\sum_i m_i (W_i \cdot Z^*) = (W \cdot Z^*) = \frac{\deg W \cdot \deg Z^*}{\deg X}$ by (11). Summing (12) we get that

$$\big((d-1)^r \deg Z\big) \cdot \mathrm{mult}_Z W \le (W \cdot Z^*) = \frac{\deg W \cdot (d-1)^r \deg Z}{\deg X}. \tag{13}$$

Canceling $(d-1)^r \deg Z$ gives (6). $\qquad\square$

Next we construct the subvariety Z^* used in the above proof.

Proposition 22 *Let* $X \subset \mathbb{P}^{n+1}$ *be a smooth hypersurface of degree* d. *Let* $Z \subset X$ *be a subvariety of dimension* $r \le \frac{n}{2}$ *and* $W_i \subset X$ *a finite set of subvarieties. Then there is a subvariety* Z^* *of dimension* r *such that*

(1) $\deg Z^* = (d-1)^r \deg Z$,
(2) $Z \cap Z^*$ *consists of at least* $(d-1)^r \deg Z$ *distinct points, and*
(3) $\dim \big(Z^* \cap (W_i \setminus Z)\big) \le \dim Z + \dim W_i - \dim X$ *for every* i.

The proof relies on the study of certain residual intersections.

23 (Residual Intersection with Cones) Let $X \subset \mathbb{P}^{n+1}$ be a hypersurface of degree d and $Z \subset X$ a subvariety. Pick a point $\mathbf{v} \in \mathbb{P}^{n+1}$ and let $\langle \mathbf{v}, Z \rangle$ denote the cone over Z with vertex $\mathbf{v}$, that is, the union of all lines $\langle \mathbf{v}, z \rangle : z \in Z$.

If $\dim Z \le n-1$ and $\mathbf{v}$ is general then $\langle \mathbf{v}, Z \rangle$ has the same degree as Z but 1 larger dimension. If $\langle \mathbf{v}, Z \rangle$ is not contained in X then $X \cap \langle \mathbf{v}, Z \rangle$ is a subscheme of X of degree $= d \cdot \deg Z$. This subscheme contains Z, thus we can write

$$X \cap \langle \mathbf{v}, Z \rangle = Z \cup Z_{\mathbf{v}}^{\mathrm{res}}, \tag{14}$$

where $Z_{\mathbf{v}}^{\text{res}}$ is called the *residual intersection* of the cone with X. Note that

$$\deg Z_{\mathbf{v}}^{\text{res}} = (d-1) \cdot \deg Z. \tag{15}$$

We are a little sloppy here; if X is singular along Z then $Z_{\mathbf{v}}^{\text{res}}$ is well defined as a cycle but not well defined as a subscheme. We will always consider the case when X is smooth at general points $z \in Z$ and $\mathbf{v}$ is not contained in the tangent plane of X at z. If these hold then $\langle \mathbf{v}, Z \rangle$ is also smooth at z and hence $Z \not\subset Z_{\mathbf{v}}^{\text{res}}$. Our aim is to understand the intersection $Z \cap Z_{\mathbf{v}}^{\text{res}}$.

Note that $Z \cap Z_{\mathbf{v}}^{\text{res}}$ can be quite degenerate. For example, let X be the cone $(x^n + y^n = z^n) \subset \mathbb{P}^3$ with vertex at $(0:0:0:1)$ and Z the line $(x - z = y = 0)$. Then $\langle \mathbf{v}, Z \rangle$ is a plane that contains Z, hence it contains the vertex of the cone. Thus $X \cap \langle \mathbf{v}, Z \rangle$ is a union of n lines through $(0:0:0:1)$. Thus $Z_{\mathbf{v}}^{\text{res}}$ is a union of $n-1$ lines and $Z \cap Z_{\mathbf{v}}^{\text{res}} = (0:0:0:1)$, a single point.

We see below that similar bad behavior does not happen for smooth hypersurfaces.

24 (Ramification Linear System) Let $X = (G = 0) \subset \mathbb{P}^{n+1}$ be a hypersurface. The tangent plane $T_{\mathbf{p}}X$ at a smooth point $(p_0: \cdots : p_{n+1})$ is given by the equation

$$\sum_i x_i \frac{\partial G}{\partial x_i}(\mathbf{p}) = 0. \tag{16}$$

Let $\mathbf{v} := (v_0: \cdots : v_{n+1}) \in \mathbb{P}^{n+1}$ be a point and $\pi_{\mathbf{v}} : \mathbb{P}^{n+1} \dashrightarrow \mathbb{P}^n$ the projection from $\mathbf{v}$. The *ramification divisor* $R_{\mathbf{v}}$ of $\pi_{\mathbf{v}}|_X$ is the set of points whose tangent plane passes through $\mathbf{v}$. Thus

$$R_{\mathbf{v}} = \left(\sum_i v_i \frac{\partial G}{\partial x_i} = 0 \right) \cap X. \tag{17}$$

Thus the $|R_{\mathbf{v}}|$ form a linear system, called the *ramification linear system,* which is the restriction of the linear system of all first derivatives of G. We denote it by $|R_X|$. The base locus of $|R_X|$ is exactly the singular locus $\text{Sing } X$.

Note that $|R_X| \subset |(\deg X - 1)H|_X$, where H is the hyperplane class.

Lemma 25 *Let $X \subset \mathbb{P}^{n+1}$ be a smooth hypersurface of degree d. Let $Z \subsetneq X$ be a subvariety of dimension r and $W_i \subset X$ a finite set of subvarieties. Then, for general $\mathbf{v} \in \mathbb{P}^{n+1}$,*

(1) $Z_{\mathbf{v}}^{\text{res}} \cap Z = R_{\mathbf{v}} \cap Z$ *(set theoretically) and*
(2) $Z_{\mathbf{v}}^{\text{res}} \cap (W_i \setminus Z)$ *has dimension* $\leq \dim Z + \dim W_i - n$.

Proof Set $\tau := \pi_{\mathbf{v}}|_X$. If τ is unramified at $\mathbf{x} \in X$ then it is a local isomorphism near $\mathbf{x}$, thus $\langle \mathbf{v}, Z \rangle \cap X = \tau^{-1}(\tau(Z))$ equals Z near X. Thus $Z \cap Z_{\mathbf{v}}^{\text{res}} \subset Z \cap R_{\mathbf{v}}$. To see the converse, it is enough to prove that $Z \cap Z_{\mathbf{v}}^{\text{res}}$ contains a dense open subset of $Z \cap R_{\mathbf{v}}$. Thus choose a point $\mathbf{x} \in Z$ that is smooth both on X, Z and such that τ ramifies at $\mathbf{x}$ but $\tau|_Z$ does not. Then the vector pointing from $\mathbf{x}$ to $\mathbf{v}$ is also a tangent vector of $\langle \mathbf{v}, Z \rangle \cap X$, hence $\mathbf{x}$ is a singular point of $\langle \mathbf{v}, Z \rangle \cap X$. So $\mathbf{x} \in Z_{\mathbf{v}}^{\text{res}}$, proving (1).

Note that $\mathbf{p} \in Z_{\mathbf{v}}^{\text{res}} \cap (W_i \setminus Z)$ iff a secant line connecting $\mathbf{p}$ with some point of Z passes through $\mathbf{v}$. The union of all secant lines connecting a point of Z with a different point of W_i has dimension $\dim Z + \dim W_i + 1$. Thus only a $\dim Z + \dim W_i + 1 - (n+1)$ dimensional family of secant lines passes through a general point of $\mathbb{P}^{n+1}$, proving (2). □

26 (Proof of Proposition 22) Set $r = \dim Z$ and $Z_0 := Z$. We inductively define

$$Z_{i+1} := (Z_i)_{\mathbf{v}_i}^{\text{res}} \quad \text{for general} \quad \mathbf{v}_i \in \mathbb{P}^{n+1}. \tag{18}$$

We claim that $Z^* := Z_r$ has the right properties. First note that Proposition 22.1 follows from (15).

Using Lemma 25.1 r times we see that $Z \cap Z^*$ consists of the intersection points

$$Z \cap R_{\mathbf{v}_1} \cap \cdots \cap R_{\mathbf{v}_r} \tag{19}$$

for general $\mathbf{v}_i$. (If $r = \frac{n}{2}$, we may also get finitely many other points $Z_{i+1} \cap (Z \setminus Z_i)$; these we can ignore.) Since X is smooth, $|R_X|$ is base point free, thus (19) consists of $(d-1)^r \deg Z$ points in general position. (We use characteristic 0 at the last step.) □

4 Multiplicity and Canonical Singularities

One can usually compute or at least estimate the multiplicity of a divisor or a linear system at a point quite easily, thus it would be useful to be able decide using multiplicities whether a pair $(X, c|M|)$ is canonical or log canonical. This turns out to be possible for surfaces, less so for threefolds, but the notions diverge more and more as the dimension grows.

If a pair $(X, c|M|)$ is not canonical, then there is a non-canonical exceptional divisor. We start with an example where this divisor is obtained by just one blow-up. Note that every exceptional divisor can be obtained by repeatedly blowing up subvarieties, but the more blow-ups we need, the harder it is to connect the multiplicity with being canonical.

Example 27 Let X be a smooth variety, $Z \subset X$ a smooth subvariety of codimension r and $|M|$ a linear system. Let $\pi : X' \to X$ denote the blow-up of Z with exceptional divisor E. Then

$$K_{X'} = \pi^* K_X + (r-1)E \quad \text{and} \quad \pi^* |M| = |M'| + \text{mult}_Z |M| \cdot E.$$

Thus

$$K_{X'} + c|M'| \sim_{\mathbb{Q}} \pi^* \big(K_X + c|M|\big) + \big(r - 1 - c \cdot \text{mult}_Z |M|\big)E. \tag{20}$$

Note that we can apply this to any subvariety, after we replace X by $X \setminus \operatorname{Sing} Z$. We have thus proved the following.

Claim 27.1 Let X be a smooth variety, $|M|$ a linear system and $Z \subset X$ a subvariety. Then the following hold.

(a) If $(X, c|M|)$ is canonical then $c \cdot \operatorname{mult}_Z |M| \le \operatorname{codim}_X Z - 1$.
(b) If $(X, c|M|)$ is log canonical then $c \cdot \operatorname{mult}_Z |M| \le \operatorname{codim}_X Z$. $\square$

The problem we have is that the converse holds only for $n = 2$ and only for part (a). Thus here our aim is to get some weaker converse statements in dimensions 2 and 3. In order to do this, we need a good series of examples.

Claim 27.2 $\left(\mathbb{A}^n, c| \sum \lambda_i x_i^{m_i}|\right)$ is log canonical iff

$$c \le \frac{1}{m_1} + \cdots + \frac{1}{m_n}.$$

A very useful way to think about this is the following. If we assign weights to the variables $w(x_i) = \frac{1}{m_i}$ then the linear system becomes weighted homogeneous of weight 1. Thus, our condition says that

$$c \cdot w\left(\sum \lambda_i x_i^{m_i}\right) \le w(x_1 \cdots x_n). \tag{21}$$

The claim is easy to prove if all the m_i are the same or if you know how to use weighted blow-ups, but can be very messy otherwise. The case $n = 2$ and $m_1 = 2$ is quite instructive and worth trying.

See [37, Sec. 6.5] for details in general (using weighted blow-ups).

The following lemma, which is a partial converse to Claim 27.1.a, proves Step 10.3.a.

Lemma 28 *Let X be a smooth variety and $|M|$ a linear system. Assume that $c \cdot \operatorname{mult}_p |M| \le 1$ for every point $p \in X$ and $\dim X \ge 2$. Then $(X, c|M|)$ is canonical.*

Proof For one blow-up $\pi : X' \to X$ as in (20) we have the formula

$$K_{X'} + c|M'| \sim_{\mathbb{Q}} \pi^*\left(K_X + c|M|\right) + \left(r - 1 - c \cdot \operatorname{mult}_Z |M|\right)E.$$

Since $r \ge 2$, our assumption $c \cdot \operatorname{mult}_p |M| \le 1$ implies that $r - 1 - c \cdot \operatorname{mult}_Z |M| \ge 0$.
If $\tau : X'' \to X'$ is any birational morphism and

$$K_{X''} + c|M''| \sim_{\mathbb{Q}} \tau^*\left(K_{X'} + c|M'|\right) + E'',$$

then we get that

$$K_{X''} + c|M''| \sim_{\mathbb{Q}} (\tau \circ \pi)^*\left(K_X + c|M|\right) + E'' + \left(r - 1 - c \cdot \operatorname{mult}_Z |M|\right)\tau^*E.$$

If $(X', c|M'|)$ is canonical then E'' is effective and so is

$$E'' + (r - 1 - c \cdot \text{mult}_Z |M|)\tau^* E.$$

Thus $(X, c|M|)$ is also canonical. If $p' \in X'$ is any point and $p = \pi(p')$ then $\text{mult}_{p'} |M'| \leq \text{mult}_p |M|$, thus $c \cdot \text{mult}_{p'} |M'| \leq 1$ and we can use induction.

The problem is that this seems to be an infinite induction, since we can keep blowing up forever. There are two ways of fixing this.

The easiest is to use a log resolution as in Definition 7 and stop when the birational transform of $|M|$ becomes base point free, hence canonical.

Theoretically it is better to focus on one divisor at a time and use a lemma of Zariski and Abhyankar, which is a very weak form of resolution; see [36, 2.45] or [37, 4.26]. □

Remark 28.1 Another proof is the following. Let $p \in B \subset X$ be a general complete intersection curve. Then $c \cdot (|M| \cdot B) \leq 1$, hence $(B, c|M|_B)$ is log canonical. By Theorem 38 this implies that $(X, c|M|)$ is canonical. □

The following partial converse to Claim 27.1.b) is a reformulation of [65], see also [37, 6.40] for a proof.

Theorem 29 *Let S be a smooth surface and $|M|$ a linear system such that $p \in S$ is a non-log-canonical center of $(S, c|M|)$. Then one can choose local coordinates (x, y) at p and weights $w(x) = a$ and $w(y) = b$ such that*

$$|M| \subset |x^i y^j : w(x^i y^j) > \tfrac{1}{c}w(xy) = \tfrac{1}{c}(a + b)|. \tag{22}$$

□

Example 30 It can be quite hard to find the right coordinate system that works; it is frequently given by complicated power series. For example, [66] writes down a degree 6 polynomial $g(x, y)$ that, in suitable local coordinates becomes $x^2 + y^{20}$. (I do not doubt the claim but I have been unable to find a clear, non-computational explanation.) Taking $a = 10$ and $b = 1$ shows that $(\mathbb{A}^2, c(g = 0))$ is log canonical for $c \leq \frac{11}{20}$. Related bounds and examples are given in [25].

The following consequence proves Step 10.3.2.b, we derive Step 10.3.b from it in Paragraph 39.

Corollary 31 ([6]) *Let S be a smooth surface and $|M|$ a movable linear system such that $p \in S$ is a non-log-canonical point of $(S, c|M|)$. Then $(M \cdot M)_p > \frac{4}{c^2}$.*

Remark Unlike for Lemma 28, a direct induction does not seem to work, but [6] sets up a more complicated inductive assumption and proves it one blow up at a time. The following argument, relying on Theorem 29, easily generalizes to all dimensions. (Unfortunately, this is less useful since Theorem 29 does not generalize to higher dimensions.)

Proof Assume first that in (22) we have $a = b$. Then every member of $|M|$ is a curve that has multiplicity $> \frac{2}{c}$ at p and the intersection multiplicity is at least the product of the multiplicities. (This is a special case of (10), but it is much simpler; see [60, IV.3.2].) Hence the intersection multiplicity is $> \frac{4}{c^2}$.

In general we get that members of $|M|$ have multiplicity $> \frac{1}{c}\left(1 + \min\{\frac{a}{b}, \frac{b}{a}\}\right)$ at p and this only gives that $(M \cdot M)_p > \frac{1}{c^2}$. Thus we need to equalize a and b. The best way to do this is by a weighted blow-up, see [37, Sec. 6.5], but here the following trick works.

After multiplying with the common denominator, we may assume that a, b are integers. Set $x = s^a$ and $y = t^b$. These define a degree ab morphism $\tau : \mathbb{A}^2_{st} \to \mathbb{A}^2_{xy}$. The inclusion

$$|M| \subset \left|x^i y^j : ai + bj > \tfrac{1}{c}(a + b)\right|$$

of (22) is now transformed into

$$\tau^*|M| \subset \left|s^{ai} t^{bj} : ai + bj > \tfrac{1}{c}(a + b)\right| \subset \left|s^m t^n : m + n > \tfrac{1}{c}(a + b)\right|.$$

That is, $\tau^*|M|$ has multiplicity $> \frac{1}{c}(a + b)$, hence $(\tau^*M \cdot \tau^*M)_p > \frac{1}{c^2}(a + b)^2$. Intersection multiplicities get multiplied by the degree of the map under pull-back, thus we conclude that $(M \cdot M)_p > \frac{1}{c^2} \cdot \frac{(a+b)^2}{ab} \geq \frac{4}{c^2}$. $\qquad\square$

A three-dimensional analog of Theorem 29 was conjectured in [6]. The method of [5] shows that it is a consequence of a result of Kawakita [26]. See also [37, Chap. 5] for more details.

Theorem 32 *Let X be a smooth threefold and $|M|$ a linear system such that $p \in X$ is a non-canonical center of $\left(X, c|M|\right)$. Then one can choose local coordinates (x, y, z) at p and weights $w(x) = a$, $w(y) = b$ and $w(z) = 1$ such that*

$$|M| \subset \left|x^i y^j z^k : w(x^i y^j z^k) > \tfrac{1}{c} w(xy) = \tfrac{1}{c}(a + b)\right|.$$

$\qquad\square$

33 (Summary) Let $|M| := |\sum \lambda_i g_i|$ be a linear system on $\mathbb{A}^n$.

- If $n = 2$ then we can decide whether $\left(\mathbb{A}^2, c|M|\right)$ is canonical at the origin just by looking at the degrees of the monomials that occur in the g_i.
- If $n = 2$ then we can decide whether $\left(\mathbb{A}^2, c|M|\right)$ is log canonical at the origin by looking at the monomials that occur in the g_i, provided we use the right coordinate system.
- If $n = 3$ then we can decide whether $\left(\mathbb{A}^3, c|M|\right)$ is canonical at the origin by looking at the monomials that occur in the g_i, provided we use the right coordinate system.
- If $n \geq 4$ then the situation is more complicated, see Example 34. However, as we discuss in Sect. 8, there is the following partial replacement.

We can frequently show that $\left(\mathbb{A}^n, c \mid \sum \lambda_i g_i \mid\right)$ is not log canonical at the origin by looking at the monomials that occur in a Gröbner basis of the ideal (g_i).

Example 34 [37, 6.45] For $r \geq 5$ consider the linear system

$$|M_r| := \left| (x^2 + y^2 + z^2)^2, x^r, y^r, z^r \right|.$$

Show that $\left(\mathbb{C}^3, c \mid M_r \mid\right)$ is log canonical iff $c \leq \frac{1}{2} + \frac{1}{r}$. However, using coordinate changes and weights only shows that $c \leq \frac{3}{4}$.

5 Hyperplane Sections and Canonical Singularities

We start with the proof of Step 10.4.a.

35 (Bertini Type Theorems) The classical Berti theorem—for differentiable maps also known as Sard's theorem—says that a general member of a base point free linear system on a smooth variety is also smooth. This has numerous analogs, all saying that if a variety has certain types of singularities then a general member of a base point free linear system also has only the same type of singularities. Thus it is not surprising that the same holds for canonical and log canonical singularities. The log canonical case of the following proves Step 10.4.a.

Proposition 35.1 Let $H \subset X$ be a general member of a base point free linear system $|H|$. If $(X, c|M|)$ is canonical (resp. log canonical) then so is $(H, c|M|_H)$.

Proof Choose a log resolution $\pi : X' \to X$ as in Definition 7 and write

$$K_{X'} = \pi^* K_X + \sum e_i E_i \quad \text{and} \quad \pi^* |M| = |M'| + \sum a_i E_i, \tag{23}$$

where $|M'|$ is base point free and $\sum E_i$ has simple normal crossing singularities only. Thus

$$K_{X'} + c|M'| \sim_{\mathbb{Q}} \pi^* \left(K_X + c|M| \right) + \sum (e_i - ca_i) E_i. \tag{24}$$

Note that $|H|$ gives us base point free linear systems $|H'|$ on X' and $|H'||_{E_i}$ on each E_i. The adjunction formula (stated only for curves but proved in general in [60, VI.1.4]) says that $K_H = \left(K_X + H \right)|_H$ and $K_{H'} = \left(K_{X'} + H' \right)|_{H'}$. Adding $H' = \pi^* H$ to (24) and restricting to H and H' we get that

$$K_{H'} + c|M'||_{H'} \sim_{\mathbb{Q}} \pi^* \left(K_H + c|M|_H \right) + \sum (e_i - ca_i)(E_i \cap H'), \tag{25}$$

where H' is smooth and $\sum (E_i \cap H')$ has simple normal crossing singularities only. If $(X, c|M|)$ is canonical (resp. log canonical) then $e_i - ca_i \geq 0$ (resp. ≥ -1) for every i. The same $e_i - ca_i$ are involved in (25), except that some of the $E_j \cap H'$

may be empty, in which case $e_j - ca_j$ does not matter for $(H, c|M|_H)$. In any case, $(H, c|M|_H)$ is also canonical (resp. log canonical). □

Let us next see what happens if we try to use the same method to prove Step 10.4.b.

36 Here we have a non-canonical center $p \in X$ and we take an $H \in |H|$ that passes through the point p. If there is an exceptional divisor $E_j \subset X'$ such that $\pi(E_j) = \{p\}$, then $\pi^*H \supset E_j$. Hence π^*H is not smooth, it is not even irreducible. In this case we write

$$\pi^*H = H' + \sum m_i E_i. \tag{26}$$

Adding $H' = \pi^*H - \sum m_i E_i$ to (23) we get

$$K_{X'} + H' + c|M'| \sim_\mathbb{Q} \pi^*(K_X + H + c|M|) + \sum(e_i - m_i - ca_i)E_i. \tag{27}$$

Thus restricting (27) to H' and H we get that

$$K_{H'} + c|M'|_{H'} \sim_\mathbb{Q} \pi^*(K_H + c|M|_H) + \sum(e_i - m_i - ca_i)(E_i \cap H'). \tag{28}$$

At first sight we are done. If p is a non-canonical center of $(X, c|M|)$ then there is an E_j such that $\pi(E_j) = \{p\}$ and $e_j - ca_j < 0$. Since H passes through p, $m_j \geq 1$ also holds, so $e_j - m_j - ca_j < -1$. Thus $E_j \cap H'$ shows that p is a non-log-canonical center of $(H, c|M|_H)$.

However, all this falls apart if $E_j \cap H' = \emptyset$. This can easily happen for some E_j, but it is enough to show that it can not happen for every E_j for which $e_j - m_j - ca_j < -1$. This is what we discuss next.

The following 2 interconnected theorems have many names. In [36] and [35] it is called *inversion of adjunction*, while [52] uses *Shokurov-Kollár connectedness principle*. A closely related result in complex analysis is the *Ohsawa-Takegoshi extension theorem*, proved in [47]. The theorems were conjectured in [61] and proved in [30, Sec. 17]. The sharpest form was established in [27], see also [35, Sec. 4.1] for other generalizations.

For simplicity I state it only for smooth varieties, though the singular case is needed for most applications. The proof is actually a quite short application of Theorem 47.3; see [36, Sec. 5.4] or [37, Chap. 6] for detailed treatments.

Theorem 37 *Let X be a smooth variety and Δ an effective $\mathbb{Q}$-divisor on X. Let $\pi : X' \to X$ be a proper, birational morphism and write*

$$K_{X'} \sim_\mathbb{Q} \pi^*(K_X + \Delta) + \sum b_i B_i,$$

where the B_i are either π-exceptional or lie over Supp Δ. *Then every fiber of*

$$\pi : \text{Supp}\left(\sum_{b_i \leq -1} B_i\right) \to X \quad \text{is connected.} \qquad \square$$

The following consequence is especially important. The first part of it directly implies Step 10.4.b–c, the second part is also used in Sect. 8.

Theorem 38 *Let X be a smooth variety and $|M|$ a linear system on X. Let $H \subset X$ be a smooth divisor. Assume that*

(1) *either H contains a non-canonical center Z^c of $(X, c|M|)$,*
(2) *or H has nonempty intersection with a non-log-canonical center Z^{lc} of $(X, c|M|)$.*

Then $(H, c|M|_H)$ is not log-canonical.

Proof Choose a log resolution $\pi : X' \to X$ and write $\pi^* H = H' + \sum m_i E_i$. Choosing a general member $M \in |M|$ gives $M' \in |M'|$. We can rearrange (27) as

$$K_{X'} \sim_{\mathbb{Q}} \pi^*(K_X + H + cM) - H' - cM' + \sum(e_i - m_i - ca_i)E_i. \qquad (29)$$

Pick a point $p \in Z^c$ (resp. $p \in H \cap Z^{lc}$). We may harmlessly assume that $c < 1$ (this is always the case for us) and then $-cM'$ does not contribute to the $\sum_{b_i \leq -1} B_i$ in Theorem 37. Thus we get that

$$F_p := \pi^{-1}(p) \cap \left(H' \cup \sum_{e_i - m_i - ca_i \leq -1} E_i\right) \quad \text{is connected.}$$

If (1) holds then there is an E_j such that $p \in \pi(E_j)$, $e_j - ca_j < 0$ and $m_j \geq 1$. If (2) holds then there is an E_j such that $p \in \pi(E_j)$ and $e_j - ca_j < -1$. Thus, in both cases, $e_j - m_j - ca_j < -1$ and $\pi^{-1}(p) \cap \sum_{e_i - m_i - ca_i \leq -1} E_i$ is not empty. Since F_p is connected, we obtain that

$$\pi^{-1}(p) \cap H' \cap \sum_{e_i - m_i - ca_i \leq -1} E_i \neq \emptyset.$$

Thus there is at least one divisor E_{j_0} such that

$$e_{j_0} - m_{j_0} - ca_{j_0} \leq -1 \quad \text{and} \quad p \in \pi(E_{j_0} \cap H').$$

Hence $E_{j_0} \cap H'$ gives the non-empty divisor that we needed in (28). A small problem is that we would like a strict inequality $e_{j_0} - m_{j_0} - ca_{j_0} < -1$. To achieve this, run the same argument with some $c' < c$. Then we get a j_0 such that

$$e_{j_0} - m_{j_0} - ca_{j_0} < e_{j_0} - m_{j_0} - c'a_{j_0} \leq -1 \quad \text{and} \quad p \in \pi(E_{j_0} \cap H').$$

Thus $(H, c|M|_H)$ is not log canonical. This completes the proof of Step 10.4.b. $\quad\square$

39 (End of the Proof of Step 10.3.b) Let $|M|$ be a movable linear system on a smooth variety Y and $\mathbf{p} \in Y$ a non-log-canonical point of $(Y, c|M|)$. The multiplicity at $\mathbf{p}$ is preserved by general hyperplane cuts through $\mathbf{p}$ and so is being non-log-canonical by Theorem 38. Thus eventually we get a surface $S = H_1 \cap \cdots \cap H_{n-2}$ containing $\mathbf{p}$ such that $(S, c|M|_S)$ is non-log-canonical at $\mathbf{p}$. Thus $\mathrm{mult}_{\mathbf{p}}(M \cdot M) = \mathrm{mult}_{\mathbf{p}}(|M|_S \cdot |M|_S) > 4/c^2$ by Corollary 31. $\qquad\square$

6 Global Sections from Isolated Singularities

The proof of Step 10.6 is a combination of 4 lemmas, which are either quite easy to prove (Lemma 41) or have been well known (Lemmas 40, 42 and 43). Nonetheless, the power of their combination was not realized before [67].

We define the upper multiplier ideals $\mathcal{J}^+(c|N|)$ and $\mathcal{J}^+(I^c)$ for a linear system $|N|$ and an ideal I in Definition 49. We use the following of its properties.

Lemma 40 *Let Y be a smooth variety and $|N|$ a linear system. Then the support of $\mathcal{O}_Y/\mathcal{J}^+(c|N|)$ is the union of all non-log-canonical centers of $(Y, c|N|)$.*

Lemma 41 *Let Y be a smooth variety and $|N|$ a linear system. Assume that $(Y, \frac{c}{2}|N|)$ is not log canonical. Then*

$$\mathcal{J}^+\big(\mathcal{J}^+(c|N|)\big) \neq \mathcal{O}_Y.$$

Lemma 42 *Let Y be a smooth variety and $I \subset \mathcal{O}_Y$ an ideal sheaf that vanishes only at finitely many points. Assume that $\mathcal{J}^+(I) \neq \mathcal{O}_Y$. Then*

$$\dim\big(\mathcal{O}_Y/I\big) \geq \tfrac{1}{2}3^{\dim Y}.$$

Lemma 43 *Let Y be a smooth, projective variety, H an ample divisor on Y and $|N|$ a linear system such that $H \sim_{\mathbb{Q}} c|N|$. Assume that $(Y, c|N|)$ is log canonical outside finitely many points. Then*

$$H^0\big(Y, \mathcal{O}_Y(K_Y + H)\big) \geq \dim\big(\mathcal{O}_Y/\mathcal{J}^+(c|N|)\big).$$

44 (Proof of Step 10.6 Using Lemmas 40–43) We apply Lemma 41 to $|N| := |2M|$ and $c = \frac{1}{m}$. We get the ideal sheaf $I := \mathcal{J}^+(\frac{1}{m}|2M|)$ such that $\mathcal{J}^+(I) \neq \mathcal{O}_Y$. By Lemma 40 I vanishes only at finitely many points. Thus, by Lemma 42,

$$\dim\big(\mathcal{O}_Y/\mathcal{J}^+(\tfrac{1}{m}|2M|)\big) \geq \tfrac{1}{2}3^{\dim Y}.$$

Finally using Lemma 43 for $H := 2L$ says that

$$H^0\big(Y, \mathcal{O}_Y(K_Y + 2L)\big) \geq \dim\big(\mathcal{O}_Y/\mathcal{J}^+(\tfrac{1}{m}|2M|)\big) \geq \tfrac{1}{2}3^{\dim Y}. \qquad\square$$

We prove Lemma 40 in Paragraph 50. Lemma 42 is local at the points where I vanishes, in fact, it is a quite general algebra statement about ideals. We discuss it in detail in Sect. 8. Lemma 43 is a restatement of Corollary 52; we explain its proof in Sect. 7.

45 (Proof of Lemma 41) Take a log resolution $\pi : Y' \to Y$ as in Definition 7 and write

$$K_{Y'} \sim \pi^* K_Y + \sum e_i E_i \quad \text{and} \quad \pi^* |N| = |N'| + \sum_i a_i E_i,$$

where $|N'|$ is base point free. By Definition 48,

$$\mathcal{J}^+(c|N|) = \pi_* \mathcal{O}_{Y'}\left(\sum_i \lceil e_i - c' a_i \rceil E_i\right),$$

where $0 < c - c' \ll 1$. Thus if

$$\pi^* \mathcal{J}^+(c|N|) = \mathcal{O}_{Y'}\left(-\sum_i b_i E_i\right),$$

then $-b_i \leq \lceil e_i - c' a_i \rceil$. (We have to be a little careful here. We need to use a $\pi : Y' \to Y$ that is a log resolution for both $|N|$ and $\mathcal{J}^+(c|N|)$.) Therefore

$$\mathcal{J}^+(\mathcal{J}^+(c|N|)) = \pi_* \mathcal{O}_{Y'}\left(\sum_i \lceil e_i - (1-\epsilon)b_i \rceil E_i\right), \tag{30}$$

for $0 < \epsilon \ll 1$. If $b_i = 0$ then $\lceil e_i - (1-\epsilon)b_i \rceil = e_i - b_i$ and if $b_i > 0$ then

$$\lceil e_i - (1-\epsilon)b_i \rceil = e_i - b_i + \lceil \epsilon b_i \rceil = e_i - b_i + 1,$$

since b_i is an integer. Thus, in both cases

$$\lceil e_i - (1-\epsilon)b_i \rceil \leq e_i - b_i + 1 \leq 2e_i + \lceil -c' a_i \rceil + 1 \leq 2e_i - ca_i + 2.$$

Since $\left(Y, \frac{c}{2}|N|\right)$ is not log canonical, there is an index j such that $e_j - \frac{c}{2}a_j < -1$. Then

$$2e_j - ca_j + 2 = 2\left(e_j - \frac{c}{2}a_j\right) + 2 < -2 + 2 = 0.$$

Thus $\mathcal{J}^+(\mathcal{J}^+(c|N|))$ vanishes along $\pi(E_j)$. □

7 Review of Vanishing Theorems

Here we prove Lemma 43. For this we need to use the cohomology of coherent sheaves. We use that the groups $H^i(Y, F)$ exist and that a short exact sequence of sheaves leads to a long exact sequence of the cohomology groups. For the uninitiated, [55, Chap. B] is a very good introduction.

We also need a vanishing theorem which says that under certain assumptions the cohomology group $H^1(Y, F)$ is 0. The reader who is willing to believe Theorem 51 need not get into any further details. However, at first sight, the definition of the multiplier ideal may appear rather strange, so I include an explanation of where these definitions and results come from.

46 (Vanishing and Global Sections) Let F be a coherent sheaf on a projective variety Y. One way to estimate the dimension of $H^0(Y, F)$ from below is to identify a subsheaf $\mathcal{S}(F) \hookrightarrow F$ and the corresponding quotient $F \twoheadrightarrow \mathcal{Q}(F)$, write down the short exact sequence

$$0 \to \mathcal{S}(F) \to F \to \mathcal{Q}(F) \to 0,$$

and the beginning of its long exact sequence

$$0 \to H^0(Y, \mathcal{S}(F)) \to H^0(Y, F) \to H^0(Y, \mathcal{Q}(F)) \to H^1(Y, \mathcal{S}(F)).$$

If the last term vanishes then

$$\dim H^0(Y, F) \geq \dim H^0(Y, \mathcal{Q}(F)).$$

In our case we have a divisor L and a linear system $|M|$ such that $c|M| \sim L$ for some c. We will use these data to construct the subsheaf

$$\mathcal{S}(\mathcal{O}_Y(K_Y + L)) \subset \mathcal{O}_Y(K_Y + L),$$

such that a generalization of Kodaira's vanishing theorem applies to $\mathcal{S}(\mathcal{O}_Y(K_Y + L))$. These vanishing theorems form a powerful machine which gives us a vanishing involving $\mathcal{O}_Y(K_Y + L)$. (For our purposes in Sect. 6, a vanishing involving pretty much any other $\mathcal{O}_Y(aK_Y + bL)$ would be good enough, as long as a, b are much smaller than $\dim Y$.)

47 (Generalizations of Kodaira's Vanishing Theorem) Kodaira's classical vanishing theorem says that if Y is a smooth, projective variety over $\mathbb{C}$ and L an ample divisor on X then $H^i(Y, \mathcal{O}_Y(K_Y + L)) = 0$ for $i > 0$. It has various generalizations when L is only close to being ample.

47.1 (Close-to-Ample Divisors) It turns out that Kodaira's vanishing theorem also works for a divisor L if one can write it as $L \sim_{\mathbb{Q}} cA + \Delta$ where

(a) $c > 0$ and A is nef and big (that is, $(A \cdot C) \geq 0$ for every curve $C \subset Y$ and $(A^{\dim Y}) > 0$), and
(b) $\Delta := \sum d_i D_i$, where $d_i \in [0, 1)$ and $\sum D_i$ has simple normal crossing singularities only.

In practice the condition that $\sum D_i$ be a simple normal crossing divisor is very rarely satisfied, but log resolution (as in Definition 7) allows us to reduce almost everything to this case. The basic vanishing theorem is the following, see [36, Sec. 2.5] or [38, 9.1.18] for proofs.

Theorem 47.2 (Kawamata-Viehweg Version) Let X be a smooth, projective variety and L a divisor as in Subparagraph 47.1. Then

$$H^i\big(X, \mathcal{O}_X(K_X + L)\big) = 0 \quad \text{for} \quad i > 0. \qquad \qquad \square$$

The following versions are easy to derive from Subparagraph 47.1; see [36, 2.68].

Theorem 47.3 (Grauert-Riemenschneider Version) Let X be a smooth, projective variety, $\pi : X \to Y$ a birational morphism and L a divisor as in Subparagraph 47.1. Then

$$R^i \pi_* \mathcal{O}_X(K_X + L) = 0 \quad \text{for} \quad i > 0. \qquad \qquad \square$$

Corollary 47.4 Let X be a smooth, projective variety, $\pi : X \to Y$ a birational morphism and L a divisor as in Subparagraph 47.1. Then

$$H^i\big(Y, \pi_* \mathcal{O}_X(K_X + L)\big) = 0 \quad \text{for} \quad i > 0. \qquad \qquad \square$$

Next we show how we get vanishing theorems starting with a linear system.

48 Let Y be a smooth, projective variety over $\mathbb{C}$ and $|M|$ a linear system on Y. Following Subparagraph 47.1 we would like to get a nef and big divisor plus a divisor with simple normal crossing support.

Thus let $\pi : Y' \to Y$ be a log resolution as in Definition 7. Write

$$K_{Y'} = \pi^* K_Y + \sum e_i E_i \quad \text{and} \quad \pi^*|M| = |M'| + \sum a_i E_i. \qquad (31)$$

Thus the E_i are either π-exceptional or belong to the base locus of $\pi^*|M|$, and we allow e_i or a_i to be 0. Let L be an ample divisor such that $L \sim_{\mathbb{Q}} c|M|$. Then we can write the pull-back of $K_Y + L$ as

$$\pi^*(K_Y + L) \sim_{\mathbb{Q}} K_{Y'} + c|M'| + \sum (-e_i + ca_i) E_i. \qquad (32)$$

The right hand side starts to look like we could apply Theorem 47.2 to it, but there are 2 problems. The coefficient $(-e_i + ca_i)$ need not lie in the interval $[0, 1)$ and, although $|M'|$ is nef, it need not be big. The latter can be arranged by keeping a little bit of L unchanged. That is, pick $0 < c' < c$ and write the pull-back of $K_Y + L$ as

$$\pi^*(K_Y + L) \sim_{\mathbb{Q}} K_{Y'} + (c - c')\pi^* L + c'|M'| + \sum (-e_i + c'a_i) E_i. \qquad (33)$$

Now $\pi^*L + c'|M'|$ is nef and big, but the first problem remains. Here we use that any number a can be uniquely written as $a = \lfloor a \rfloor + \{a\}$ where $\lfloor a \rfloor$ is an integer and $\{a\} \in [0, 1)$. Furthermore, let $\lceil a \rceil := -\lfloor -a \rfloor$ denote the rounding up of a. Applying this and rearranging we get that

$$
\begin{aligned}
\pi^*(K_Y + L) + \sum \lceil e_i - c'a_i \rceil E_i \\
\sim_{\mathbb{Q}} K_{Y'} + (c - c')\pi^*L + c'|M'| + \sum \{-e_i + c'a_i\}E_i.
\end{aligned}
\tag{34}
$$

Now the vanishing Corollary 47.4 applies to the right hand side of (34). Note also that

$$
\pi_*\mathcal{O}_{Y'}\big(\pi^*(K_X+L)+\sum \lceil e_i - c'a_i \rceil E_i\big) = \mathcal{O}_Y(K_Y+L)\otimes\pi_*\mathcal{O}_{Y'}\big(\sum \lceil e_i - c'a_i \rceil E_i\big).
$$

This suggests that the basic object is $\pi_*\mathcal{O}_{Y'}\big(\sum \lceil e_i - c'a_i \rceil E_i\big)$. Note that this does not depend on c' as long as $c - c'$ is small enough. Indeed, then

$$
\lceil e_i - c'a_i \rceil = \begin{cases} \lceil e_i - ca_i \rceil & \text{if } e_i - ca_i \notin \mathbb{Z} \quad \text{and} \\ \lceil e_i - ca_i \rceil + 1 & \text{if } e_i - ca_i \in \mathbb{Z}. \end{cases}
$$

Definition 49 (Multiplier Ideal) Let X be a smooth, projective variety over $\mathbb{C}$ and $|M|$ a linear system. The *upper multiplier ideal* of $c|M|$ is

$$
\mathcal{J}^+\big(c|M|\big) := \pi_*\mathcal{O}_{Y'}\big(\sum \lceil e_i - c'a_i \rceil E_i\big)
$$

for any c' satisfying $0 < c - c' \ll 1$. It is not hard to see that this does not depend on the choice of $\pi : Y' \to Y$. Note that [38, Sec. 9.2] calls

$$
\mathcal{J}\big(c|M|\big) := \pi_*\mathcal{O}_{Y'}\big(\sum \lceil e_i - ca_i \rceil E_i\big)
$$

the multiplier ideal. Clearly $\mathcal{J}^+\big(c|M|\big) = \mathcal{J}\big(c'|M|\big)$ for $0 < c - c' \ll 1$.

For an ideal sheaf $I(g_i)$ let $|M| := |\sum \lambda_i g_i|$ and set $\mathcal{J}^+\big(I^c\big) := \mathcal{J}^+\big(c|M|\big)$.

50 (Proof of Lemma 40) Set $W := \operatorname{Supp}\mathcal{O}_Y/\mathcal{J}^+\big(c|M|\big)$. Then W is exactly the π-image of the support of the negative part of $\sum \lceil e_i - c'a_i \rceil E_i$. If $e_i - ca_i < -1$ then $e_i - c'a_i < -1$ so $\lceil e_i - c'a_i \rceil \leq -1$. If $e_i - ca_i \geq -1$ and $a_i > 0$ then $e_i - c'a_i > -1$ so $\lceil e_i - c'a_i \rceil \geq 0$. If $a_i = 0$ then $e_i - ca_i = e_i \geq 0$.   □

Now we can apply Corollary 47.4 and get the following, see [38, Sec. 9.4].

Theorem 51 (Nadel Vanishing) *Let Y be a smooth, projective variety, L an ample divisor and $|M|$ a linear system such that $c|M| \sim_{\mathbb{Q}} L$. Then*

$$
H^i\big(Y, \mathcal{O}_Y(K_Y + L) \otimes \mathcal{J}^+\big(c|M|\big)\big) = 0 \quad \text{for} \quad i > 0.
$$
   □

As we discussed in Paragraph 46, this immediately implies the following.

Corollary 52 *Let Y be a smooth, projective variety, L an ample divisor on Y and $|M|$ a linear system such that $L \sim_{\mathbb{Q}} c|M|$. Assume that $(Y, c|M|)$ is log canonical outside finitely many points. Then*

$$H^0(Y, \mathcal{O}_Y(K_Y + L)) \geq \dim (\mathcal{O}_Y/\mathcal{J}^+(c|M|)).$$ □

8 Review of Monomial Ideals

In this section we prove Lemma 42. Its claim is local at the points where I vanishes, we can thus work using local coordinates at a point. Though not important, it is notationally simpler to pretend that we work at the origin of $\mathbb{A}^n$. (This is in fact completely correct, one needs to argue that Y and $\mathbb{A}^n$ have the same completions, [60, Sec. II.2.2].)

As a general rule, an ideal is log canonical iff it contains low multiplicity polynomials. In this section we give a precise version of this claim. Key special cases of the following are proved by Reid [54] and Corti [6]. More general versions are in [12, 19]. An excellent detailed treatment of this topic is given in [38, Chap. 9], so I concentrate on the definitions and explanations, leaving the details to [38].

Theorem 53 *Let $I \subset R := k[x_1, \ldots, x_n]$ be an ideal vanishing only at the origin. Assume that I is not log canonical. Then*

$$\dim (R/I) \geq \min_{a_1,\ldots,a_n \geq 0} \#\left\{ \mathbb{N}^n \cap \left(\sum a_i r_i \leq \sum a_i \right) \right\}.$$

The proof is given in two steps. We first reduce to the case of monomial ideals in Corollary 55.2 and then to counting lattice points in a simplex Corollary 56.2. Following the proof shows that the lower bound is sharp, but I do not know a closed formula for it. However, a simple argument, given in Paragraph 57, gives the following.

Corollary 54 *Let $I \subset k[x_1, \ldots, x_n]$ be a non-log-canonical ideal that vanishes only at the origin. Then*

$$\dim (R/I) \geq \tfrac{1}{2} 3^n.$$

55 (Deformation to Monomial Ideals) (See [8, Chap. 2] for details.) Let $I \subset R := k[x_1, \ldots, x_n]$ be an ideal. Write every $g \in R$ as $g = \mathrm{in}(g) + \mathrm{rem}(g)$ where $\mathrm{in}(g) := a_g \prod x_i^{r_i}$ is the lexicographically lowest monomial that appears in g with nonzero coefficient. Define the *initial ideal* of I (with respect to the lexicographic ordering) as

$$\mathrm{in}(I) := (\mathrm{in}(g) : g \in I). \tag{35}$$

Thus in(I) is generated by monomials and it is not hard to see that

$$\dim(R/I) = \dim\big(R/\operatorname{in}(I)\big). \tag{36}$$

A key property is the following.

Proposition 55.1 If in(I) is log canonical then so is I.

Comments on the Proof Choose integers $1 \le w_1 \ll \cdots \ll w_n$. For $g \in R$ let $w(g)$ denote the largest t power that divides $g(t^{w_1}x_1, \ldots, t^{w_n}x_n)$. Then

$$t^{-w(g)}g(t^{w_1}x_1, \ldots, t^{w_n}x_n) = \operatorname{in}(g)(x_1, \ldots, x_n) + t(\text{other terms}). \tag{37}$$

Any finite collection of these defines a linear system $|M|$ on $Y := \mathbb{A}^{n+1}$ with coordinates $(x_1, \ldots, x_n, t)$.

If we choose w_i that work for a Gröbner basis $g_i \in I$, then we get $|M|$ whose restriction to $(t = 0)$ gives $I_0 = \operatorname{in}(I)$ and to $(t = \lambda)$ gives $I_\lambda \cong I$ for $\lambda \ne 0$.

If I_0 is log canonical then $\big(Y, |M|\big)$ is also log canonical by Theorem 38.2, and so is $I_\lambda \cong I$ by Proposition 35.1. $\qquad\square$

Combining (36) and Proposition 55.1 gives the first reduction step of the proof of Theorem 53.

Corollary 55.2 If Theorem 53 holds for monomial ideals then it also holds for all ideals. $\qquad\square$

56 (Monomial Ideals) Let $I \subset k[_1, \ldots, x_n]$ be a *monomial ideal,* that is, an ideal generated by monomials. A very good description of I is given by its *Newton polytope.*

For $\prod x_i^{r_i} \in I$ we mark the point $(r_1, \ldots, r_n)$ with a big dot for elements of $I \setminus (x_1, \ldots, x_n)I$ and with an invisible dot for elements of $(x_1, \ldots, x_n)I$. The Newton polytope is the boundary of the convex hull of the marked points, as in the next example.

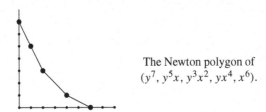

The Newton polygon of
$(y^7, y^5x, y^3x^2, yx^4, x^6).$

A face of the Newton polytope is called *central* if it contains a point all of whose coordinates are equal.

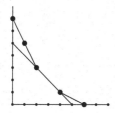

The Newton polygon of
$(y^7, y^5x, y^3x^2, yx^4, x^6)$,
with central face extended.

The next version of Claim 27.2 follows from [54]; see [19] for various generalizations and [38, Sec. 9.3.C] for proof.

Proposition 56.1 A monomial ideal I is log canonical iff its Newton polytope contains the point $(1, \ldots, 1)$. □

Thus I is not log canonical iff a central face of its Newton polytope contains a point $(d, \ldots, d)$ with $d > 1$. The equation of this face can then be written as

$$\sum a_i r_i = d \sum a_i \quad \text{for some} \quad a_i \geq 0.$$

In particular, a monomial $\prod_i x_i^{r_i}$ is not contained in I if $\sum a_i r_i \leq \sum a_i$. We have thus proved the following.

Corollary 56.2 A monomial ideal I is not log canonical iff there is a simplex

$$\Delta(\mathbf{a}) := \left(0 \leq r_i, \ \sum a_i r_i \leq \sum a_i\right)$$

that is disjoint from the Newton polytope of I. If this holds then

$$\dim\left(R/I\right) \geq \left(\begin{array}{c}\text{number of lattice points} \\ \text{in the simplex } \Delta(\mathbf{a})\end{array}\right).$$ □

57 (Lattice Points in Simplices) We thus need to estimate from below the number of lattice points in the n-simplex $\left(0 \leq r_i, \sum a_i r_i \leq \sum a_i\right)$, independent of the a_i. I could not find the optimal values.

The lower bound $\frac{1}{2}3^n$ comes from the observation that if $r_i \in \{0, 1, 2\}$ then either $(r_1, \ldots, r_n)$ or $(2 - r_1, \ldots, 2 - r_n)$ satisfies $\sum a_i r_i \leq \sum a_i$. We can do a little better by adding the points with coordinates $3, \ldots, n$ on at least 1 of the coordinate axes.

Another lower bound is $\frac{1}{n}\binom{2n}{n}$, which is asymptotically $4^n/(n\sqrt{\pi n})$. This comes from the observation that if $\sum r_i \leq n$ then at least one of the cyclic permutations of $(r_1, \ldots, r_n)$ satisfies $\sum a_i r_i \leq \sum a_i$.

The first bound is better for $n \leq 5$, the second for $n \geq 6$.

We can also combine the two bounds to get

$$\frac{1}{n}\left[\binom{2n}{n} - \frac{1}{2}3^n\right] + \frac{1}{2}3^n = \frac{1}{n}\binom{2n}{n} + \frac{n-1}{2n}3^n.$$

Acknowledgements These notes are based on lectures given at the AGNES Conference at Rutgers and at Northwestern University in April and May, 2018. The hospitality of Northwestern University, and especially of M. Popa, gave an ideal time to write them up in expanded form. A received many helpful comments and references from N. Alon, V. Cheltsov, A. Corti, L. Ein and Y. Liu. Several e-mails from T. de Fernex helped me to clarify some of the key points and to improve the paper considerably.

Partial financial support was provided by the NSF under grant number DMS-1362960 and by the Nemmers Prize of Northwestern University.

References

1. F. Ambro, Quasi-log varieties. Tr. Mat. Inst. Steklova **240**, 220–239 (2003). No. Biratsion. Geom. Linein. Sist. Konechno Porozhdennye Algebry. MR 1993751
2. I.A. Cheltsov, On a smooth four-dimensional quintic. Mat. Sb. **191**(9), 139–160 (2000). MR 1805602
3. I.A. Cheltsov, Birationally rigid Fano varieties. Usp. Mat. Nauk **60**(5)(365), 71–160 (2005). MR 2195677
4. I.A. Cheltsov, Nonexistence of elliptic structures on general Fano complete intersections of index one. Vestnik Moskov. Univ. Ser. I Mat. Mekh. (3), 50–53 (2005). Translation in Moscow Univ. Math. Bull. **60** (2005), no. 3, 30–33 (2006)
5. A. Corti, Factoring birational maps of threefolds after Sarkisov. J. Algebraic Geom. **4**(2), 223–254 (1995). MR 1311348
6. A. Corti, Singularities of linear systems and 3-fold birational geometry, in *Explicit Birational Geometry of 3-Folds*. London Mathematical Society Lecture Note Series, vol. 281, pp. 259–312. Cambridge University Press, Cambridge (2000). MR MR1798984 (2001k:14041)
7. A. Corti, M. Reid (eds.), *Explicit Birational Geometry of 3-Folds*. London Mathematical Society Lecture Note Series, vol. 281. Cambridge University Press, Cambridge (2000). MR MR1798978 (2001f:14004)
8. D. Cox, J. Little, D. O'Shea, *Ideals, Varieties, and Algorithms: An Introduction to Computational Algebraic Geometry and Commutative Algebra*. Undergraduate Texts in Mathematics (Springer, New York, 1992). MR 1189133
9. T. de Fernex, Erratum to: Birationally rigid hypersurfaces. Invent. Math. **203**(2), 675–680 (2016). MR 3455160
10. T. de Fernex, Birational rigidity of singular Fano hypersurfaces. Ann. Sc. Norm. Super. Pisa Cl. Sci. **17**, 911–929 (2017)
11. T. de Fernex, L. Ein, M. Mustaţă, Bounds for log canonical thresholds with applications to birational rigidity. Math. Res. Lett. **10**(2–3), 219–236 (2003). MR 1981899 (2004e:14060)
12. T. de Fernex, L. Ein, M. Mustaţă, Multiplicities and log canonical threshold. J. Algebraic Geom. **13**(3), 603–615 (2004). MR 2047683 (2005b:14008)
13. G. Fano, Sopra alcune varieta algebriche a tre dimensioni aventi tutti i generi nulli. Atti. Ac. Torino **43**, 973–977 (1908)
14. G. Fano, Osservazioni sopra alcune varieta non razionali aventi tutti i generi nulli. Atti. Ac. Torino **50**, 1067–1072 (1915)
15. O. Fujino, Foundations of the minimal model program. Mathematical Society of Japan Memoirs, vol. 35 (2017). MR 3643725
16. W. Fulton, *Intersection Theory*, 2nd edn. (Springer, Berlin, 1998). MR 99d:14003
17. P. Griffiths, J. Harris, *Principles of Algebraic Geometry*. (Wiley, New York, 1978). Pure and Applied Mathematics. MR MR507725 (80b:14001)
18. C.D. Hacon, J. McKernan, The Sarkisov program. J. Algebraic Geom. **22**(2), 389–405 (2013). MR 3019454

19. J.A. Howald, Multiplier ideals of monomial ideals. Trans. Amer. Math. Soc. **353**(7), 2665–2671 (2001). MR 1828466
20. V.A. Iskovskikh, *Birational Automorphisms of Three-Dimensional Algebraic Varieties*. Current Problems in Mathematics, vol. 12 (Russian) (VINITI, Moscow, 1979), pp. 159–236, 239 (loose errata). MR 537686
21. V.A. Iskovskikh, Minimal models of rational surfaces over arbitrary fields. Izv. Akad. Nauk SSSR Ser. Mat. **43**(1), 19–43, 237 (1979). MR 80m:14021
22. V.A. Iskovskikh, Birational rigidity of Fano hypersurfaces in the framework of Mori theory. Usp. Mat. Nauk **56**(2)(338), 3–86 (2001). MR 1859707
23. V.A. Iskovskikh, Ju.I. Manin, Three-dimensional quartics and counterexamples to the Lüroth problem. Mat. Sb. (N.S.) **86**(128), 140–166 (1971). MR 0291172
24. V.A. Iskovskikh, Yu.G. Prokhorov, Fano varieties, in *Algebraic Geometry, V*. Encyclopaedia of Mathematical Sciences, vol. 47, pp. 1–247 (Springer, Berlin, 1999)
25. J.M. Johnson, J. Kollár, How small can a polynomial be near infinity? Amer. Math. Monthly **118**(1), 22–40 (2011). MR 2795944
26. M. Kawakita, Divisorial contractions in dimension three which contract divisors to smooth points. Invent. Math. **145**(1), 105–119 (2001). MR MR1839287
27. M. Kawakita, Inversion of adjunction on log canonicity. Invent. Math. **167**(1), 129–133 (2007). MR 2264806
28. J. Kollár, Flops. Nagoya Math. J. **113**, 15–36 (1989). MR 986434
29. J. Kollár, Flips, flops, minimal models, etc, in *Surveys in Differential Geometry* (Cambridge, MA, 1990), pp. 113–199 (Lehigh University, Bethlehem, 1991). MR 1144527
30. J. Kollár (ed.), Flips and abundance for algebraic threefolds, Société Mathématique de France, Papers from the Second Summer Seminar on Algebraic Geometry held at the University of Utah, Salt Lake City, Utah, August 1991, Astérisque No. 211 (1992)
31. J. Kollár, Nonrational hypersurfaces. J. Am. Math. Soc. **8**(1), 241–249 (1995)
32. J. Kollár, *Rational Curves on Algebraic Varieties*. Ergebnisse der Mathematik und ihrer Grenzgebiete. 3, Folge., vol. 32 (Springer, Berlin, 1996)
33. J. Kollár, Singularities of pairs, in *Algebraic Geometry—Santa Cruz 1995*. Proceedings of Symposium in Pure Mathematics, vol. 62, pp. 221–287 (American Mathematical Society, Providence, 1997)
34. J. Kollár, *Lectures on resolution of singularities*, Annals of Mathematics Studies, vol. 166 (Princeton University Press, Princeton, NJ, 2007)
35. J. Kollár, *Singularities of the Minimal Model Program*. Cambridge Tracts in Mathematics, vol. 200 (Cambridge University Press, Cambridge, 2013). With the collaboration of Sándor Kovács
36. J. Kollár, S. Mori, *Birational Geometry of Algebraic Varieties*. Cambridge Tracts in Mathematics, vol. 134 (Cambridge University Press, Cambridge, 1998). With the collaboration of C. H. Clemens and A. Corti, Translated from the 1998 Japanese original
37. J. Kollár, K.E. Smith, A. Corti, *Rational and Nearly Rational Varieties*. Cambridge Studies in Advanced Mathematics, vol. 92 (Cambridge University Press, Cambridge, 2004)
38. R. Lazarsfeld, *Positivity in Algebraic Geometry. I-II*. Ergebnisse der Mathematik und ihrer Grenzgebiete. 3. Folge., vol. 48–49 (Springer, Berlin, 2004). MR 2095471 (2005k:14001a)
39. S. Lefschetz, *L'analysis situs et la géométrie algébrique* (Gauthier-Villars, Paris, 1950). MR 0033557
40. Y. Liu, The volume of singular Kähler-Einstein Fano varieties. Compos. Math. **154**, 1131–1158 (2018)
41. Y. Liu, Z. Zhuang, Birational superrigidity and K-stability of singular Fano complete intersections, ArXiv e-prints (2018)
42. Yu.I. Manin, Rational surfaces over perfect fields. Inst. Hautes Études Sci. Publ. Math. **30**(1), 55–113 (1966). MR 37 #1373
43. T. Matsusaka, D. Mumford, Two fundamental theorems on deformations of polarized varieties. Am. J. Math. **86**, 668–684 (1964). MR 0171778 (30 #2005)
44. S. Mori, *Lectures on Extremal Rays and Fano Threefolds* (Nagoya University, Nagoya, 1983)

45. D. Mumford, *Algebraic Geometry. I.* Complex projective varieties, Grundlehren der Mathematischen Wissenschaften, No. 221 (Springer, Berlin, 1976)

46. M. Noether, Ueber Flächen, welche Schaaren rationaler Curven besitzen. Math. Ann. **3**(2), 161–227 (1870). MR 1509694

47. T. Ohsawa, K. Takegoshi, On the extension of L^2 *holomorphic functions*. Math. Z. **195**(2), 197–204 (1987). MR MR892051 (88g:32029)

48. A.V. Pukhlikov, Birational isomorphisms of four-dimensional quintics. Invent. Math. **87**(2), 303–329 (1987). MR 870730

49. A.V. Pukhlikov, A remark on the theorem of V. A. Iskovskikh and Yu. I. Manin on a three-dimensional quartic. Trudy Mat. Inst. Steklov. **208**, 278–289 (1995). No. Teor. Chisel, Algebra i Algebr. Geom., Dedicated to Academician Igor Rostislavovich Shafarevich on the occasion of his seventieth birthday (Russian). MR 1730270

50. A.V. Pukhlikov, Birational automorphisms of Fano hypersurfaces. Invent. Math. **134**(2), 401–426 (1998). MR 1650332 (99i:14046)

51. A.V. Pukhlikov, Birationally rigid Fano hypersurfaces. Izv. Ross. Akad. Nauk Ser. Mat. **66**(6), 159–186 (2002). MR 1970356

52. A.V. Pukhlikov, *Birationally Rigid Varieties*. Mathematical Surveys and Monographs, vol. 190 (American Mathematical Society, Providence, 2013). MR 3060242

53. A.V. Pukhlikov, Birational geometry of Fano hypersurfaces of index two. Math. Ann, **366**, 721–782 (2016)

54. M. Reid, Canonical 3-folds. Journées de Géometrie Algébrique d'Angers, Juillet 1979/Algebraic Geometry, Angers, 1979, Sijthoff & Noordhoff, Alphen aan den Rijn, 1980, pp. 273–310. MR 605348 (82i:14025)

55. M. Reid, Chapters on algebraic surfaces, in *Complex Algebraic Geometry* (Park City, UT, 1993), IAS/Park City Mathematical Series, vol. 3, pp. 3–159 (American Mathematical Society, Providence, RI, 1997). MR 1442522

56. P. Samuel, *Méthodes d'algèbre abstraite en géométrie algébrique*, Ergebnisse der Mathematik und ihrer Grenzgebiete (N.F.), Heft 4 (Springer, Berlin, 1955). MR 0072531

57. V.G. Sarkisov, Birational automorphisms of conic bundles. Izv. Akad. Nauk SSSR Ser. Mat. **44**(4), 918–945, 974 (1980). MR 587343

58. S. Schreieder, Stably irrational hypersurfaces of small slopes, ArXiv e-prints (2018)

59. B. Segre, A note on arithmetical properties of cubic surfaces. J. Lond. Math. Soc **18**, 24–31 (1943). MR 0009471

60. I.R. Shafarevich, *Basic Algebraic Geometry* (Springer, New York, 1974). Die Grundlehren der mathematischen Wissenschaften, Band 213

61. V.V. Shokurov, Three-dimensional log perestroikas. Izv. Ross. Akad. Nauk Ser. Mat. **56**(1), 105–203 (1992). MR 1162635 (93j:14012)

62. C. Stibitz, Z. Zhuang, K-stability of birationally superrigid Fano varieties, ArXiv e-prints (2018)

63. F. Suzuki, Birational rigidity of complete intersections. Math. Z. **285**(1-2), 479–492 (2017). MR 3598821

64. K. Takeuchi, Some birational maps of Fano 3-folds. Compos. Math. **71**(3), 265–283 (1989). MR 1022045

65. A.N. Varčenko, Newton polyhedra and estimates of oscillatory integrals. Funkcional. Anal. i Priložen. **10**(3), 13–38 (1976). MR 54 #10248

66. H. Yoshihara, On plane rational curves. Proc. Jpn. Acad. Ser. A Math. Sci. **55**(4), 152–155 (1979). MR 533711

67. Z. Zhuang, Birational superrigidity and K-stability of Fano complete intersections of index one (with an appendix written jointly with Charlie Stibitz), ArXiv e-prints (2018)

Hodge Theory of Cubic Fourfolds, Their Fano Varieties, and Associated K3 Categories

Daniel Huybrechts

Abstract These are notes of lectures given at the school 'Birational Geometry of Hypersurfaces' in Gargnano in March 2018. The main goal was to discuss the Hodge structures that come naturally associated with a cubic fourfold. The emphasis is on the Hodge and lattice theoretic aspects with many technical details worked out explicitly. More geometric or derived results are only hinted at.

The primitive Hodge structure of a smooth cubic fourfold $X \subset \mathbb{P}^5$ is concentrated in degree four and it is of a very particular type. Once a Tate twist is applied and the sign of the intersection form is changed, it reveals its true nature. It very much looks like the Hodge structure of a K3 surface. In his thesis Hassett [12] studied this curious relation and the intricate lattice theory behind it in greater detail. He established a transcendental correspondence between polarized K3 surfaces of certain degrees and special cubic fourfolds, some aspects of which are reminiscent of the Kuga–Satake construction. The geometric nature of the Hassett correspondence is still not completely understood but it seems that derived categories are central for its understanding. Work of Addington and Thomas [2] represents an important step in this direction, combining Hassett's Hodge theory with Kuznetsov's categorical approach to hypersurfaces.

The aim of the lectures was to discuss the Hodge structures $H^4(X, \mathbb{Z})$, $H^4(X, \mathbb{Z})_{\mathrm{pr}}$, $\widetilde{H}(X, \mathbb{Z})$, and $H^2(F(X), \mathbb{Z})$, all naturally associated with a cubic fourfold X, and their relation to the Hodge structures $H^2(S, \mathbb{Z})$, $H^2(S, \mathbb{Z})_{\mathrm{pr}}$, $\widetilde{H}(S, \mathbb{Z})$, and $\widetilde{H}(S, \alpha, \mathbb{Z})$ that come with a (polarized, twisted) K3 surface S. For a discussion of more motivic aspects, partially covered by the original lectures, and of derived aspects, not touched upon at all, we have to refer to the existing literature. Most of the content of the lectures is also covered by Huybrechts [19].

D. Huybrechts (✉)
Mathematisches Institut, Universität Bonn, Bonn, Germany
e-mail: huybrech@math.uni-bonn.de

© Springer Nature Switzerland AG 2019
A. Hochenegger et al. (eds.), *Birational Geometry of Hypersurfaces*,
Lecture Notes of the Unione Matematica Italiana 26,
https://doi.org/10.1007/978-3-030-18638-8_5

1   Lattice and Hodge Theory for Cubic Fourfolds and K3 Surfaces

In the first section, we collect all facts from Hodge and lattice theory relevant for the study of cubic fourfolds. The curious relation between the lattice theory of cubic fourfolds and K3 surfaces has been systematically studied first by Hassett [12]. Earlier results in this direction are due to Beauville and Donagi [5].

1.1 As abstract lattices, the middle cohomology and the primitive cohomology of a smooth cubic fourfold $X \subset \mathbb{P}^5$ are described by

$$H^4(X, \mathbb{Z}) \simeq I_{21,2} \simeq E_8^{\oplus 2} \oplus U^{\oplus 2} \oplus I_{3,0},$$

$$H^4(X, \mathbb{Z})_{\mathrm{pr}} \simeq E_8^{\oplus 2} \oplus U^{\oplus 2} \oplus A_2,$$

where the square of the hyperplane class h is given as $h^2 = (1, 1, 1) \in I_{3,0}$. Here, we use the common notation E_8 and U for the unique, unimodular, even lattices of signature $(8, 0)$ and $(1, 1)$, respectively, and $I_{m,n}$ for the unique, unimodular, odd lattice of signature (m, n), see [19, Sec. 1.1.5] for details and references. It will be convenient to change the sign and introduce the *cubic lattice* and the *primitive cubic lattice* as

$$\bar{\Gamma} := I_{2,21} \simeq E_8(-1)^{\oplus 2} \oplus U^{\oplus 2} \oplus I_{0,3} \simeq H^4(X, \mathbb{Z})(-1),$$

$$\Gamma := E_8(-1)^{\oplus 2} \oplus U^{\oplus 2} \oplus A_2(-1) \simeq H^4(X, \mathbb{Z})_{\mathrm{pr}}(-1).$$

In particular, from now on $(h^2)^2 = -3$. The twist should not be confused with the Tate twist of the Hodge structure. It turns out that $E_8(-1)^{\oplus 2}$, certainly the most interesting part of these lattices, will hardly play any role in our discussion. We shall henceforth abbreviate it by

$$E := E_8(-1)^{\oplus 2}$$

and consequently write

$$\bar{\Gamma} \simeq E \oplus U^{\oplus 2} \oplus I_{0,3} \text{ and } \Gamma \simeq E \oplus U^{\oplus 2} \oplus A_2(-1).$$

Although there is a priori no geometric reason why K3 surfaces should enter the picture at all, their intersection form will play a central role in our discussion. We will first address this first purely on the level of abstract lattice theory and later add Hodge structures.

Recall that for a complex K3 surface S, its middle cohomology with the intersection form is the lattice

$$H^2(S, \mathbb{Z}) \simeq E \oplus U^{\oplus 3} \simeq E \oplus U_1 \oplus U_2 \oplus U_3 =: \Lambda,$$

see [16, Ch. 14]. The summands U_i, $i = 1, 2, 3$, are copies of the hyperbolic plane U. Indexing them will make the discussion more explicit and will help us to avoid ambiguities later on.

The full cohomology $H^*(S, \mathbb{Z})$ is also endowed with a unimodular intersection form. It is customary to introduce a sign in the pairing on $(H^0 \oplus H^4)(S, \mathbb{Z})$, which, however, does not change the abstract isomorphism type, for $U \simeq U(-1)$. The resulting lattice is the *Mukai lattice*

$$\widetilde{H}(S, \mathbb{Z}) := H^2(S, \mathbb{Z}) \oplus (H^0 \oplus H^4)(S, \mathbb{Z}) \simeq E \oplus U^{\oplus 3} \oplus U_4$$

$$\simeq E \oplus U_1 \oplus U_2 \oplus U_3 \oplus U_4 =: \widetilde{\Lambda}.$$

The standard basis of U consists of isotropic vectors e, f with $(e.f) = 1$. We shall denote the standard bases in the first three copies of U as e_i, $f_i \in U_i$, $i = 1, 2, 3$. However, in order to take into account the sign change in the Mukai pairing, we shall use the convention that $(e_4.f_4) = -1$ and that $e_4 = [S] \in H^0(S, \mathbb{Z})$ and $f_4 = [x] \in H^4(S, \mathbb{Z})$ with $x \in S$ a point.

Next, we introduce an explicit embedding $A_2 \hookrightarrow \widetilde{\Lambda}$. Here, $A_2 = \mathbb{Z}\lambda_1 \oplus \mathbb{Z}\lambda_2$ is the lattice of rank two given by the intersection form $\begin{pmatrix} 2 & -1 \\ -1 & 2 \end{pmatrix}$ and we define

$$A_2 \hookrightarrow U_3 \oplus U_4 \subset \widetilde{\Lambda} \tag{1}$$

by $\lambda_1 \mapsto e_4 - f_4$ and $\lambda_2 \mapsto e_3 + f_3 + f_4$. The orthogonal complement $\langle \lambda_1, \lambda_2 \rangle^\perp = A_2^\perp \subset \widetilde{\Lambda}$ is the lattice

$$A_2^\perp = E \oplus U_1 \oplus U_2 \oplus A_2(-1),$$

where $A_2(-1) \subset U_3 \oplus U_4$ is spanned by $\mu_1 := e_3 - f_3$ and $\mu_2 := -e_3 - e_4 - f_4$ satisfying $(\mu_i)^2 = -2$ and $(\mu_1.\mu_2) = 1$.

Remark 1.1 We observe that $\lambda_1^\perp = E \oplus U_1 \oplus U_2 \oplus U_3 \oplus \mathbb{Z}(-2)$, where the last direct summand is generated by $e_4 + f_4$. Hence, $\lambda_1^\perp \simeq \Lambda \oplus \mathbb{Z}(-2)$, which is a lattice of discriminant[1] $\mathrm{disc}(\lambda_1^\perp) = 2$ and which contains $A_2^\perp \oplus \mathbb{Z}(\lambda_1 + 2\lambda_2)$ as a sublattice of index three. As $H^2(S, \mathbb{Z}) \simeq \Lambda$ and $H^2(S^{[2]}, \mathbb{Z}) \simeq H^2(S, \mathbb{Z}) \oplus \mathbb{Z}(-2)$ for the Hilbert scheme $S^{[2]}$ of any K3 surface S, this can be read as a lattice isomorphism $\lambda_1^\perp \simeq H^2(S^{[2]}, \mathbb{Z})$.

The discussion so far leads to the fundamental observation that there exists an isomorphism

$$\bar{\Gamma} \supset \Gamma \simeq A_2^\perp \subset \widetilde{\Lambda}$$

---

[1]The sign of the discriminant will be of no importance in our discussion, we tacitly work with its absolute value.

between the primitive cubic lattice Γ and the lattice $A_2^\perp$ inside the Mukai lattice $\widetilde{\Lambda}$.

For later use, we record that (1) induces inclusions of index three:

$$A_2 \oplus A_2(-1) \subset U_3 \oplus U_4 \text{ and } A_2 \oplus A_2^\perp \subset \widetilde{\Lambda},$$

where, for example, the quotient of the latter is generated by the image of the class $(1/3)(\mu_1 - \mu_2 - \lambda_1 + \lambda_2) = e_3 + f_4$.

Another technical result that will be crucial at some point later, is the following elementary statement which is surprisingly difficult to prove, cf. [2, Prop. 3.2].

Lemma 1.2 *Consider* $A_2 \subset \widetilde{\Lambda}$ *as before, let* $U \hookrightarrow \widetilde{\Lambda}$ *be an isometric embedding of a copy of the hyperbolic plane, and denote by* $\overline{A_2 + U}$ *the saturation of* $A_2 + U \subset \widetilde{\Lambda}$. *Then there exists an isometric embedding of a copy of the hyperbolic plane* $U' \hookrightarrow \overline{A_2 + U}$ *such that* $\mathrm{rk}(A_2 + U') = 3$.

Proof See [2] for the proof.   □

Remark 1.3 To motivate the notion of Noether–Lefschetz (or Heegner) divisors for cubic fourfolds, let us recall the corresponding concept for K3 surfaces: For a primitive class $\ell \in \Lambda$ with $(\ell)^2 = d$, we write

$$\Lambda_d := \ell^\perp \subset \Lambda.$$

As ℓ is in the same $O(\Lambda)$-orbit as the class $e_2 + (d/2) f_2$, cf. [16, Cor. 14.1.10], it can abstractly be described as

$$\Lambda_d \simeq E \oplus U^{\oplus 2} \oplus \mathbb{Z}(-d).$$

It is important to note that the lattices Λ_d are in general not contained in $A_2^\perp \subset \widetilde{\Lambda}$.

We shall call any primitive vector $v \in \Gamma \simeq A_2^\perp$ with $(v)^2 < 0$ a *Noether–Lefschetz* vector. With such Noether–Lefschetz vector one naturally associates two lattices. On the cubic side, one defines

$$\mathbb{Z}h^2 \oplus \mathbb{Z}v \subset K_v \subset \overline{\Gamma}$$

as the saturation of $\mathbb{Z}h^2 \oplus \mathbb{Z}v \subset \overline{\Gamma}$. On the K3 side, we introduce the saturation

$$A_2 \oplus \mathbb{Z}v \subset L_v \subset \widetilde{\Lambda}.$$

Note that L_v is of rank three and signature $(2, 1)$, while K_v is of rank two and signature $(0, 2)$. Clearly, their respective orthogonal complements are isomorphic:

$$\overline{\Gamma} \supset K_v^\perp \simeq L_v^\perp \subset \widetilde{\Lambda},$$

as they are both described as $v^\perp \subset \Gamma \simeq A_2^\perp$. In particular, for the discriminants we have

$$d := \mathrm{disc}(L_v) = \mathrm{disc}(K_v).$$

The situation has been studied in depth in [12, Prop. 3.2.2]:

Lemma 1.4 (Hassett) *Only the following two cases can occur:*

(i) *Either* $\mathbb{Z}h^2 \oplus \mathbb{Z}v = K_v$, $A_2 \oplus \mathbb{Z}v = L_v$, *and*

$$d = \mathrm{disc}(K_v) = \mathrm{disc}(L_v) = -3\,(v)^2 \equiv 0\,(6)$$

(ii) *or* $\mathbb{Z}h^2 \oplus \mathbb{Z}v \subset K_v$, $A_2 \oplus \mathbb{Z}v \subset L_v$ *are both of index three, and*

$$d = \mathrm{disc}(K_v) = \mathrm{disc}(L_v) = -\frac{1}{3}(v)^2 \equiv 2\,(6).$$

Proof The main ingredient is the standard formula, see e.g. [16, Sec. 14.0.2],

$$\mathrm{disc}(K_v) \cdot [K_v : \mathbb{Z}h^2 \oplus \mathbb{Z}v]^2 = \mathrm{disc}(\mathbb{Z}h^2 \oplus \mathbb{Z}v) = -3\,(v)^2.$$

Any $y \in K_v$ is of the form $y = s\,h^2 + t\,v$, with $s, t \in \mathbb{Q}$. From $(h.y) \in \mathbb{Z}$ one concludes $s \in (1/3)\,\mathbb{Z}$ and hence also $t \in (1/3)\,\mathbb{Z}$. This shows that $[K_v : \mathbb{Z}h^2 \oplus \mathbb{Z}v] = 1, 3$, or $= 9$, but the last possibility is excluded as $(1/3)\,h^2 \notin \Gamma$.

In the first case, i.e. $K_v = \mathbb{Z}h^2 \oplus \mathbb{Z}v$, one finds $d = \mathrm{disc}(K_v) = -3\,(v)^2 \equiv 0\,(6)$. In the second case, so when the index is three, then $3\,d = -(v)^2 \equiv 0, 2, 4\,(6)$. On the other hand, K_v admits a basis consisting of h^2 and another class x. Indeed, pick any class $x \in K_v$ whose image generates the quotient $K_v/(\mathbb{Z}h^2 \oplus \mathbb{Z}v) \simeq \mathbb{Z}/3\mathbb{Z}$. We may assume $3\,x = s\,h^2 + t\,v$ with $s, t = \pm 1$ and, therefore, $K_v = \mathbb{Z}h^2 \oplus \mathbb{Z}x$. Hence, its discriminant satisfies $d = -3\,(x)^2 - (x.h^2)^2 \equiv 0, 2, 3, 5\,(6)$. Altogether this shows that $d \equiv 0, 2\,(6)$.

We claim that $d \equiv 0\,(6)$ holds if and only if $K_v = \mathbb{Z}h^2 \oplus \mathbb{Z}v$. The 'if'-direction was proven already. For the 'only if'-direction, assume that $d \equiv 0\,(6)$ but $[K_v : \mathbb{Z}h^2 \oplus \mathbb{Z}v] = 3$. Pick $x \in K_v$ as above. Then, write $v = s\,h^2 + t\,x$, $s, t \in \mathbb{Z}$, and use $(v.h^2) = 0$ and the primitivity of v to show $v = r\,((x.h^2)\,h^2 + 3\,x)$ with $r = \pm 1, \pm(1/3)$ as v is primitive. However, $(x.h^2) \equiv 0\,(3)$ under the assumption that $d \equiv 0\,(6)$. Hence, $\pm v = m\,h^2 + x$, $m \in \mathbb{Z}$, and, therefore, $x \in \mathbb{Z}h^2 \oplus \mathbb{Z}v$. This yields a contradiction and thus proves the assertion.

The assertions for the lattice L_v follows directly from the ones for K_v.   □

Remark 1.5 Depending on the perspective, it may be useful to study the various cases from the point of view of d or, alternatively, of $(v)^2$. To have the results handy for later use, we restate the above discussion as

$$d \equiv 0\,(6) \;\Rightarrow\; (v)^2 = -d/3 \equiv 0\,(6)\ \text{or}\ \equiv \pm 2\,(6),$$
$$d \equiv 2\,(6) \;\Rightarrow\; (v)^2 = -3\,d \equiv 0\,(6)$$

and

$$(v)^2 \equiv \pm 2\,(6) \implies d = -3\,(v)^2 \equiv 0\,(6),$$
$$(v)^2 \equiv 0\,(6) \quad \implies d = -3\,(v)^2 \equiv 0\,(6) \text{ or } d = -(1/3)\,(v)^2 \equiv 2\,(6).$$

In particular, d determines $(v)^2$ uniquely, but not vice versa unless $(v)^2 \equiv \pm 2\,(6)$.

Proposition 1.6 (Hassett) *Let $v, v' \in \Gamma$ be two primitive vectors and assume that $\mathrm{disc}(L_v) = \mathrm{disc}(L_{v'})$ or, equivalently, $\mathrm{disc}(K_v) = \mathrm{disc}(K_{v'})$. Then there exist an orthogonal transformations $g \in \tilde{O}(\Gamma)$ such that $g(v) = \pm v'$ and, in particular,*

$$L_{v'} \simeq L_{g(v)} \text{ and } K_{v'} \simeq K_{g(v)}.$$

The definition of $\tilde{O}(\Gamma)$ will be recalled below.

Proof We apply Eichler's criterion, cf. [11, Prop. 3.3]. If an even lattice N is of the form $N \simeq N' \oplus U^{\oplus 2}$, then a primitive vector $v \in N$ with prescribed $(v)^2 \in \mathbb{Z}$ and $(1/n)\,\bar{v} \in A_N$, with n determined by $(v.N) = n\,\mathbb{Z}$, is unique up to the action of $\tilde{O}(N)$. Apply this to $v \in \Gamma \simeq A_2^{\perp} \simeq E \oplus U^{\oplus 2} \oplus A_2(-1)$ and use that for any primitive $v \in \Gamma$, either $(v.\Gamma) = \mathbb{Z}$ or $3\,\mathbb{Z}$. This follows from $[\bar{\Gamma} : \Gamma \oplus \mathbb{Z}\,h^2] = 3$ and the unimodularity of $\bar{\Gamma}$.

(i) If $(v)^2 \equiv 0\,(6)$, there are two cases: Assume first that $d \equiv 2\,(6)$ or, equivalently, that $\mathbb{Z}\,v \oplus \mathbb{Z}\,h^2$ is not saturated. Then, one finds an element of the form $\alpha := (1/3)\,v + t\,h^2 \in \bar{\Gamma}$. As $(\alpha.w) \in \mathbb{Z}$ for all $w \in \Gamma$, this shows $(v.\Gamma) \subset 3\,\mathbb{Z}$. Hence, $n = 3$ and $(1/3)\,\bar{v} = \pm 1 \in A_\Gamma \simeq \mathbb{Z}/3\,\mathbb{Z}$.

Assume now that $d \equiv 0\,(6)$ and write $v = n_1 v_1 + n_2 v_2$ with $v_1 \in E \oplus U_1 \oplus U_2$ and $v_2 \in A_2(-1)$, both primitive, and $n_1, n_2 \in \mathbb{Z}$. If $n_1 \not\equiv 0\,(3)$, then there exists a class w in the unimodular lattice $E \oplus U_1 \oplus U_2 \subset \Gamma$ with $(v.w) \notin 3\,\mathbb{Z}$ and hence $(v.\Gamma) = \mathbb{Z}$. If $n_1 \equiv 0\,(3)$, then $n_2 \not\equiv 0\,(3)$, as v is primitive. However, in this case $(1/3)\,(v \pm h^2) = (n_1/3)v_1 + (1/3)\,(n_2 v_2 \pm h^2) \in \bar{\Gamma}$ and so $\mathbb{Z}\,v \oplus \mathbb{Z}\,h^2$ is not saturated, contradicting $d \equiv 0\,(6)$.

(ii) If $(v)^2 \equiv \pm 2\,(6)$ and hence $(v)^2 \not\equiv 0\,(3)$, then $(v.\Gamma) = \mathbb{Z}$, $n = 1$, and $\bar{v} \in A_\Gamma$ is trivial.

Hence, in case (i) and (ii), if indeed d and not only $(v)^2$ is fixed, then $(v)^2 = (v')^2$ and $(1/n)\,\bar{v} = (1/n)\,\bar{v}' \in A_\Gamma$ (up to sign). $\qquad\square$

Remark 1.7 Due to the uniqueness, no information is lost when explicit classes $v \in \Gamma \simeq A_2^{\perp}$ are chosen for any given d. In the sequel, we will work with the following ones.

(i) For $d \equiv 0\,(6)$, one may choose $v_d := e_1 - (d/6)\,f_1 \in U_1 \subset \Gamma$. Observe that indeed, as explained in the general context above, $(v_d)^2 = -d/3$ and that the lattice $A_2 \oplus \mathbb{Z}\,v_d$ is saturated (use $A_2 \subset U_2 \oplus U_3$ and $v_d \in U_1$), i.e.

$$L_d := L_{v_d} = A_2 \oplus \mathbb{Z}\,v_d.$$

Similarly,

$$K_d := K_{v_d} = \mathbb{Z}\,h^2 \oplus \mathbb{Z}\,v_d,$$

which again shows $(v_d)^2 = -d/3$. Their orthogonal complement is

$$\Gamma_d := L_d^\perp \simeq K_d^\perp \simeq E \oplus U_2 \oplus A_2(-1) \oplus \mathbb{Z}\,(e_1 + (d/6)\,f_1)$$

and their discriminant group

$$A_{K_d^\perp} \simeq A_{K_d} \simeq \mathbb{Z}/3\mathbb{Z} \oplus \mathbb{Z}/(d/3)\mathbb{Z}$$

is cyclic if and only if $9 \nmid d$.

(ii) For $d \equiv 2\,(6)$, one sets $v_d := 3\,(e_1 - ((d-2)/6)\,f_1) + \mu_1 - \mu_2 \in U_1 \oplus A_2(-1)$. Then both inclusions

$$A_2 \oplus \mathbb{Z}\,v_d \subset L_d := L_{v_d} \text{ and } \mathbb{Z}\,h^2 \oplus \mathbb{Z}\,v_d \subset K_d := K_{v_d}$$

are of index three, for example $v_d - \lambda_1 + \lambda_2$ and $v_d - h^2$ are divisible by 3. Use $\lambda_1 = e_4 - f_4, \lambda_2 = e_3 + f_3 + f_4, \mu_1 = e_3 - f_3$, and $\mu_2 = -e_3 - e_4 - f_4$, the latter corresponding to $(1, -1, 0), (0, 1, -1) \in \mathbb{Z}^{\oplus 3}$. In this case, see [1, 12, 31],

$$\Gamma_d := L_d^\perp \simeq K_d^\perp \simeq E \oplus U_2 \oplus (\mathbb{Z}^{\oplus 3}, (\,.\,)_A) \text{ with } A := \begin{pmatrix} -2 & 1 & 0 \\ 1 & -2 & 1 \\ 0 & 1 & (d-2)/3 \end{pmatrix}$$

and L_d and K_d are given by the matrices $-A$ and

$$\begin{pmatrix} -3 & 1 \\ 1 & -(d+1)/3 \end{pmatrix},$$

respectively. The discriminant groups for $d \equiv 2\,(6)$ are cyclic, indeed $A_{K_d^\perp} \simeq A_{K_d} \simeq \mathbb{Z}/d\mathbb{Z}$.

In addition to the orthogonal group

$$\tilde{O}(\Gamma) := \{\, g \in O(\bar{\Gamma}) \mid g(h^2) = h^2 \,\}, \tag{2}$$

which we will also think of as $\tilde{O}(\Gamma) = \{\, g \in O(\Gamma) \mid \bar{g} \equiv \text{id on } A_\Gamma \,\}$, we need to consider

$$\tilde{O}(\Gamma, K_d) := \{\, g \in \tilde{O}(\Gamma) \mid g(K_d) = K_d, \text{ i.e. } g(v_d) = \pm v_d \,\}$$
$$\cup$$
$$\tilde{O}(\Gamma, v_d) := \{\, g \in \tilde{O}(\Gamma) \mid g|_{K_d} = \text{id, i.e. } g(v_d) = v_d \,\}.$$

Observe that $\tilde{O}(\Gamma, v_d)$ can be identified with the subgroup of all $g \in O(\Gamma_d)$ with trivial action on the discriminant group $A_{\Gamma_d} \simeq A_{K_d}$. Also, by definition, $\tilde{O}(\Gamma, v_d) \subset \tilde{O}(\Gamma, K_d)$ is a subgroup of index one or two. Note that the natural homomorphism $\tilde{O}(\Gamma, K_d) \longrightarrow O(K_d)$ is neither surjective (let alone injective) nor is its image contained in the subgroup of transformations acting trivially on the discriminant $\tilde{O}(K_d)$.

Lemma 1.8 (Hassett) *The subgroup $\tilde{O}(\Gamma, v_d) \subset \tilde{O}(\Gamma, K_d)$ is of index at most two. More precisely, one distinguishes the following cases:*

(i) *If $d \equiv 0\,(6)$, then*

$$\tilde{O}(\Gamma, v_d) \subset \tilde{O}(\Gamma, K_d)$$

has index two.

(ii) *If $d \equiv 2\,(6)$, then*

$$\tilde{O}(\Gamma, v_d) = \tilde{O}(\Gamma, K_d).$$

Proof

(i) According to Lemma 1.4, $d \equiv 0\,(6)$ if and only if $\mathbb{Z}h^2 \oplus \mathbb{Z}v_d = K_d$, which is contained in $I_{0,3} \oplus U_1$. Let $g \in \tilde{O}(\Gamma)$ be the orthogonal transformation defined by $g = \mathrm{id}$ on $E \oplus U_2 \oplus I_{0,3}$ and by $g = -\mathrm{id}$ on U_1. Then g is an element in $\tilde{O}(\Gamma, K_d) \setminus \tilde{O}(\Gamma, v_d)$.

(ii) Now, $d \equiv 2\,(6)$ if and only if $\mathbb{Z}h^2 \oplus \mathbb{Z}v_d \subset K_d$ has index three and then $v_d = 3\,(e_1 - ((d-2)/6)\,f_1) + \mu_1 - \mu_2$ with $\mu_1 = (1, -1, 0)$, $\mu_2 = (0, 1, -1) \in A_2(-1) \subset I_{0,3}$ and $h^2 = (1, 1, 1)$. Now observe that $(1/3)\,(v_d - h^2) \in K_d$, but $(1/3)\,(-v_d - h^2) \notin K_d$.

$\square$

1.2 It turns out that certain geometric properties of cubic fourfolds are encoded by lattice-theoretic properties of Noether–Lefschetz vectors $v \in \Gamma$. The following ones are relevant for our purposes. It is a matter of choice, whether they are read as conditions on d or on the primitive $v \in \Gamma$. For $d \in \mathbb{Z}$ one considers the conditions:

(∗)   ⇔ There exists an L_d.

(∗∗′) ⇔ There exists an L_d and an embedding $U(n) \hookrightarrow L_d$ for some $n \neq 0$.

(∗∗)   ⇔ There exists an L_d and a primitive embedding $U \hookrightarrow L_d$.

(∗∗∗) ⇔ There exists an L_d and a primitive embedding $U \hookrightarrow L_d$ with $\lambda_1 \in U$.

Remark 1.9

(i) The following implications trivially hold

$$(***) \Rightarrow (**) \Rightarrow (**') \Rightarrow (*).$$

(ii) Each of the conditions in fact splits in two, distinguishing between $d \equiv 0\,(6)$ and $d \equiv 2\,(6)$. We shall write accordingly $(*)_0$, $(*)_2$, $(**)_0$, $(**')_2$, etc.

Lemma 1.10 *Condition* $(**)$ *holds if and only if there exists an isomorphism of lattices*

$$\varepsilon : \Gamma_d \xrightarrow{\ \sim\ } \Lambda_d.$$

In this case, one also has an isomorphism of groups

$$\tilde{O}(\Gamma, v_d) \simeq \tilde{O}(\Lambda_d).$$

Proof Assume that there exists a (primitive) hyperbolic plane $U \hookrightarrow L_d$. As the composition with the inclusion $L_d \subset \tilde{\Lambda}$ can be identified with $U_4 \hookrightarrow \tilde{\Lambda}$ up to the action of $O(\tilde{\Lambda})$, see [16, Thm. 14.1.12], one has $U^\perp \simeq \Lambda$. Hence, $\Gamma_d = L_d^\perp \subset U^\perp \simeq \Lambda$ is a primitive sublattice of corank one, signature $(2, 19)$, discriminant d, and is, therefore, isomorphic to Λ_d. Conversely, if $L_d^\perp = \Gamma_d \simeq \Lambda_d \subset \Lambda \subset \tilde{\Lambda}$, then $U_4 \subset L_d$. Here, one again uses that up to $O(\tilde{\Lambda})$, there exists only one primitive embedding $\Lambda_d \hookrightarrow \tilde{\Lambda}$.

For the isomorphism between the two orthogonal groups, just recall that they are both described as the subgroup of all orthogonal transformations of $\Gamma_d \simeq \Lambda_d$ acting trivially on the discriminant $A_{\Gamma_d} \simeq A_{\Lambda_d} \simeq \mathbb{Z}/d\mathbb{Z}$. □

Remark 1.11 As any isometric embedding $U \hookrightarrow L_d$ splits, see [16, Ex. 14.0.3], one concludes that for d satisfying $(**)_0$ and $(**)_2$, respectively, that

$(**)_0$: $A_2 \oplus \mathbb{Z}\, v_d \simeq L_d \simeq U \oplus \mathbb{Z}(d)$ and $(v_d)^2 = -(1/3)\,d$

$(**)_2$: $A_2 \oplus \mathbb{Z}\, v_d \hookrightarrow L_d \simeq U \oplus \mathbb{Z}(d)$ index three and $(v_d)^2 = -3\,d$.

Remark 1.12 For a numerical description of these conditions one needs the following classical facts determining which numbers are represented by A_2, see [10, 23].

(i) For a given even, positive integer d there exists a vector $w \in A_2$ with $(w)^2 = d$ if and only if the prime factorization of $d/2$ satisfies

$$\tfrac{d}{2} = \prod p^{n_p} \text{ with } n_p \equiv 0\,(2) \text{ for all } p \equiv 2\,(3). \tag{3}$$

(ii) For a given even, positive integer d there exists a primitive vector $w \in A_2$ with $(w)^2 = d$ if and only if

$$\tfrac{d}{2} = \prod p^{n_p} \text{ with } n_p = 0 \text{ for all } p \equiv 2 \ (3) \text{ and } n_3 \leq 1. \tag{4}$$

Proposition 1.13 *Numerically,* $(*)$, $(**')$, $(**)$, *and* $(***)$ *are described by:*

(i) $(*)$ $\Leftrightarrow$ $d \equiv 0, 2 \ (6)$.
(ii) $(**')$ $\Leftrightarrow$ $\exists \, w \in A_2 \colon (w)^2 = d$ $\Leftrightarrow$ (3).
(iii) $(**)$ $\Leftrightarrow$ $\exists \, w \in A_2$ primitive: $(w)^2 = d$ $\Leftrightarrow$ (4) $\Leftrightarrow$ $\exists \, a, n \in \mathbb{Z} \colon d = \frac{2n^2+2n+2}{a}$.
(iv) $(***)$ $\Leftrightarrow$ $\exists \, a, n \in \mathbb{Z} \colon d = \frac{2n^2+2n+2}{a^2}$.

Proof The first assertion follows from Lemma 1.4.

To prove (ii), one has to distinguish between the two cases $d \equiv 0 \ (6)$ and $d \equiv 2 \ (6)$. Assume first that $(**')_0$ holds. Then $L_d = A_2 \oplus \mathbb{Z} \, v_d$, which contains the isotropic vector $e \in U(n) \subset L_d$. Writing $e = w_0 + a \, v_d$ for some $w_0 \in A_2$ and $a \in \mathbb{Z}$, one has $(w_0)^2 = a^2 d/3$. Hence, $a^2 d/6$ satisfies (3) and, therefore, $d/2$ does. The latter then yields the existence of some $w \in A_2$ with $(w)^2 = d$. Assume now we are in case $(**')_2$, then the standard basis vector $e \in U \subset L_d$ itself might not be contained in $A_2 \oplus \mathbb{Z} \, v_d$, but $3 \, e$ is and replacing e by $3 \, e$ and $(1/3)$ by 3, one can argue as before.

Conversely, if $d/2$ satisfies (3), then we can pick $w \in A_2$ with $(w)^2 = d/3$ for $d \equiv 0 \ (6)$ and with $(w)^2 = 3d$ for $d \equiv 2 \ (6)$. Then $e := w + v_d$ is isotropic. Furthermore, there exists $w' \in A_2$ with $m := (e.w') = (w.w') \neq 0$. Then $f := m \, w' - ((w')^2/2) \, e$ satisfies $(f)^2 = 0$ and $(e.f) = (e.w')^2 =: n$, which yields an embedding $U(n) \hookrightarrow A_2 \oplus \mathbb{Z} \, v_d \subset L_d$ proving $(**')$.

Turning to (iii) and using the notation in (ii), observe that in case $(**)_0$, which implies $(**')_0$, the class w_0 has to be primitive. Indeed, if $w_0 = p \, w_1$ for some prime p, then $p \mid a$ or $p \mid d/3$. On the other hand, writing $f = w'_0 + a' \, v_d$ yields the contradiction $1 = (e.f) = p \, (w_1.w'_0) + aa'/3 \equiv 0 \ (p)$. Hence, $a^2 d/6$ satisfies (4) and, therefore, $d/2$ does, i.e. there exists a primitive $w \in A_2$ with $(w)^2 = d/2$. The argument for $(**)_2$ is similar: If $e = w_0 + a \, v_d$, one argues as before. If not, then $3 \, e = w_0 + a \, v_d$ and if $w_0 = p \, w_1$, then $p \neq 3$. All other primes are excluded as before.

For the converse in this situation, we use the arguments above and pick a primitive $w \in A_2$ with $(w)^2 = d/3$ or $= 3d$, respectively. As $A_{A_2} \simeq \mathbb{Z}/3\mathbb{Z}$, either $(w.A_2) = \mathbb{Z}$ or $= 3 \, \mathbb{Z}$. If $(w)^2 = d/3$, then the former holds (because $3^2 \nmid d$) and, therefore, w' above can be chosen such that $m = 1$. Hence, there exists $U \hookrightarrow L_d$. If $(w.A_2) = 3 \, \mathbb{Z}$, so in particular $(w)^2 = 3 \, d$ and $d \equiv 2 \ (6)$, then the class $e := w \pm v_d$ is of the form $e = 3 \, e'$ with $e' \in L_d$. Therefore, the two classes e' and $f' := w' - ((w')^2/2) \, e'$, where $w' \in A_2$ is chosen such that $(w.w') = 3$, define an embedding $U \hookrightarrow L_d$.

As we will not use the presentation of d as $(2n^2+2n+2)/a$ and $(2n^2+2n+2)/a^2$, respectively, we leave the proof of the other equivalences to the reader, see [12, Prop. 6.1.3] and [1, Sec. 3]. □

The following table lists the first special discriminants, highlighting the difference between the four conditions.

(∗∗∗)			14				26				38	42
(∗∗)			14				26				38	42
(∗∗′)	8		14	18		24	26		32		38	42
(∗)	8	12	14	18	20	24	26	30	32	36	38	42

(∗∗∗)							62					
(∗∗)							62				74	78
(∗∗′)			50				62		68		74	78
(∗)	44	48	50	54	56	60	62	66	68	72	74	78

1.3 In the theory of K3 surfaces, there are good reasons to pass from the K3 lattice $\Lambda \simeq H^2(S, \mathbb{Z})$ to the Mukai lattice $\widetilde{\Lambda} \simeq \widetilde{H}(S, \mathbb{Z}) \simeq H^2(S, \mathbb{Z}) \oplus U_4$, see [16, Ch. 16] for a survey and references. A similar extension of lattices, though slightly more technical due to the non-triviality of the canonical bundle, turns out to be useful for cubics and their comparison with K3 surfaces.

We have already constructed and fixed an isomorphism $\Gamma \simeq E \oplus U_1 \oplus U_2 \oplus A_2(-1) \simeq A_2^{\perp}$, where $A_2(-1) \oplus A_2 \hookrightarrow U_3 \oplus U_4$. On the cubic side, one also finds a natural sublattice isomorphic to $U_3 \oplus U_4$, namely $H^{*\neq 4}(X, \mathbb{Z})$. However, the distinguished $A_2(-1) \subset \Gamma$ sits in $H^4(X, \mathbb{Z})$, so this has to be modified. Moreover, we will embed A_2 into rational cohomology $H^*(X, \mathbb{Q})$ and the intersection product on $H^*(X, \mathbb{Q})$ is modified by more than a mere sign.

Definition 1.14 The *Mukai pairing* on $H^*(X, \mathbb{Q})$ is defined as

$$(\alpha.\alpha') := -\int e^{\frac{c_1(X)}{2}} \cdot \alpha^* \cdot \alpha'. \tag{5}$$

Here, $(\alpha_0 + \alpha_2 + \alpha_4 + \alpha_6 + \alpha_8)^* := \alpha_0 - \alpha_2 + \alpha_4 - \alpha_6 + \alpha_8$ and

$$e^{\frac{c_1(X)}{2}} = e^{\frac{3h}{2}} = 1 + \frac{3}{2}h + \frac{9}{8}h^2 + \frac{27}{48}h^3 + \frac{81}{384}h^4.$$

Warning Unlike the Mukai pairing for K3 surfaces, the pairing (5) is not symmetric.

Definition 1.15 The *Mukai vector* of a coherent sheaf $E \in \mathrm{Coh}(X)$, or a complex $E \in \mathrm{D}^b(X)$, or simply a class $E \in K_{\mathrm{top}}(X)$ is defined as

$$v(E) := \mathrm{ch}(E) \cdot \sqrt{\mathrm{td}(X)}.$$

One easily computes

$$\sqrt{\mathrm{td}(X)} = 1 + \frac{3}{4}h + \frac{11}{32}h^2 + \frac{15}{128}h^3 + \frac{121}{6144}h^4.$$

Using the general fact $\sqrt{\mathrm{td}}^* = e^{-\frac{c_1(X)}{2}} \cdot \sqrt{\mathrm{td}}$ and the Grothendieck–Riemann–Roch formula, one expresses the Euler–Poincaré pairing of two coherent sheaves as

$$\chi(E, E') = -(v(E).v(E')). \tag{6}$$

Note that the left hand side is not symmetric, as ω_X is not trivial. This confirms the observation that (5) is not symmetric.

Example 1.16 For our purposes the following classes are of importance:

$$w_0 := v(\mathcal{O}_X) = \sqrt{\mathrm{td}(X)}, \quad w_1 := v(\mathcal{O}_X(1)) = e^h \cdot \sqrt{\mathrm{td}(X)},$$

$$\text{and } w_2 := v(\mathcal{O}_X(2)) = e^{2h} \cdot \sqrt{\mathrm{td}(X)}.$$

In a sense to be made more precise, these classes are responsible for (.) not being symmetric. Explicitly, they are

$$w_0 = 1 + \frac{3}{4}h + \frac{11}{32}h^2 + \frac{15}{128}h^3 + \frac{121}{6144}h^4, \quad w_1 = 1 + \frac{7}{4}h + \frac{51}{32}h^2 + \frac{385}{384}h^3 + \frac{2921}{6144}h^4,$$

$$\text{and } w_2 = 1 + \frac{11}{4}h + \frac{132}{32}h^2 + \frac{1397}{384}h^3 + \frac{16025}{6144}h^4.$$

In addition to the classes w_0, w_1, w_2, one also needs the following ones

$$v(\lambda_1) := 3 + \frac{5}{4}h - \frac{7}{32}h^2 - \frac{77}{384}h^3 + \frac{41}{2048}h^4.$$

$$v(\lambda_2) := -3 - \frac{1}{4}h + \frac{15}{32}h^2 + \frac{1}{384}h^3 - \frac{153}{2048}h^4.$$

Remark 1.17 The notation suggests that the $v(\lambda_i)$, $i = 1, 2$, are Mukai vectors of some natural (complexes of) sheaves. This is almost true, as we explain next.

Consider an arbitrary line $L \subset X$ and the two natural sheaves $\mathcal{O}_L(i)$, $i = 1, 2$, on X. Their Mukai vectors are

$$u_i := v(\mathcal{O}_L(i)) = \begin{cases} \frac{1}{3}h^3 + \frac{5}{12}h^4 & \text{if } i = 1 \\ \frac{1}{3}h^3 + \frac{9}{12}h^4 & \text{if } i = 2. \end{cases}$$

Under the right orthogonal projection $H^*(X, \mathbb{Q}) \twoheadrightarrow \{w_0, w_1, w_2\}^\perp$ they are mapped to λ_i. Explicitly,

$$v(\lambda_1) = u_1 - w_1 + 4w_0 \text{ and } v(\lambda_2) = u_2 - w_2 + 4w_1 - 6w_0. \tag{7}$$

Here, one uses $(u_i.u_j) = 0$ for all i, j and

$$\begin{aligned} (w_i.w_j) &= \chi(\mathcal{O}_X(i), \mathcal{O}_X(j)) = \chi(X, \mathcal{O}_X(j - i)), \\ (w_i.u_j) &= \chi(\mathcal{O}_X(i), \mathcal{O}_L(j)) = \chi(\mathbb{P}^1, \mathcal{O}_{\mathbb{P}^1}(j - i)), \\ (u_i.w_j) &= \chi(\mathcal{O}_L(i), \mathcal{O}_X(j)) = \chi(\mathbb{P}^1, \mathcal{O}_{\mathbb{P}^1}(i - j - 3)), \end{aligned}$$

Lemma 1.18 *If $H^*(X, \mathbb{Q})$ is considered with the negative Mukai pairing, then*

$$A_2 \hookrightarrow H^*(X, \mathbb{Q}), \quad \lambda_i \mapsto v(\lambda_i)$$

defines an isometric embedding. Furthermore,

(i) $v(\lambda_1), v(\lambda_2) \in \{w_0, w_1, w_2\}^\perp$.
(ii) $w_0, w_1, w_2, v(\lambda_1), v(\lambda_2) \in \mathbb{Q}[h]$ *are linearly independent.*
(iii) $\{w_0, w_1, w_2, v(\lambda_1), v(\lambda_2)\}^\perp = H^4(X, \mathbb{Q})_{\mathrm{pr}} = {}^\perp\{w_0, w_1, w_2, v(\lambda_1), v(\lambda_2)\}$, *on which the Mukai pairing coincides with the intersection product (up to sign).*
(iv) *The Mukai pairing (.) is symmetric on the right orthogonal complement*

$$\{w_0, w_1, w_2\}^\perp \subset H^*(X, \mathbb{Q}).$$

Proof The first assertion can be verified by a computation or using (7). Similarly, (i) follows from the observation that $v(\lambda_i)$ is the orthogonal projection of u_i and (ii) is again proven by a computation. Finally, (ii) implies (iii) and (iv) can be deduced from (iii). □

Corollary 1.19 *The lattices $A_2^\perp \simeq \Gamma \simeq H^4(X, \mathbb{Z})_{\mathrm{pr}} \subset H^*(X, \mathbb{Q})$ and $A_2 \simeq \mathbb{Z}\,v(\lambda_1) \oplus \mathbb{Z}\,v(\lambda_2) \subset H^*(X, \mathbb{Q})$ are orthogonal with respect to the Mukai pairing (5). The induced embedding of their direct sum $A_2^\perp \oplus A_2$ extends to*

$$A_2^\perp \oplus A_2 \subset \widetilde{\Lambda} \hookrightarrow H^*(X, \mathbb{Q}). \tag{8}$$

A more conceptual understanding of these calculations is provided by the discussion in [2]. In particular, cohomology with rational coefficients $H^*(X, \mathbb{Q})$

is replaced by integral topological K-theory. Denote by $K_{\text{top}}(X)$ the topological K-theory of all complex vector bundles. Traditionally, the Chern character is used to identify $K_{\text{top}}(X) \otimes \mathbb{Q}$ with $H^*(X, \mathbb{Q}) = H^{2*}(X, \mathbb{Q})$. For our purposes the Mukai vector is better suited

$$v \colon K_{\text{top}}(X) \hookrightarrow K_{\text{top}}(X) \otimes \mathbb{Q} \xrightarrow{\ \sim\ } H^*(X, \mathbb{Q}).$$

Note that the torsion freeness of $K_{\text{top}}(X)$ follows from the torsion freeness of $H^*(X, \mathbb{Z})$ and the Atiyah–Hirzebruch spectral sequence. Then $K_{\text{top}}(X)$ is equipped with a non-degenerate but non-symmetric linear form with values in $\mathbb{Q}$. Due to (6), it takes values in $\mathbb{Z}$ on the image of the highly non-injective map $K(X) \longrightarrow K_{\text{top}}(X)$. Clearly, the classes $[\mathcal{O}_X(i)]$, $i = 0, 1, 2$, and $[\mathcal{O}_L(i)]$, $i = 1, 2$, are all contained in the image. We shall be interested in the right orthogonal complement of the former three classes and introduce the notation:

$$K'_{\text{top}}(X) := \{ [\mathcal{O}_X], [\mathcal{O}_X(1)], [\mathcal{O}_X(2)] \}^{\perp} \subset K_{\text{top}}(X).$$

Proposition 1.20 (Addington–Thomas) *The restriction of the Mukai pairing* $(.) = -\chi(\, , \,)$ *to* $K'_{\text{top}}(X)$ *is symmetric and integral Moreover, as abstract lattices*

$$\widetilde{\Lambda} \simeq K'_{\text{top}}(X).$$

Proof Note that $v \colon K'_{\text{top}}(X) \otimes \mathbb{Q} \xrightarrow{\ \sim\ } \{w_0, w_1, w_2\}^{\perp}$. Hence, Lemma 1.18 implies the first assertion. The original proof [2] of the second assertion uses derived categories. Here is a sketch of a more direct, purely topological argument. Consider the right orthogonal projection $p \colon K_{\text{top}}(X) \twoheadrightarrow K'_{\text{top}}(X)$. It really is defined over $\mathbb{Z}$, as $(w_i)^2 = 1$. Analogously to (7), one has $p[\mathcal{O}_L(1)] = [\mathcal{O}_L(1)] - [\mathcal{O}_X(1)] + 4[\mathcal{O}_X]$ and $p[\mathcal{O}_L(2)] = [\mathcal{O}_L(2)] - [\mathcal{O}_X(2)] + 4[\mathcal{O}_X(1)] - 6[\mathcal{O}_X]$. Hence, $\lambda_i \mapsto p[\mathcal{O}_L(i)]$ defines an isometric embedding $A_2 \hookrightarrow K'_{\text{top}}(X)$.

First, $H^4(X, \mathbb{Z})_{\text{pr}} \subset H^*(X, \mathbb{Q})$ is contained in $v(K'_{\text{top}}(X))$. Indeed, $H^4(X, \mathbb{Z})_{\text{pr}}$ is spanned by classes of all vanishing spheres and those lift to $K_{\text{top}}(X)$. After fixing an isometry $E \oplus U^{\oplus 2} \oplus A_2(-1) \simeq A_2^{\perp} \simeq \Gamma \simeq H^4(X, \mathbb{Z})_{\text{pr}} \subset K'_{\text{top}}(X)$, this yields an isometric embedding $\Gamma \oplus A_2 \hookrightarrow K'_{\text{top}}(X)$ and allows one to view $\mu_1, \mu_2 \in A_2(-1)$ as classes in $K'_{\text{top}}(X)$.

Second, one needs to show that the class $(1/3)(\mu_1 - \mu_2 - \lambda_1 + \lambda_2) \in (A_2(-1) \oplus A_2) \otimes \mathbb{Q} \subset K_{\text{top}}(X) \otimes \mathbb{Q}$ is integral, i.e. contained in $K_{\text{top}}(X)$. This presumably can be achieved algebraically on some particular cubic fourfold.[2] Hence, the embedding in step one extends to an isometric embedding $\widetilde{\Lambda} \hookrightarrow K'_{\text{top}}(X)$ of finite index. The unimodularity of $\widetilde{\Lambda}$ then implies the second assertion. $\qquad\square$

---

[2] This mysterious class would need to satisfy the two equations $(\lambda_1 . \alpha) = -1$ and $(\lambda_2 . \alpha) = 1$.

1.4 We now endow the various lattices considered above with natural Hodge structures. Let us first briefly recall the well known theory for K3 surfaces, see [16, Ch. 16] for further details and references.

For any complex K3 surface S its second cohomology $H^2(S, \mathbb{Z})$, which as a lattice is isomorphic to Λ, comes with a natural Hodge structure of weight two given by the $(2, 0)$-part $H^{2,0}(S)$. The full Hodge structure is then determined by additionally requiring $H^{1,1}(S) \perp H^{2,0}(S)$ with respect to the intersection pairing.

The global Torelli theorem for complex K3 surfaces asserts that two K3 surfaces S and S' are isomorphic if and only if there exists a Hodge isometry $H^2(S, \mathbb{Z}) \simeq H^2(S', \mathbb{Z})$, i.e. an isomorphism of integral Hodge structures that is compatible with the intersection pairing:

$$S \simeq S' \Leftrightarrow \exists \, H^2(S, \mathbb{Z}) \simeq H^2(S', \mathbb{Z}) \text{ Hodge isometry.}$$

Let (S, L) be a polarized K3 surface. Then the primitive cohomology $H^2(S, \mathbb{Z})_{L-\mathrm{pr}} \subset H^2(S, \mathbb{Z})$ is endowed with the induced structure. Its $(2, 0)$-part is again $H^{2,0}(S)$ and its $(1, 1)$-part is the primitive part of $H^{1,1}(S)$, i.e. the kernel of $(L. \ ) \colon H^{1,1}(S) \longrightarrow \mathbb{C}$. The polarized version of the global Torelli theorem is the the statement that two polarized K3 surfaces (S, L) and (S', L') are isomorphic if and only if there exists a Hodge isometry $H^2(S, \mathbb{Z}) \simeq H^2(S', \mathbb{Z})$ inducing $H^2(S, \mathbb{Z})_{L-\mathrm{pr}} \simeq H^2(S', \mathbb{Z})_{L-\mathrm{pr}}$:

$$(S, L) \simeq (S', L) \Leftrightarrow \exists \, H^2(S; \mathbb{Z}) \simeq H^2(S', \mathbb{Z}), \ L \longmapsto L', \text{ Hodge isometry}$$

The result will be stated again in moduli theoretic terms in Theorem 2.9.

Warning A Hodge isometry $H^2(S, \mathbb{Z})_{L-\mathrm{pr}} \simeq H^2(S', \mathbb{Z})_{L-\mathrm{pr}}$ does not necessarily extend to a Hodge isometry between the full cohomology. Hence, in general, the existence of a Hodge isometry between the primitive Hodge structures of two polarized K3 surfaces does not imply that (S, L) and (S', L') are isomorphic. In fact, even the unpolarized K3 surfaces S and S' may be non-isomorphic.

Next comes the Mukai Hodge structure $\widetilde{H}(S, \mathbb{Z})$. The underlying lattice is $H^*(S, \mathbb{Z})$ with the sign change in $U_4 = (H^0 \oplus H^4)(S, \mathbb{Z})$. The Hodge structure of weight two is again given by the $(2, 0)$-part being $\widetilde{H}^{2,0}(S) := H^{2,0}(S)$ and the condition that $\widetilde{H}^{1,1}(S) \perp H^{2,0}(S)$ with respect to the Mukai pairing. In particular, $U \simeq U_4 = (H^0 \oplus H^4)(S, \mathbb{Z})$ is contained in $\widetilde{H}^{1,1}(S, \mathbb{Z})$. The derived global Torelli theorem is the statement that for two projective K3 surfaces S and S' there exists an exact, $\mathbb{C}$-linear equivalence $\mathrm{D}^b(S) \simeq \mathrm{D}^b(S')$ between their bounded derived categories of coherent sheaves if and only of there exists a Hodge isometry $\widetilde{H}(S, \mathbb{Z}) \simeq \widetilde{H}(S', \mathbb{Z})$:

$$\mathrm{D}^b(S) \simeq \mathrm{D}^b(S') \Leftrightarrow \exists \, \widetilde{H}(S, \mathbb{Z}) \simeq \widetilde{H}(S', \mathbb{Z}) \text{ Hodge isometry.}$$

A twisted K3 surface (S, α) consists of a K3 surface S together with a Brauer class $\alpha \in \mathrm{Br}(S) \simeq H^2(S, \mathcal{O}_S^*)$ (we work in the analytic topology). Choosing a lift $B \in H^2(S, \mathbb{Q})$ of α under the natural morphism $H^2(S, \mathbb{Q}) \longrightarrow \mathrm{Br}(S)$ induced by the exponential sequence allows one to introduce a natural Hodge structure $\widetilde{H}(S, \alpha, \mathbb{Z})$ of weight two associated with (S, α). As a lattice, this is just $\widetilde{H}(S, \mathbb{Z})$, but the $(2, 0)$-part is now given by $\widetilde{H}^{2,0}(S, \alpha) := \mathbb{C} (\sigma + \sigma \wedge B)$, where $0 \neq \sigma \in H^{2,0}(S)$. This defines a Hodge structure by requiring, as before, that $\widetilde{H}^{1,1}(S, \alpha) \perp \widetilde{H}^{2,0}(S, \alpha)$ with respect to the Mukai pairing. Although the definition depends on the choice of B, the Hodge structures induced by two different lifts B and B' of the same Brauer class α are Hodge isometric albeit not canonically, see [21].

The twisted version of the derived global Torelli theorem is the statement that the bounded derived categories of twisted coherent sheaves on (S, α) and (S', α') are equivalent if and only if there exists a Hodge isometry $\widetilde{H}(S, \alpha, \mathbb{Z}) \simeq \widetilde{H}(S', \alpha', \mathbb{Z})$ preserving the natural orientation of the four positive directions, cf. [16, Ch. 16.4] and [30]:

$$\mathrm{D}^b(S, \alpha) \simeq \mathrm{D}^b(S', \alpha') \Leftrightarrow \exists\, \widetilde{H}(S, \alpha, \mathbb{Z}) \simeq \widetilde{H}(S', \alpha', \mathbb{Z}) \text{ oriented Hodge isometry.}$$

Next consider $H^4(X, \mathbb{Z})$ and $H^4(X, \mathbb{Z})_{\mathrm{pr}}$ of a smooth cubic fourfold X. These are Hodge structures of weight four determined by the one-dimensional $H^{3,1}(X)$ and the condition that $H^{3,1}(X) \perp H^{2,2}(X)$ with respect to the intersection product.

The global Torelli theorem for smooth cubic fourfolds, which we will state again as Theorem 2.12 in moduli theoretic terms, is the statement that two smooth cubic fourfolds X and X' are isomorphic (as abstract complex varieties) if and only if there exists a Hodge isometry $H^4(X, \mathbb{Z})_{\mathrm{pr}} \simeq H^4(X', \mathbb{Z})_{\mathrm{pr}}$:

$$X \simeq X' \Leftrightarrow \exists\, H^4(X, \mathbb{Z})_{\mathrm{pr}} \simeq H^4(X', \mathbb{Z})_{\mathrm{pr}} \text{ Hodge isometry.}$$

Note that any such Hodge isometry can be extended to a Hodge isometry $H^4(X, \mathbb{Z}) \simeq H^4(X', \mathbb{Z})$ that maps h_X^2 to $\pm h_{X'}^2$. The situation here is easier compared to the case of polarized K3 surfaces as the discriminant of $H^4(X, \mathbb{Z})_{\mathrm{pr}}$ is just $\mathbb{Z}/3\mathbb{Z}$.[3]

To relate $H^4(X, \mathbb{Z})$ of a cubic fourfolds to K3 surfaces one has to change the sign of the intersection product, so that as abstract lattices $H^4(X, \mathbb{Z}) \simeq \overline{\Gamma}$ and $H^4(X, \mathbb{Z})_{\mathrm{pr}} \simeq \Gamma$ (with an implicit sign change), and Tate shift the Hodge structure to obtain $H^4(X, \mathbb{Z})(1)$ and $H^4(X, \mathbb{Z})_{\mathrm{pr}}(1)$, which are now Hodge structures of weight two.

Definition 1.21 The integral Hodge structure $\widetilde{H}(X, \mathbb{Z})$ of K3 type associated with a smooth cubic fourfold X is the lattice

$$\widetilde{H}(X, \mathbb{Z}) := K'_{\mathrm{top}}(X)$$

---

[3] We will encounter yet another Torelli theorem in Sect. 3.3.

with the Hodge structure of weight two given by $\widetilde{H}^{2,0}(X) := v^{-1}(H^{3,1}(X))$ and the requirement that $\widetilde{H}^{1,1}(X)$ and $\widetilde{H}^{2,0}(X)$ are orthogonal with respect to the Mukai pairing on $K_{\text{top}}(X)$.

The Mukai vector $K_{\text{top}}(X) \otimes \mathbb{Q} \xrightarrow{\sim} H^*(X, \mathbb{Q})$ induces an isometry

$$\widetilde{H}(X, \mathbb{Z}) = K'_{\text{top}}(X) \simeq \widetilde{\Lambda} \subset H^*(X, \mathbb{Q})$$

with $\widetilde{\Lambda} \subset H^*(X, \mathbb{Q})$ provided by (8). Observe that there is a natural isometric inclusion of Hodge structures

$$H^4(X, \mathbb{Z})_{\text{pr}}(1) \subset \widetilde{H}(X, \mathbb{Z}).$$

Moreover, the sublattice A_2 is algebraic, i.e. $A_2 \subset \widetilde{H}^{1,1}(X, \mathbb{Z})$, and its orthogonal Hodge structure is $A_2^{\perp} \simeq H^4(X, \mathbb{Z})_{\text{pr}}(1)$. Also note that according to Remark 1.1 $\lambda_1^{\perp} \subset \widetilde{H}(X, \mathbb{Z})$ is a sub Hodge structure with underlying lattice isomorphic to $\Lambda \oplus \mathbb{Z}(-2)$.

Remark 1.22 Once the Kuznetsov category $\mathcal{A}_X \subset D^b(X)$ is introduced, one also writes $\widetilde{H}(\mathcal{A}_X, \mathbb{Z}) = \widetilde{H}(X, \mathbb{Z})$. The notation $\widetilde{H}(X, \mathbb{Z})$ is analogous to the notation $\widetilde{H}(S, \mathbb{Z})$ for K3 surfaces and the Hodge structure plays a similar role. In fact, as a consequence of the above discussion we know that as lattices $\widetilde{H}(X, \mathbb{Z}) \simeq \widetilde{H}(S, \mathbb{Z})$ and the analogy goes further: For a K3 surface, the algebraic part naturally contains a hyperbolic plane:

$$U \simeq (H^0 \oplus H^4)(S, \mathbb{Z}) \hookrightarrow \widetilde{H}^{1,1}(S, \mathbb{Z}).$$

Similarly, for a smooth cubic fourfold the algebraic part naturally contains a copy of A_2:

$$v \colon A_2 \simeq \mathbb{Z}\, p[\mathcal{O}_L(1)] \oplus \mathbb{Z}\, p[\mathcal{O}_L(2)] \hookrightarrow \widetilde{H}^{1,1}(X, \mathbb{Z}).$$

Here, $p \colon K_{\text{top}}(X) \twoheadrightarrow K'_{\text{top}}(X)$ is the projection as in the proof of Proposition 1.20, so the composition maps $\lambda \mapsto p[\mathcal{O}_L(i)] \mapsto v(\lambda_i)$. Their respective orthogonal complements are

$$H^2(S, \mathbb{Z}) = U^{\perp} \hookrightarrow \widetilde{H}(S, \mathbb{Z}) \quad \text{and} \quad H^4(X, \mathbb{Z})_{\text{pr}}(1) = A_2^{\perp} \hookrightarrow \widetilde{H}(X, \mathbb{Z}),$$

in terms of which the global Torelli theorem is formulated in both instances. Also, $e_4 - f_4 = (1, 0, -1) \in \widetilde{H}^{1,1}(S, \mathbb{Z})$ and $v(\lambda_1) \in \widetilde{H}^{1,1}(X, \mathbb{Z})$ are both algebraic classes satisfying $(e_4 - f_4)^2 = 2 = (v(\lambda_1))^2$. Their orthogonal complements are isometric.

Definition 1.23 Let (S, L) be a polarized K3 surface and X a smooth cubic fourfold.

(i) We say (S, L) and X are *associated*, $(S, L) \sim X$, if there exists an isometric embedding of Hodge structures

$$H^2(S, \mathbb{Z})_{L-\mathrm{pr}} \hookrightarrow H^4(X, \mathbb{Z})_{\mathrm{pr}}(1). \tag{9}$$

(ii) We say S and X are *associated*, $S \sim X$, if there exists a Hodge isometry

$$\widetilde{H}(S, \mathbb{Z}) \simeq \widetilde{H}(X, \mathbb{Z}).$$

(iii) For $\alpha \in \mathrm{Br}(S)$ we say that the twisted K3 surface (S, α) and X are *associated*, $(S, \alpha) \sim X$, if there exists a Hodge isometry

$$\widetilde{H}(S, \alpha, \mathbb{Z}) \simeq \widetilde{H}(X, \mathbb{Z}).$$

First observe the immediate implication:

$$(S, L) \sim X \ \Rightarrow \ S \sim X.$$

Indeed, any isometric embedding (9) can be extended to an isometry $\widetilde{H}(S, \mathbb{Z}) \simeq \widetilde{H}(X, \mathbb{Z})$. This follows from the existence of the hyperbolic plane $U \subset H^2(S, \mathbb{Z})^{\perp}_{L-\mathrm{pr}}$, cf. [16, Rem. 14.1.13].

As an aside, observe that a K3 surface S that is associated with a cubic fourfold in any sense is necessarily projective. Indeed, if for example $S \sim X$, then $\widetilde{H}^{1,1}(S, \mathbb{Z}) \simeq \widetilde{H}^{1,1}(X, \mathbb{Z})$ contains the positive plane A_2 and, therefore, $H^{1,1}(S, \mathbb{Z})$ contains at least one class of positive square.

The key to link $S \sim X$, (S, L), and $(S, \alpha) \sim X$ to the properties $(**)$ and $(**')$ is the following result in [2] generalized to the twisted case in [17].

Proposition 1.24 (Addington–Thomas, Huybrechts) *Assume X is a smooth cubic fourfold.*

(i) *There exists a K3 surface S with $S \sim X$ if and only if there exists a (primitive) embedding $U \hookrightarrow \widetilde{H}^{1,1}(X, \mathbb{Z})$.*

(ii) *There exists a twisted K3 surface (S, α) with $(S, \alpha) \sim X$ if and only if there exists an embedding $U(n) \hookrightarrow \widetilde{H}^{1,1}(X, \mathbb{Z})$ for some $n \neq 0$.*

Proof Any Hodge isometry $\widetilde{H}(S, \mathbb{Z}) \simeq \widetilde{H}(X, \mathbb{Z})$ yields a hyperbolic plane $U \simeq (H^0 \oplus H^4)(S, \mathbb{Z}) \subset \widetilde{H}^{1,1}(S, \mathbb{Z}) \simeq \widetilde{H}^{1,1}(X, \mathbb{Z})$. Conversely, if $U \subset \widetilde{H}^{1,1}(X, \mathbb{Z}) \subset \widetilde{H}(X, \mathbb{Z})$, then as a lattice $U^{\perp} \simeq \Lambda$. Moreover, the Hodge structure of $\widetilde{H}(X, \mathbb{Z})$ induces a Hodge structure on $U^{\perp} \simeq \Lambda$ which due to the surjectivity of the period map [16, Thm. 7.4.1] is Hodge isometric to $H^2(S, \mathbb{Z})$ for some K3 surface S. However, as before, $U^{\perp} \simeq H^2(S, \mathbb{Z})$ extends to $\widetilde{H}(X, \mathbb{Z}) \simeq \widetilde{H}(S, \mathbb{Z})$. This proves (i).

For (ii), again one direction is easy, as $\widetilde{H}^{1,1}(S,\mathbb{Z})$ contains the B-field shift of $(H^0 \oplus H^4)(S,\mathbb{Z})$, cf. [16, Ch. 14]. More precisely, $\widetilde{H}^{1,1}(S,\alpha,\mathbb{Z}) = (\exp(B)\,\widetilde{H}^{1,1}(S,\mathbb{Q})) \cap \widetilde{H}(S,\mathbb{Z})$, which clearly contains the lattice $(\langle 1, B, B^2/2 \rangle \cap \widetilde{H}(S,\mathbb{Z})) \oplus H^4(S,\mathbb{Z}) \simeq U(n)$, where n is minimal with $n\,(1, B, B^2) \in \widetilde{H}(S,\mathbb{Z})$. The other direction needs a surjectivity statement for twisted K3 surfaces which is an easy consequence of the surjectivity of the untwisted period map.   □

Proposition 1.25 *Assume a smooth cubic fourfold X is associated with some K3 surface S, so $S \sim X$. Then there exists a polarized K3 surface $(S', L') \sim X$:*

$$\widetilde{H}(S,\mathbb{Z}) \simeq \widetilde{H}(X,\mathbb{Z}) \;\Rightarrow\; H^2(S',\mathbb{Z})_{L'-\mathrm{pr}} \hookrightarrow H^4(X,\mathbb{Z})_{\mathrm{pr}}.$$

Proof Assume $S \sim X$. Then there exists a Hodge isometry $\widetilde{H}(S,\mathbb{Z}) \simeq \widetilde{H}(X,\mathbb{Z})$. On the left hand side, one finds $U \simeq (H^0 \oplus H^4)(S,\mathbb{Z}) \subset \widetilde{H}^{1,1}(S,\mathbb{Z})$ and, on the right hand side, $A_2 \subset \widetilde{H}^{1,1}(X,\mathbb{Z})$. Consider the saturation of the sum of both as a lattice $\overline{U + A_2} \subset \widetilde{H}^{1,1}(S,\mathbb{Z})$. According to Lemma 1.2, there exists another hyperbolic plane $U' \subset \overline{U + A_2}$ with $\mathrm{rk}(U' + A_2) = 3$. Using the surjectivity of the period map, one finds another K3 surface S' and a Hodge isometry

$$\widetilde{H}(S',\mathbb{Z}) \simeq \widetilde{H}(S,\mathbb{Z}) \simeq \widetilde{H}(X,\mathbb{Z}) \tag{10}$$

inducing $H^2(S',\mathbb{Z}) \simeq U'^{\perp}$. But then $H^2(S',\mathbb{Z}) \cap A_2^{\perp} \subset H^2(S',\mathbb{Z})$ is of corank one and we can assume it to be of the form $H^2(S',\mathbb{Z})_{L'\text{-pr}}$. However, being contained in $A_2^{\perp}$ implies that under (10) $H^2(S',\mathbb{Z})_{L'-\mathrm{pr}}$ embeds into $H^4(X,\mathbb{Z})_{\mathrm{pr}}(1)$, which ensures $(S', L') \sim X$.   □

Corollary 1.26 *A smooth cubic fourfold X is associated with some polarized K3 surface, $(S, L) \sim X$, if and only if there exists an isometric embedding $U \hookrightarrow \widetilde{H}^{1,1}(X,\mathbb{Z})$.*   □

2   Period Domains and Moduli Spaces

The comparison of the Hodge theory of K3 surfaces and cubic fourfolds is now considered in families. Via period maps, this leads to an algebraic correspondence between the moduli space of polarized K3 surfaces of certain degrees and the moduli space of cubic fourfolds. The approach has been initiated by Hassett [12] and has turned out to be very valuable indeed.

2.1 Here is a very brief reminder on some results, mostly due to Borel and Baily–Borel, on arithmetic quotients of orthogonal type. Let $(N, (\,.\,))$ be a lattice of signature $(2, n_-)$ and set $V := N \otimes \mathbb{R}$. Then the period domain D_N associated with

N is the Grassmannian of positive, oriented planes $W \subset V$, which alternatively can be described as

$$D_N \simeq \{ x \mid (x)^2 = 0, \ (x.\bar{x}) > 0 \} \subset \mathbb{P}(N \otimes \mathbb{C})$$
$$\simeq O(2, n_-)/(O(2) \times O(n_-)).$$

By definition, the period domain D_N associated with N has the structure of a complex manifold. This is turned into an algebraic statement by the following fundamental result [3]. It uses the fact that under the assumption on the signature of N the orthogonal group $O(N)$ acts properly discontinuously on D_N.

Theorem 2.1 (Baily–Borel) *Assume $G \subset O(N)$ is a torsion free subgroup of finite index. Then the quotient*

$$G \setminus D_N$$

has the structure of a smooth, quasi-projective complex variety.

As G acts properly discontinuously as well, the stabilizers are finite and hence trivial. This already proves the smoothness of the quotient $G \setminus D_N$. The difficult part of the theorem is to find a Zariski open embedding into a complex projective variety.

Finite index subgroups $G \subset O(N)$ with torsion are relevant, too. In this situation, one uses Minkowski's theorem stating that the map $\pi_p \colon \mathrm{Gl}(n, \mathbb{Z}) \longrightarrow \mathrm{Gl}(n, \mathbb{F}_p)$, $p \geq 3$, is injective on finite subgroups or, equivalently, that its kernel is torsion free. Hence, for every finite index subgroup $G \subset O(N)$ there exists a normal and torsion free subgroup $G_0 := G \cap \mathrm{Ker}(\pi_p) \subset G$ of finite index.

Corollary 2.2 *Assume $G \subset O(N)$ is a subgroup of finite index. Then the quotient $G \setminus D_N$ has the structure of a normal, quasi-projective complex variety with finite quotient singularities.* $\qquad\Box$

We remark that not only these arithmetic quotients, but also holomorphic maps into them are algebraic. This is the following remarkable GAGA style result, see [7].

Theorem 2.3 (Borel) *Assume $G \subset O(N)$ is a torsion free subgroup of finite index. Then any holomorphic map $\varphi \colon Z \longrightarrow G \setminus D_N$ from a complex variety Z is regular.*

Remark 2.4 Often, the result is applied to holomorphic maps to singular quotients $G \setminus D_N$, i.e. in situations when G is not necessarily torsion free. This is covered by the above only when $Z \longrightarrow G \setminus D_N$ is induced by a holomorphic map $Z' \longrightarrow G_0 \setminus D_N$, where $Z' \longrightarrow Z$ is a finite quotient and $G_0 \subset G$ is a normal, torsion free subgroup of finite index.

2.2 We shall be interested in (at least) three different types of period domains: For polarized K3 surfaces and for (special) smooth cubic fourfolds. These are the period domains associated with the lattices Γ, Γ_d, and Λ_d:

$$D \subset \mathbb{P}(\Gamma \otimes \mathbb{C}), \ D_d \subset \mathbb{P}(\Gamma_d \otimes \mathbb{C}), \text{ and } Q_d \subset \mathbb{P}(\Lambda_d \otimes \mathbb{C}).$$

These period domains are endowed with the natural action of the corresponding orthogonal groups $O(\Gamma)$, $O(\Gamma_d)$, and $O(\Lambda_d)$ and we will be interested in the following quotients by distinguished finite index subgroups of those:

$$\mathcal{C} := \tilde{O}(\Gamma) \setminus D = O(\Gamma) \setminus D,$$

$$\tilde{\mathcal{C}}_d := \tilde{O}(\Gamma, K_d) \setminus D_d, \qquad \check{\mathcal{C}}_d := \tilde{O}(\Gamma, v_d) \setminus D_d, \text{ and}$$

$$\mathcal{M}_d := \tilde{O}(\Lambda_d) \setminus Q_d.$$

For the first equality note that $\tilde{O}(\Gamma) \subset O(\Gamma)$ is of index two, but $-\mathrm{id} \in O(\Gamma) \setminus \tilde{O}(\Gamma)$ acts trivially on D. The subgroup $\tilde{O}(\Lambda_d) \subset O(\Lambda_d)$ is defined analogously to (2).

Due to Theorems 2.1 and 2.3, see also Remark 2.4, the induced maps $\tilde{\mathcal{C}}_d \dashrightarrow \check{\mathcal{C}}_d \to \mathcal{C}$ are regular morphisms between normal quasi-projective varieties. The image in $\mathcal{C}$ shall be denoted by $\mathcal{C}_d$, so that

$$\tilde{\mathcal{C}}_d \dashrightarrow \check{\mathcal{C}}_d \dashrightarrow \mathcal{C}_d \subset \mathcal{C}.$$

The condition ($*$) will in the sequel be interpreted as the condition that $\mathcal{C}_d \neq \emptyset$.

Corollary 2.5 (Hassett) *Assume d satisfies ($*$). The naturally induced maps*

$$\tilde{\mathcal{C}}_d \dashrightarrow \check{\mathcal{C}}_d \dashrightarrow \mathcal{C}_d$$

are surjective, finite, and algebraic.

Furthermore, $\check{\mathcal{C}}_d \dashrightarrow \mathcal{C}_d$ is the normalization of $\mathcal{C}_d$ and $\tilde{\mathcal{C}}_d \to \check{\mathcal{C}}_d$ is a finite morphism between normal varieties, which is an isomorphism if $d \equiv 2\ (6)$ and of degree two if $d \equiv 0\ (6)$.

Proof Clearly, if d satisfies $(*)_2$, then $\tilde{O}(\Gamma, K_d) = \tilde{O}(\Gamma, v_d)$ by Lemma 1.8 and, therefore, $\tilde{\mathcal{C}}_d \simeq \check{\mathcal{C}}_d$. Otherwise, $\tilde{\mathcal{C}}_d \dashrightarrow \check{\mathcal{C}}_d$ is the quotient by the involution $g \in \tilde{O}(\Gamma)$ defined by $g = \mathrm{id}$ on $E \oplus U_2 \oplus I_{0,3}$ and $g = -\mathrm{id}$ on U_1, which indeed acts non-trivially on $\tilde{\mathcal{C}}_d$.

To prove that $\check{\mathcal{C}}_d \dashrightarrow \mathcal{C}_d$ is quasi-finite, use that $\check{\mathcal{C}}_d \to \mathcal{C}$ is algebraic with discrete and hence finite fibres. For a very general $x \in D_d$ such that there does not exist any proper primitive sublattice $N \subset \Gamma_d$ with $x \in N \otimes \mathbb{C}$, any $g \in \tilde{O}(\Gamma)$ with $g(x) = x$ also satisfies $g(\Gamma_d) = \Gamma_d$ and, therefore, $g(K_d) = K_d$, i.e. $g \in \tilde{O}(\Gamma, K_d)$. This proves that $\check{\mathcal{C}}_d \to \mathcal{C}$ is generically injective. Thus, once $\check{\mathcal{C}}_d \to \mathcal{C}$ is shown to be finite, and not only quasi-finite, it is the normalization of its image $\mathcal{C}_d$. We refer to [8, 12] for more details on this point. $\qquad\square$

Remark 2.6 Note that while the fibre of $\widetilde{C}_d \longrightarrow \check{C}_d$ consists of at most two points, the fibres of $\check{C}_d \longrightarrow C_d$ may contain more points, depending on the singularity type of the points in C_d. For fixed d, the cardinality of the fibres is bounded. However, it is unbounded when d is allowed to grow.

Lemma 1.10 immediately yields the following result which eventually leads to the mysterious relation between K3 surfaces and cubic fourfolds.

Corollary 2.7 *Assume d satisfies (**). We choose an isomorphism $\varepsilon \colon \Gamma_d \xrightarrow{\sim} \Lambda_d$.*

(i) *If d satisfies (*)$_0$, then ε naturally induces an isomorphism $\mathcal{M}_d \simeq \widetilde{C}_d$. Therefore, $\mathcal{M}_d$ comes with a finite morphism onto C_d generically of degree two:*

$$\Phi_\varepsilon \colon \ \mathcal{M}_d \simeq \widetilde{C}_d \ \xrightarrow{\ 2:1\ } \ \check{C}_d \ \xrightarrow{\ \text{norm}\ } \ C_d \subset \mathcal{C}.$$

(ii) *If d satisfies (*)$_2$, then ε naturally induces an isomorphism $\mathcal{M}_d \simeq \widetilde{C}_d \simeq \check{C}_d$. Therefore, $\mathcal{M}_d$ can be seen as the normalization of $C_d \subset \mathcal{C}$:*

$$\Phi_\varepsilon \colon \mathcal{M}_d \simeq \widetilde{C}_d \simeq \check{C}_d \xrightarrow{\ \text{norm}\ } C_d \subset \mathcal{C}. \qquad \qquad \square$$

Remark 2.8 As indicated by the notation, the morphism $\Phi_\varepsilon \colon \mathcal{M}_d \longrightarrow C_d \subset \mathcal{C}$, which will be seen to link polarized K3 surfaces (S, L) of degree d with special cubic fourfolds X, depends on the choice of $\varepsilon \colon \Gamma_d \xrightarrow{\sim} \Lambda_d$. There is no distinguished choice for ε and, therefore, one should not expect to find a distinguished morphism $\mathcal{M}_d \longrightarrow C_d$ that can be described by a geometric procedure associating a cubic fourfold X to a polarized K3 surface (S, L).[4]

To avoid any dependance on ε, one could think of defining a morphism from the finite quotient

$$\pi_d \colon \mathcal{M}_d = \tilde{\mathrm{O}}(\Lambda_d) \setminus \mathcal{Q}_d \longrightarrow \bar{\mathcal{M}}_d := \mathrm{O}(\Lambda_d) \setminus \mathcal{Q}_d$$

to some meaningful quotient of $\mathcal{C}$. But, as the degree of π_d grows with d, there definitely is no reasonable quotient of $\mathcal{C}$ that would receive all of them. However, it seems plausible that a quotient $C_d \longrightarrow \bar{C}_d$ can be constructed that allows for a morphism $\bar{\mathcal{M}}_d \longrightarrow \bar{C}_d$. The derived point of view to be explained later will shed more light on this.

2.3 We start by recalling the central theorem in the theory of K3 surfaces: the global Torelli theorem. In the situation at hand, it is due to Pjateckiĭ-Šapiro and Šafarevič, see [16] for details, generalizations, and references.

---

[4]I wish to thank E. Brakkee and P. Magni for discussions concerning this point.

Consider the coarse moduli space M_d of polarized K3 surfaces (S, L) with $(L)^2 = d$, which can be constructed as a quasi-projective variety either by (not quite) standard GIT methods, by using the theorem below, or as a Deligne–Mumford stack.

The period map associates with any $[(S, L)] \in M_d$ a point in $\mathcal{M}_d$. For this, choose an isometry $H^2(S, \mathbb{Z}) \simeq \Lambda$, called a marking, that maps $c_1(L)$ to $\ell = e_2 + (d/2)f_2$ and, therefore, induces an isometry $H^2(S, \mathbb{Z})_{L-\mathrm{pr}} \simeq \Lambda_d$. Then the $(2, 0)$-part $H^{2,0}(S) \subset H^2(S, \mathbb{C}) \simeq \Lambda \otimes \mathbb{C}$ defines a point in the period domain Q_d. The image point in the quotient $\tilde{O}(\Lambda_d) \setminus Q_d$ is then independent of the choice of any marking. This defines the period map $\mathcal{P} \colon M_d \longrightarrow \mathcal{M}_d$ which Hodge theory reveals to be holomorphic. Note that both spaces, M_d and $\mathcal{M}_d$, are quasi-projective varieties with quotient singularities.

Theorem 2.9 (Pjateckiĭ-Šapiro and Šafarevič) *The period map is an algebraic, open embedding*

$$\mathcal{P} \colon M_d \hookrightarrow \mathcal{M}_d = \tilde{O}(\Lambda_d) \setminus Q_d. \tag{11}$$

Remark 2.10 Coming back to Remark 2.8, one might wonder how the image of M_d under the finite quotient $\pi_d \colon \mathcal{M}_d \longrightarrow \bar{\mathcal{M}}_d$, can be interpreted geometrically in terms of the polarized K3 surfaces (S, L) parametrized by M_d. There is no completely satisfactory answer to this, i.e. the image $\pi_d(M_d)$ is not known (and should probably not be expected) to be the coarse moduli space of a nice geometric moduli functor. The best one can say is that for $(S, L) \in M_d$ with $\rho(S) = 1$, the fibre $\pi_d^{-1}(\pi_d(S, L))$ can be viewed as the set of all Fourier–Mukai partners of S, which come with a unique polarization, cf. [13, 18].

To understand the complement of the open embedding (11), note first that any $x \in Q_d$ is the period of some K3 surface S. This surface then comes with a natural line bundle L (up to the action of the Weyl group) corresponding to $\ell = e_2 + (d/2)f_2 \in \Lambda$. Furthermore, L is ample (again, possibly after applying the Weyl group action) if and only if there exists no $\delta \in \Lambda_d$ with $(\delta)^2 = -2$ orthogonal to x, i.e. $x \in Q_d \setminus \bigcup \delta^\perp$ with $\delta \in \Delta_d := \Delta(\Lambda_d)$, the set of all (-2)-classes in Λ_d. Hence, the complement of $M_d \subset \mathcal{M}_d$ can be described as the quotient

$$\tilde{O}(\Lambda_d) \setminus \bigcup \delta^\perp \subset \mathcal{M}_d. \tag{12}$$

Note that $\tilde{O}(\Lambda_d)$ acts on Δ_d and that the quotient (12) really is a finite union. In fact, it consists of at most two components due to the following result.[5]

Proposition 2.11 *The complement $\mathcal{M}_d \setminus M_d$ consists of either one or two irreducible Noether–Lefschetz divisors depending on d:*

(i) *If $d/2 \not\equiv 1$ (4), then the complement (12) of $M_d \subset \mathcal{M}_d$ is irreducible.*

---

[5]Thanks to O. Debarre for pointing this out to me.

(ii) *If $d/2 \equiv 1\,(4)$, then the complement (12) of $M_d \subset \mathcal{M}_d$ has of two irreducible components.*

Proof This is again an application of Eichler's criterion, see the proof of Proposition 1.6. For $\delta \in \Lambda_d$ with $(\delta)^2 = -2$, one has $(\delta.\Lambda_d) = n\,\mathbb{Z}$ with $n = 1$ or $n = 2$. In the first case, the residue class $(1/n)\,\bar\delta \in A_{\Lambda_d} \simeq \mathbb{Z}/d\mathbb{Z}$ is trivial. In the second case, $(1/2)\,\bar\delta \equiv 0$ or $\equiv d/2\,(d)$ in $\mathbb{Z}/d\mathbb{Z}$. However, the second case is only possible if $d/2 \equiv 1\,(4)$. Indeed, write $\delta = \delta' + \delta'' \in U_2^\perp \oplus U_2$ with $\delta'' \in \ell^\perp \cap U_2 = \mathbb{Z}\,(e_2 - (d/2)f_2)$. Then $(1/2)\,\delta' + (1/2)\,\delta'' + (m/2)\,\ell \in \Lambda$ for some $m \in \mathbb{Z}$. Hence, $(1/2)\,\delta' \in \Lambda$ and, therefore, $-2 = (\delta)^2 \equiv (\delta'')^2\,(8)$. Combine this with $(1/2)\,\delta'' + (m/2)\,\ell \in U_2$, which implies $(\delta'')^2 \equiv -m^2 d\,(8)$. $\qquad\square$

To be more explicit, one can write

$$
M_d = \begin{cases} \mathcal{M}_d \setminus \delta_0^\perp & \text{if } \frac{d}{2} \not\equiv 1\,(4) \\ \mathcal{M}_d \setminus (\delta_0^\perp \cup \delta_1^\perp) & \text{if } \frac{d}{2} \equiv 1\,(4), \end{cases}
$$

where δ_0, δ_1 are chosen explicitly as $\delta_0 = e_1 - f_1$ and $\delta_1 = 2e_1 + \frac{d/2-1}{2}f_1 + e_2 - (d/2)\,f_2$.

2.4 We now switch to the cubic side. The moduli space M of smooth cubic fourfolds can be constructed by means of standard GIT methods as the quotient

$$
M = |\mathcal{O}_{\mathbb{P}^5}(3)|_{\mathrm{sm}}/\!/\mathrm{PGl}(6).
$$

As in the case of K3 surfaces, mapping a smooth cubic fourfold X to its period $H^{3,1}(X) \subset H^4(X, \mathbb{C})_{\mathrm{pr}} \simeq \Gamma \otimes \mathbb{C}$, which is a point in the period domain $D \subset \mathbb{P}(\Gamma \otimes \mathbb{C})$, defines a holomorphic map $\mathcal{P} \colon M \longrightarrow \mathcal{C}$. In analogy to the situation for K3 surfaces, the following global Torelli theorem has been proven [9, 20, 27, 34, 35].

Theorem 2.12 (Voisin, Looijenga, ..., Charles, Huybrechts–Rennemo, ...) *The period map is an algebraic, open embedding*

$$
\mathcal{P} \colon M \hookrightarrow \mathcal{C} = \mathrm{O}(\Gamma) \setminus D.
$$

This central result is complemented by a result of Laza and Looijenga, which can be seen as an analogue of Proposition 2.11, see [26, 27]. First note that for $d = 2$ and $d = 6$ the lattice K_d is given by the matrices $\begin{pmatrix} -3 & 1 \\ 1 & -1 \end{pmatrix}$ and $\begin{pmatrix} -3 & 0 \\ 0 & -2 \end{pmatrix}$, respectively, see Remark 1.7. Hence, if a smooth cubic fourfold X defined a point in $\mathcal{C}_6$, then $H^{2,2}(X, \mathbb{Z})_{\mathrm{pr}}$ would contain a class δ with $(\delta)^2 = 2$ contradicting [34, §4, Prop. 1]. In [12] one finds an argument using limiting mixed Hodge structures to also exclude the case $[X] \in \mathcal{C}_2$. So, $M \subset \mathcal{C} \setminus (\mathcal{C}_2 \cup \mathcal{C}_6)$.

Theorem 2.13 (Laza, Looijenga) *The period map identifies the moduli space M of smooth cubic fourfolds with the complement of $C_2 \cup C_6$:*

$$\mathcal{P}: M \xrightarrow{\sim} \mathcal{C} \setminus (C_2 \cup C_6).$$

To complete the picture, we state the following result. We refrain from giving a proof, but refer to similar results in the theory of K3 surfaces [16, Prop. 6.2.9].

Proposition 2.14 *The union $\bigcup C_d \subset \mathcal{C}$ of all C_d with d satisfying $(\ast\ast\ast)$ is analytically dense in $\mathcal{C}$. Consequently, the union of all C_d for satisfying $(\ast\ast')$ (or $(\ast\ast)$ or $(\ast)$) is analytically dense.*

Remark 2.15 On the level of moduli spaces, the theory of K3 surfaces is linked with the theory of cubic fourfolds in terms of the morphism

$$\Phi_\varepsilon: \mathcal{M}_d \subset \mathcal{M}_d \longrightarrow C_d \subset \mathcal{C},$$

cf. Corollary 2.7. Note that the image of a point $[(S, L)] \in \mathcal{M}_d$ corresponding to a polarized K3 surface (S, L) can a priori be contained in the boundary $\mathcal{C} \setminus M = C_2 \cup C_6$. However, unless $d = 2$ or $d = 6$, generically this is not the case and the map defines a rational map

$$\Phi_\varepsilon: \mathcal{M}_d \dashrightarrow M,$$

which is of degree one or two.

2.5 In Sect. 1.4 we have linked Hodge theory of K3 surfaces and Hodge theory of cubic fourfolds. We will now cast this in the framework of period maps and moduli spaces, i.e. in terms of the maps Φ_ε.

Proposition 2.16 *A smooth cubic fourfold X and a polarized K3 surface (S, L) are associated, $(S, L) \sim X$, in the sense of Definition 1.23 if and only if $\Phi_\varepsilon[(S, L)] = [X]$ for some choice of $\varepsilon: \Gamma_d \xrightarrow{\sim} \Lambda_d$:*

$$(S, L) \sim X \Leftrightarrow \exists \varepsilon: \Phi_\varepsilon[(S, L)] = [X].$$

Proof Assume $\Phi_\varepsilon[(S, L)] = [X]$. Pick an arbitrary marking $H^2(S, \mathbb{Z}) \xrightarrow{\sim} \Lambda$ with $L \mapsto \ell$. Composing the induced isometry $H^2(S, \mathbb{Z})_{L-\mathrm{pr}} \xrightarrow{\sim} \Lambda_d$ with $\varepsilon^{-1}: \Lambda_d \xrightarrow{\sim} \Gamma_d \subset \Gamma$ yields a point in $D_d \subset D$. Then there exists a marking $H^4(X, \mathbb{Z})_{\mathrm{pr}} \simeq \Gamma$ such that X yields the same period point in D, which thus yields a Hodge isometric embedding $H^2(S, \mathbb{Z})_{L-\mathrm{pr}} \hookrightarrow H^4(X, \mathbb{Z})_{\mathrm{pr}}(1)$. Conversely, any such Hodge isometric embedding defines a sublattice of $\Gamma \simeq H^4(X, \mathbb{Z})_{\mathrm{pr}}$

isomorphic to some $v^\perp$ which after applying some element in $O(\Gamma)$ becomes Γ_d, see Proposition 1.6. Composing with a marking of (S, L) yields the appropriate ε.   □

Corollary 2.17 *Let X be a smooth cubic fourfold.*

(i) *For fixed d, there exists a polarized K3 surface (S, L) of degree d with $X \sim (S, L)$ if and only if $X \in C_d$ and d satisfies $(**)$.*
(ii) *There exists a twisted K3 surface (S, α) with $X \sim (S, \alpha)$ if and only if $X \in C_d$ for some d satisfying $(**')$.*

Proof Consider $\mathcal{M}_d$ as the moduli space of quasi-polarized K3 surfaces (S, L), i.e. with L only big and nef but not necessarily ample. One then has to show that whenever there exists a Hodge isometric embedding $H^2(S, \mathbb{Z})_{L-\mathrm{pr}} \hookrightarrow H^4(X, \mathbb{Z})_{\mathrm{pr}}(1)$, then L is not orthogonal to any algebraic class $\delta_S \in H^2(S, \mathbb{Z})$ with $(\delta_S)^2 = -2$. Indeed, in this case L would be automatically ample. However, such a class δ_S would correspond to a class $\delta \in H^{2,2}(X, \mathbb{Z})_{\mathrm{pr}}$ with $(\delta)^2 = 2$, which contradicts $[X] \in M = \mathcal{C} \setminus (\mathcal{C}_2 \cup \mathcal{C}_6)$. Of course, the argument is purely Hodge theoretic and one can easily avoid talking about quasi-polarized K3 surfaces.

To prove (ii), observe that the period of X is contained in D_d if and only if one finds $L_d \hookrightarrow \widetilde{H}^{1,1}(X, \mathbb{Z})$. If d satisfies $(**')$, then there exists $U(n) \hookrightarrow L_d$ and we can conclude by Proposition 1.24. Conversely, if $(S, \alpha) \sim X$, one finds $U(n) \hookrightarrow \widetilde{H}^{1,1}(S, \alpha, \mathbb{Z}) \simeq \widetilde{H}^{1,1}(X, \mathbb{Z})$. As there also exists a positive plane $A_2 \hookrightarrow \widetilde{H}^{1,1}(X, \mathbb{Z})$, the lattice $U(n)$ is contained in a primitive sublattice of rank three in $H^{1,1}(X, \mathbb{Z})$, which is then necessarily of the form L_d for some d satisfying $(**')$.   □

A geometric interpretation of the condition $(***)$, involving the Fano variety of lines $F(X)$, will be explained in the next section, see Proposition 3.4. The conditions $(**)$ and $(**')$ will occur there again as well.

Remark 2.18 Note that a given cubic fourfold X can be associated with more than one polarized K3 surface (S, L) and, in fact, sometimes even with infinitely many (S, L). To start, there are the finitely many choices of $\varepsilon \in O(\Lambda_d)/\tilde{O}(\Lambda_d)$, see [12, Thm. 5.2.3]. Then, Φ_ε is only generically injective for d satisfying $(**)_2$ and even of degree two for $(**)_0$. And finally, X could be contained in more than one $\mathcal{C}_d$. In fact, it can happen that $X \in \mathcal{C}_d$ for infinitely many d satisfying $(**)$. To be more precise, depending on the degree d, there may exist non-isomorphic K3 surfaces S and S' endowed with polarizations L and L', respectively, such there nevertheless exists a Hodge isometry $H^2(S, \mathbb{Z})_{L-\mathrm{pr}} \simeq H^2(S', \mathbb{Z})_{L'-\mathrm{pr}}$. Indeed, the latter may not extend to a Hodge isometry $H^2(S, \mathbb{Z}) \simeq H^2(S', \mathbb{Z})$, see Sect. 1.4.

The situation is not quite as bad as it sounds. Although there may be infinitely many polarized K3 surfaces (S, L) associated with one X, only finitely many isomorphism types of unpolarized K3 surfaces S will be involved.

Remark 2.19 In [8] a geometric interpretation for the generic fibre of the rational map $\Phi_\varepsilon \colon \mathcal{M}_d \dashrightarrow \mathcal{C}_d$ in the case $d \equiv 0\,(6)$ is described. It turns out that

$\Phi_\varepsilon[(S, L)] = [(S', L')]$ implies that S' is isomorphic to $M(3, L, d/6)$, the moduli space of stable bundles on S with the indicated Mukai vector.

3 Fano Perspective

We come back to the Hodge structure $v(\lambda_1)^\perp \subset \tilde{H}(X, \mathbb{Z})$, see Remarks 1.1 and 1.22. To give it a geometric interpretation, we consider the Fano correspondence

$$F(X) \xleftarrow{\ p\ } \mathbb{L} \xrightarrow{\ q\ } X. \tag{13}$$

Here, $F(X)$ is the Fano variety of lines contained in X, $p: \mathbb{L} \longrightarrow F(X)$ is the universal line, and q is the natural projection, cf. [19, Ch. 3] for details and references. Due to work of Beauville and Donagi [5], it is known that $F(X)$ is a four-dimensional hyperkähler manifold deformation equivalent to the Hilbert scheme $S^{[2]}$ of a K3 surface S.

3.1 The fact that $F(X)$ is of K3$^{[2]}$-type implies that $H^2(F(X), \mathbb{Z})$ with the Beauville–Bogomolov pairing is isometric to the lattice $H^2(S^{[2]}, \mathbb{Z}) \simeq \Lambda \oplus \mathbb{Z}(-2)$. But the cohomology of the Fano variety can also be compared to $\tilde{H}(X, \mathbb{Z})$ by the following combination of [1, 5].

Theorem 3.1 (Beauville–Donagi, Addington) *The Fano correspondence (13) induces two compatible Hodge isometries*

$$
\begin{array}{ccc}
H^4(X, \mathbb{Z})_{\mathrm{pr}}(1) & \xrightarrow{\sim} & H^2(F(X), \mathbb{Z})_{\mathrm{pr}} \\
\cap & & \cap \\
v(\lambda_1)^\perp & \xrightarrow{\sim} & H^2(F(X), \mathbb{Z}) \\
\cap & & \\
\tilde{H}(X, \mathbb{Z}). & &
\end{array}
$$

On the left hand side, $H^4(X, \mathbb{Z})_{\mathrm{pr}}(1) \subset v(\lambda_1)^\perp \subset \tilde{H}(X, \mathbb{Z})$ is the Hodge structure introduced earlier on the sublattice $v(\lambda_1)^\perp \simeq \lambda_1^\perp \simeq \Lambda \oplus \mathbb{Z}(-2)$. As before, the sign of the intersection pairing on $H^4(X, \mathbb{Z})_{\mathrm{pr}}$ is changed. On the right hand side, $H^2(F(X), \mathbb{Z})_{\mathrm{pr}}$ is the primitive cohomology with respect to the Plücker polarization $g \in H^2(F(X), \mathbb{Z})$. It is endowed with a natural quadratic form, the Beauville–Bogomolov form on the hyperkähler fourfold $F(X)$. We shall not attempt to prove the result but we will define the maps that are used and indicate the main steps of the argument.

First, it has been observed in [5] that

$$\varphi := p_* \circ q^* : H^4(X, \mathbb{Z})(1) \longrightarrow H^2(F(X), \mathbb{Z})$$

maps h^2 to the Plücker polarization $g \in H^2(F(X), \mathbb{Z})$ and that for four-dimensional cubics the map induces an isomorphism

$$H^4(X, \mathbb{Z})_{\mathrm{pr}}(1) \xrightarrow{\sim} H^2(F(X), \mathbb{Z})_{\mathrm{pr}}$$

of Hodge structures of weight two satisfying $(\alpha)^2 = -\frac{1}{6} \int_{F(X)} \varphi(\alpha)^2 \cdot g^2$, cf. [19, Sec. 3.4] for statements and further references.

Now, as $v(\lambda_1)^\perp \subset \widetilde{H}(X, \mathbb{Z}) \subset H^*(X, \mathbb{Q})$ is not concentrated in degree four, we need to extend the above to the full cohomology. As was observed by Mukai, the natural map $p_* \circ q^*$ needs to be modified to enjoy certain functoriality properties. More precisely, it is known that the following diagram commutes

$$
\begin{array}{ccc}
K_{\mathrm{top}}(X) & \xrightarrow{p_* \circ q^*} & K_{\mathrm{top}}(F(X)) \\
{\scriptstyle v}\downarrow & & \downarrow {\scriptstyle v} \\
H^*(X, \mathbb{Q}) & \longrightarrow & H^*(F(X), \mathbb{Q}).
\end{array}
\tag{14}
$$

Here, the top and bottom rows are given by $E \longmapsto p_*(q^*E)$ and $\alpha \longmapsto p_*(q^*\alpha \cdot v(i_*\mathcal{O}_\mathbb{L}))$, respectively, where $i : \mathbb{L} \subset X \times F(X)$ is the inclusion, see [14, Ch. 5]. The Mukai vector $i_*\mathcal{O}_\mathbb{L}$ can be computed by means of the Grothendieck–Riemann–Roch formula as

$$v(i_*\mathcal{O}_\mathbb{L}) = i_*(\mathrm{td}(p)) \cdot \left(\mathrm{td}(X)^{-1} \boxtimes \mathrm{td}(F(X))\right)^{1/2}.$$

From here it is a straightforward computation to show that the commutativity of the diagram (14) implies the commutative diagram

$$
\begin{array}{ccc}
K_{\mathrm{top}}(X) & \xrightarrow{p_* \circ q^*} & K_{\mathrm{top}}(F(X)) \\
{\scriptstyle \mathrm{ch}}\downarrow & & \downarrow {\scriptstyle \mathrm{ch}} \\
H^*(X, \mathbb{Q}) & \xrightarrow{\varphi} & H^*(F(X), \mathbb{Q}),
\end{array}
$$

where now the bottom row is defined as $\varphi : \alpha \longmapsto p_*(q^*\alpha \cdot \mathrm{td}(p))$. In particular, for any class $\gamma \in K_{\mathrm{top}}(X)$ one finds $c_1(p_*(q^*(\gamma))) = \{p_*(q^*\mathrm{ch}(\gamma) \cdot \mathrm{td}(p))\}_2$.

The restriction of $c_1 \circ p_* \circ q^* : K_{\mathrm{top}}(X) \longrightarrow H^2(F(X), \mathbb{Z})$ to the primitive part $A_2^\perp \subset K'_{\mathrm{top}}(X)$, i.e. the part mapping to $H^4(X, \mathbb{Q})_{\mathrm{pr}}$ under ch (or, equivalently, under the Mukai vector v), factors over the original isometry

$H^4(X, \mathbb{Z})_{\mathrm{pr}}(1) \xrightarrow{\sim} H^2(F(X), \mathbb{Z})_{\mathrm{pr}}$. As observed in Remark 1.1, $\lambda_1^\perp \subset K'_{\mathrm{top}}(X)$ contains $A_2^\perp \oplus \mathbb{Z}(\lambda_1 + 2\lambda_2)$ as a sublattice of index three. A computation reveals where the second summand is mapped to, cf. [1].

Lemma 3.2 (Addington) *Under the map* $c_1 \circ p_* \circ q^* \colon K_{\text{top}}(X) \longrightarrow H^2(F(X), \mathbb{Z})$ *the class* $\lambda_1 + 2\lambda_2$ *is mapped to the Plücker polarization* $g \in H^2(F(X), \mathbb{Z})$. *Furthermore,* $(\lambda_1 + 2\lambda_2)^2 = (g)^2 = 6$, *where the second square is with respect to the Beauville–Bogomolov form.* □

Therefore, there exists an isometric embedding of the sublattice

$$A_2^\perp \oplus \mathbb{Z}\,(\lambda_1 + 2\lambda_2) \lhook\joinrel\longrightarrow H^2(F(X), \mathbb{Z}), \tag{15}$$

where $A_2^\perp \oplus \mathbb{Z}\,(\lambda_1 + 2\lambda_2)$ is a sublattice of $\lambda_1^\perp$ of index three and discriminant disc $=$ 18. On the other hand, as abstract lattices $H^2(F(X), \mathbb{Z}) \simeq \lambda_1^\perp$. Using this, one then

proves that (15) indeed extends to an isometry $\lambda_1^\perp \overset{\sim}{\longrightarrow} H^2(F(X), \mathbb{Z})$. Composition

with $\lambda_1^\perp \simeq v(\lambda_1)^\perp$ yields the Hodge isometry $v(\lambda_1)^\perp \overset{\sim}{\longrightarrow} H^2(F(X), \mathbb{Z})$. Here, the orthogonal complements $\lambda_1^\perp$ and $v(\lambda_1)^\perp$ are taken in $K'_{\text{top}}(X)$ and $\widetilde{H}(X, \mathbb{Z})$, respectively.

3.2 In the sequel, we will think of $H^2(F(X), \mathbb{Z})$ as a natural sub Hodge structure of $\widetilde{H}(X, \mathbb{Z})$:

$$H^2(F(X), \mathbb{Z}) \subset \widetilde{H}(X, \mathbb{Z}),$$

orthogonal to the distinguished class $v(\lambda_1) \in \widetilde{H}^{1,1}(X, \mathbb{Z})$. This should be thought of as analogous to the inclusion

$$H^2(S^{[2]}, \mathbb{Z}) \subset \widetilde{H}(S, \mathbb{Z}),$$

which is orthogonal to $v(\mathcal{I}_x) = (1, 0, -1) \in (H^0 \oplus H^4)(S, \mathbb{Z}) \subset \widetilde{H}^{1,1}(S, \mathbb{Z})$. Note that both vectors, $v(\lambda_1)$ and $v(\mathcal{I}_x)$, are of square two, which immediately leads to the following observation.

Lemma 3.3 *Let X be a smooth cubic fourfold and S a K3 surface. Then every*

Hodge isometry $H^2(F(X), \mathbb{Z}) \overset{\sim}{\longrightarrow} H^2(S^{[2]}, \mathbb{Z})$ *extends to a Hodge isometry*

$\widetilde{H}(X, \mathbb{Z}) \overset{\sim}{\longrightarrow} \widetilde{H}(S, \mathbb{Z})$ *mapping* $v(\lambda_1)$ *to* $v(\mathcal{I}_x)$. □

The result should be compared to the observation made earlier that every Hodge isometry $H^4(X, \mathbb{Z})_{\text{pr}} \overset{\sim}{\longrightarrow} H^4(X', \mathbb{Z})_{\text{pr}}$ extends to $H^4(X, \mathbb{Z}) \overset{\sim}{\longrightarrow} H^4(X', \mathbb{Z})$ with $h_X^2 \longmapsto \pm h_{X'}^2$.

This enables one to prove the Fano analogue of Proposition 1.24, see [1, 12, 19].

Proposition 3.4 (Addington, Hassett, Huybrechts) *Assume X is a smooth cubic fourfold.*

(i) *There exist a K3 surface S and a Hodge isometry*

$$H^2(S^{[2]}, \mathbb{Z}) \simeq H^2(F(X), \mathbb{Z}) \tag{16}$$

if and only if there exists an embedding $U \hookrightarrow \widetilde{H}^{1,1}(X, \mathbb{Z})$ *with* $v(\lambda_1)$ *contained in its image.*

(ii) *There exist a K3 surface S and a Hodge isometry*

$$H^2(M_S(v), \mathbb{Z}) \simeq H^2(F(X), \mathbb{Z}) \qquad (17)$$

for some smooth, projective, four-dimensional moduli space $M_S(v)$ *of stable sheaves on S if and only if there exists a K3 surface S with* $S \sim X$ *if and only if there exists an embedding* $U \hookrightarrow \widetilde{H}^{1,1}(X, \mathbb{Z})$.

(iii) *There exist a twisted K3 surface* (S, α) *and a Hodge isometry*

$$H^2(M_{S,\alpha}(v), \mathbb{Z}) \simeq H^2(F(X), \mathbb{Z}) \qquad (18)$$

for some smooth, projective, four-dimensional moduli space $M_{S,\alpha}(v)$ *of twisted stable sheaves on S if and only if there exists a twisted K3 surface* (S, α) *with* $(S, \alpha) \sim X$ *if and only if there exists an embedding* $U(n) \hookrightarrow \widetilde{H}^{1,1}(X, \mathbb{Z})$ *for some* $n \neq 0$.

Proof Any Hodge isometry (16) extends to a Hodge isometry $\widetilde{H}(S, \mathbb{Z}) \simeq H^2(F(X), \mathbb{Z})$ with $(1, 0, -1) \mapsto v(\lambda_1)$. As $(1, 0, -1) \in U \simeq (H^0 \oplus H^4)(S, \mathbb{Z}) \subset \widetilde{H}^{1,1}(S, \mathbb{Z})$, this proves one direction in (i). For the other direction use the arguments in the proof of Proposition 1.24 to show that there exists a K3 surface S with $S \sim X$ such that the given $U \hookrightarrow \widetilde{H}^{1,1}(X, \mathbb{Z})$ corresponds to $(H^0 \oplus H^4)(S, \mathbb{Z})$.

For (ii) and (iii) recall that there exists a Hodge isometry $H^2(M_{S,\alpha}(v), \mathbb{Z}) \simeq v^{\perp} \subset \widetilde{H}(S, \alpha, \mathbb{Z})$, cf. [16, Ch. 10] for references in the untwisted case and [21] for the twisted case. Then, if a Hodge isometry $\widetilde{H}(S, \alpha, \mathbb{Z}) \simeq \widetilde{H}(X, \mathbb{Z})$ is given, let $v \in \widetilde{H}^{1,1}(S, \alpha, \mathbb{Z})$ be the vector that is mapped to $v(\lambda_1)$. Then (17) and (18) hold. The remaining assertions follow from Proposition 1.24. $\qquad \square$

This leads to the following analogue of Corollary 2.17.

Corollary 3.5 *For a smooth cubic fourfold X the condition (i) (or (ii) or (iii)) is equivalent to* $X \in \mathcal{C}_d$ *for some d satisfying* (∗∗∗) *(or* (∗∗) *or* (∗∗′)*, respectively).*

$\qquad \square$

So, at one glance:

$$H^2(S^{[2]}, \mathbb{Z}) \simeq H^2(F(X), \mathbb{Z}) \Leftrightarrow (***),$$
$$H^2(M_S(v), \mathbb{Z}) \simeq H^2(F(X), \mathbb{Z}) \Leftrightarrow (**),$$
$$H^2(M_{S,\alpha}(v), \mathbb{Z}) \simeq H^2(F(X), \mathbb{Z}) \Leftrightarrow (**').$$

3.3 The purely Hodge and lattice theoretic considerations above can now be combined with the global Torelli theorem for hyperkähler fourfolds due to Verbitsky [33] and Markman [29], see also [15]: Two hyperkähler fourfolds Y and Y' of

K3$^{[2]}$-type are birational if and only if there exists a Hodge isometry $H^2(Y, \mathbb{Z}) \simeq H^2(Y', \mathbb{Z})$:

$$Y \sim Y' \Leftrightarrow H^2(Y, \mathbb{Z}) \simeq H^2(Y', \mathbb{Z}).$$

This then implies the following reformulation of the above results:

$$S^{[2]} \sim F(X) \Leftrightarrow (***), \quad M_S(v) \sim F(X) \Leftrightarrow (**)$$

$$\text{and } M_{S,\alpha}(v) \sim F(X) \Leftrightarrow (**').$$

More precisely, one has:

Corollary 3.6 *Let X be a smooth cubic fourfold and $F(X)$ its Fano variety of lines.*

(i) *There exists a K3 surface S such that $F(X)$ is birational to $S^{[2]}$ if and only if $X \in \mathcal{C}_d$ for some d satisfying $(***)$.*

(ii) *There exists a K3 surface S such that $F(X)$ is birational to a certain smooth, projective moduli space $M_S(v)$ of stable sheaves on S if and only if $X \in \mathcal{C}_d$ for some d satisfying $(**)$.*

(iii) *There exists a twisted K3 surface (S, α) such that $F(X)$ is birational to a certain smooth, projective moduli space $M_{S,\alpha}(v)$ of twisted stable sheaves on S if and only if $X \in \mathcal{C}_d$ for some d satisfying $(**')$.*   □

Remark 3.7 For $d \equiv 0\,(6)$ and very general $(S, L) \in M_d$, i.e. $\mathrm{Pic}(S) \simeq \mathbb{Z}L$, there exists exactly one other polarized K3 surface $(S', L') \in M_d$ with $\Phi_\varepsilon[(S, L)] = \Phi_\varepsilon[(S', L')] =: [X]$. In particular, the Fano variety $F(X)$ of lines in the corresponding cubic fourfold X is a natural four-dimensional hyperkähler manifold associated with (S, L) and (S', L'). Other hyperkähler manifolds that come naturally with S and S' would be $S^{[2]}$ and $S'^{[2]}$. From Corollary 3.6 we know that for d not satisfying $(***)$ the Hilbert scheme $S^{[2]}$ and the Fano variety $F(X)$ are not isomorphic. It was recently shown in [8] that also $S^{[2]}$ and $S'^{[2]}$ need not be isomorphic (nor birational). More precisely, they are isomorphc if and only if the Pell equation $3p^2 - (d/6)q^2 = -1$ has an integral solution.

4 The Hodge Theory of Kuznetsov's Category

In this short last section we touch upon the Hodge theoretic aspects of Kuznetsov's triangulated category $\mathcal{A}_X$ naturally associated with every smooth cubic fourfold $X \subset \mathbb{P}^5$. For the more categorical aspects we refer to the original [24, 25] or the lecture notes in this volume [28]. The Hodge theoretic investigation of $\mathcal{A}_X$ was initiated by Addington and Thomas [2], the algebraic part of it played a crucial role already in [25].

4.1 We consider the bounded derived category $D^b(X) = D^b(\text{Coh}(X))$ of the abelian category $\text{Coh}(X)$ of coherent sheaves on X. The three line bundles $\mathcal{O}_X, \mathcal{O}_X(1), \mathcal{O}_X(2) \in D^b(X)$ form an exceptional collection, i.e. $\text{Hom}(\mathcal{O}_X(i), \mathcal{O}(j)[*]) = 0$ for $i > j$ and $\mathbb{C}[0]$ for $i = j$. According to a result of Bondal and Orlov [6], the derived category $D^b(X)$ determines X uniquely. More precisely, if there exists an exact, linear equivalence $D^b(X) \simeq D^b(X')$ for two smooth cubic fourfolds $X, X' \subset \mathbb{P}^5$, then $X \simeq X'$. This could be called a categorical global Torelli theorem, although the existence of such an equivalence is almost as hard as writing down an explicit isomorphism between them. However, it turns out that $D^b(X)$ contains a natural subcategory which is a much subtler invariant.

Definition 4.1 For a smooth cubic fourfold $X \subset \mathbb{P}^5$, we denote by

$$\mathcal{A}_X := \langle \mathcal{O}_X, \mathcal{O}_X(1), \mathcal{O}_X(2) \rangle^\perp \subset D^b(X)$$

the full triangulated subcategory of all objects $F \in D^b(X)$ right orthogonal to $\mathcal{O}_X, \mathcal{O}_X(1)$, and $\mathcal{O}_X(2)$, i.e. such that $\text{Hom}(\mathcal{O}_X(i), F[*]) = 0$ for $i = 0, 1, 2$.

Theorem 4.2 (Kuznetsov) *The triangulated category $\mathcal{A}_X$ is a Calabi–Yau category of dimension two, i.e. $F \longmapsto F[2]$ defines a Serre functor.*   □

In other words, for all $E, F \in \mathcal{A}_X$ there exist functorial isomorphisms

$$\text{Hom}(E, F) \simeq \text{Hom}(F, E[2])^*.$$

Other examples of such categories are provided by $D^b(S)$ and $D^b(S, \alpha)$ associated with K3 surfaces S and twisted K3 surfaces (S, α). A natural question in this context is now to determine when the Kuznetsov category $\mathcal{A}_X$ associated with a cubic fourfold is equivalent to the derived category $D^b(S)$ or $D^b(S, \alpha)$ for some (twisted) K3 surface.

4.2 The goal of [2] was to compare Hassett's condition $(**)$ with the condition $\mathcal{A}_X \simeq D^b(S)$. Building upon [2], the twisted version was later dealt with in [17].

Theorem 4.3 (Addington–Thomas, Huybrechts) *Let X be a smooth cubic fourfold and (S, α) a twisted K3 surface.*

(i) *Any exact, linear equivalence $\mathcal{A}_X \simeq D^b(S)$ induces a Hodge isometry $\widetilde{H}(X, \mathbb{Z}) \simeq \widetilde{H}(S, \mathbb{Z})$. In particular, X is contained in $\mathcal{C}_d$ with d satisfying $(**)$.*

(ii) *Any exact, linear equivalence $\mathcal{A}_X \simeq D^b(S, \alpha)$ induces a Hodge isometry $\widetilde{H}(X, \mathbb{Z}) \simeq \widetilde{H}(S, \alpha, \mathbb{Z})$. In particular, X is contained in $\mathcal{C}_d$ with d satisfying $(**')$.*

In fact, it is also known that for very general $X \in \mathcal{C}_d$ with d satisfying $(**)$ or $(**')$, respectively, the converse in (i) and (ii) hold true. The proof, however, requires a fair amount of deformation theory for Fourier–Mukai kernels developed in [2, 17, 22, 32]. For non-special cubic fourfolds one has the following result.

Proposition 4.4 (Huybrechts) *Let X and X′ be smooth cubic fourfolds. Then any Fourier–Mukai equivalence $\mathcal{A}_X \simeq \mathcal{A}_{X'}$ induces a Hodge isometry $\widetilde{H}(X, \mathbb{Z}) \simeq \widetilde{H}(X', \mathbb{Z})$. The converse holds for all non-special X and for general special ones.*

The results of the forthcoming [4] complete this picture, so that eventually we will have

$$\mathcal{A}_X \simeq D^b(S) \quad \Leftrightarrow \widetilde{H}(S, \mathbb{Z}) \simeq \widetilde{H}(X, \mathbb{Z}) \text{ Hodge isometry,}$$
$$\mathcal{A}_X \simeq D^b(S, \alpha) \Leftrightarrow \widetilde{H}(S, \alpha, \mathbb{Z}) \simeq \widetilde{H}(X, \mathbb{Z}) \text{ Hodge isometry,}$$
$$\mathcal{A}_X \simeq \mathcal{A}_{X'} \quad \Leftrightarrow \widetilde{H}(X, \mathbb{Z}) \simeq \widetilde{H}(X', \mathbb{Z}) \text{ Hodge isometry.}$$

Acknowledgements I wish to thank Andreas Hochenegger and Paolo Stellari for the organization of the school and PS for gently insisting that I should write up these notes. The many questions of the participants have been stimulating and helped me to improve the quality of the notes. Special thanks to Emma Brakkee, who also went through a first draft and pointed out many inaccuracies, and Pablo Magni. The author is supported by the SFB/TR 45 'Periods, Moduli Spaces and Arithmetic of Algebraic Varieties' of the DFG (German Research Foundation) and the Hausdorff Center for Mathematics.

References

1. N. Addington, On two rationality conjectures for cubic fourfolds. Math. Res. Lett. **23**, 1–13 (2016)
2. N. Addington, R. Thomas, Hodge theory and derived categories of cubic fourfolds. Duke Math. J. **163**, 1885–1927 (2014)
3. W. Baily, A. Borel, Compactification of arithmetic quotients of bounded symmetric domains. Ann. Math. **84**, 442–528 (1966)
4. A. Bayer, M. Lahoz, E. Macrì, H. Nuer, A. Perry, P. Stellari, Stability conditions in family. arXiv:1902.08184
5. A. Beauville, R. Donagi, La variété des droites d'une hypersurface cubique de dimension 4. C. R. Acad. Sci. Paris Sér. I Math. **301**, 703–706 (1085)
6. A. Bondal, D. Orlov, Derived categories of coherent sheaves, in *Proceedings of the ICM, Beijing*, vol. II (2002), pp. 47–56
7. A. Borel, Some metric properties of arithmetic quotients of symmetric spaces and an extension theorem. J. Diff. Geom. **6**, 543–560 (1972)
8. E. Brakkee, Two polarized K3 surfaces associated to the same cubic fourfold (2018). arXiv:1808.01179
9. F. Charles, A remark on the Torelli theorem for cubic fourfolds (2012). arXiv:1209.4509
10. D. Cox, *Primes of the form $x^2 + ny^2$* (Wiley, New York, 1989)
11. V. Gritsenko, K. Hulek, G. Sankaran, Abelianisation of orthogonal groups and the fundamental group of modular varieties. J. Algebra **322**, 463–478 (2009)
12. B. Hassett, Special cubic fourfolds. Compos. Math. **120**, 1–23 (2000)
13. K. Hulek, D. Ploog, Fourier–Mukai partners and polarised K3 surfaces. Fields Inst. Commun. **67**, 333–365 (2013)
14. D. Huybrechts, Fourier–Mukai transforms in algebraic geometry, in *Oxford Mathematical Monographs* (Oxford University, Oxford, 2006)
15. D. Huybrechts, A global Torelli theorem for hyperkähler manifolds (after Verbitsky). Séminaire Bourbaki, Exp. No. 1040, Astérisque No. 348, 375–403 (2012)

16. D. Huybrechts, *Lectures on K3 Surfaces* (Cambridge University, Cambridge, 2016). http://www.math.uni-bonn.de/people/huybrech/K3.html
17. D. Huybrechts, The K3 category of a cubic fourfold. Compos. Math. **153**, 586–620 (2017)
18. D. Huybrechts, Finiteness of polarized K3 surfaces and hyperkähler manifolds. Ann. Henri Lebesgue **1**, 227–246 (2018)
19. D. Huybrechts, The geometry of cubic hypersurfaces. Lectures Notes (in preparation). http://www.math.uni-bonn.de/people/huybrech/Notes.pdf
20. D. Huybrechts, J. Rennemo, Hochschild cohomology versus the Jacobian ring, and the Torelli theorem for cubic fourfolds. Algebr. Geom. **6**(1), 76–99 (2019)
21. D. Huybrechts, P. Stellari, Equivalences of twisted K3 surfaces. Math. Ann. **332**, 901–936 (2005)
22. D. Huybrechts, E. Macrì, P. Stellari, Derived equivalences of K3 surfaces and orientation. Duke Math. J. **149**, 461–507 (2009)
23. M. Kneser, *Quadratische Formen* (Springer, Heidelberg, 2002)
24. A. Kuznetsov, Derived categories of cubic and V_{14} threefolds. Proc. Steklov Inst. Math. **3**(246), 171–194 (2004). arXiv:math/0303037
25. A. Kuznetsov, Derived categories of cubic fourfolds, in *Cohomological and Geometric Approaches to Rationality Problems*. Progress in Mathematics, vol. 282 (Birkhaüser Verlag, Boston, 2010), pp. 219–243
26. R. Laza, The moduli space of cubic fourfolds via the period maps. Ann. Math. **172**, 673–7111 (2010)
27. E. Looijenga, The period map for cubic fourfolds. Invent. Math. **177**, 213–233 (2009)
28. E. Macrì, P. Stellari, Lectures on non-commutative K3 surfaces, Bridgeland stability, and moduli spaces, in *Birational Geometry of Hypersurfaces*, ed. by A. Hochenegger et al. Lecture Notes of the Unione Matematica Italiana, vol. 26 (Springer, Cham, 2019). https://doi.org/10.1007/978-3-030-18638-8_6
29. E. Markman, A survey of Torelli and monodromy results for holomorphic-symplectic varieties, in *Complex and Differential Geometry*. Springer Proceedings Mathematical, vol. 8 (2011), pp. 257–322
30. E. Reinecke, Autoequivalences of twisted K3 surfaces. Compos. Math. **155**(5), 912–937 (2019)
31. S. Tanimoto, A. Várilly-Alvarado, Kodaira dimension of moduli of special cubic fourfolds. J. reine angew. Math. **2019**(752), 265–300 (2019). https://doi.org/10.1515/crelle-2016-0053
32. Y. Toda, Deformations and Fourier–Mukai transforms. J. Differ. Geom. **81**, 197–224 (2009)
33. M. Verbitsky, Mapping class group and a global Torelli theorem for hyperkähler manifolds. Duke Math. J. **162**, 2929–2986 (2013)
34. C. Voisin, Théorème de Torelli pour les cubiques de $\mathbb{P}^5$. Invent. Math. **86**, 577–601 (1986)
35. C. Voisin, Erratum: "A Torelli theorem for cubics in $\mathbb{P}^5$". Invent. Math. **172**, 455–458 (2008)

Lectures on Non-commutative K3 Surfaces, Bridgeland Stability, and Moduli Spaces

Emanuele Macrì and Paolo Stellari

Abstract We survey the basic theory of non-commutative K3 surfaces, with a particular emphasis to the ones arising from cubic fourfolds. We focus on the problem of constructing Bridgeland stability conditions on these categories and we then investigate the geometry of the corresponding moduli spaces of stable objects. We discuss a number of consequences related to cubic fourfolds including new proofs of the Torelli theorem and of the integral Hodge conjecture, the extension of a result of Addington and Thomas and various applications to hyperkähler manifolds.

These notes originated from the lecture series by the first author at the school on *Birational Geometry of Hypersurfaces*, Palazzo Feltrinelli - Gargnano del Garda (Italy), March 19–23, 2018.

1 Introduction

K3 surfaces have been extensively studied during the last decades of the previous century. The techniques used to understand their geometry include Hodge theory, lattice theory and homological algebra. In fact the lattice and Hodge structures on their second cohomology groups determine completely their geometry in the following sense:

(K3.1) **Torelli Theorem:** Two K3 surfaces S_1 and S_2 are isomorphic if and only if there is a Hodge isometry $H^2(S_1, \mathbb{Z}) \cong H^2(S_2, \mathbb{Z})$.

The result is originally due to Pijateckiĭ-Šapiro and Šafarevič in the algebraic case [144] and to Burns and Rapoport in the analytic case [41] (see also [61, 114]

E. Macrì
Department of Mathematics, Northeastern University, Boston, MA, USA
e-mail: e.macri@northeastern.edu; https://web.northeastern.edu/emacri/

P. Stellari (✉)
Dipartimento di Matematica "F. Enriques", Università degli Studi di Milano, Milano, Italy
e-mail: paolo.stellari@unimi.it; http://users.unimi.it/stellari

© Springer Nature Switzerland AG 2019   199
A. Hochenegger et al. (eds.), *Birational Geometry of Hypersurfaces*,
Lecture Notes of the Unione Matematica Italiana 26,
https://doi.org/10.1007/978-3-030-18638-8_6

for other proofs). Another way to rephrase this is in terms of fibers of the period map for K3 surfaces: the period map is generically injective. Such a map turns out to be surjective as well. Roughly speaking this means that one can get complete control on which weight-2 Hodge structures on the abstract lattice correspond to an actual surface.

One can go further and construct other compact *hyperkähler manifolds* out of K3 surfaces. Following Beauville [25], one could first consider Hilbert schemes (or Douady spaces) of points on such surfaces. More generally, for a *projective* K3 surface S and a primitive vector $\mathbf{v}$ in the algebraic part of the total cohomology $H^*(S, \mathbb{Z})$, one can construct the moduli space $M_H(S, \mathbf{v})$ of stable sheaves E on S with Mukai vector $v(E) := \operatorname{ch}(E) \cdot \sqrt{\operatorname{td}(S)} = \mathbf{v}$, for the choice of an ample line bundle H which is generic with respect to $\mathbf{v}$.

(K3.2) The moduli space $M_H(S, \mathbf{v})$ is a smooth projective hyperkähler manifold of dimension $\mathbf{v}^2 + 2$ which is deformation equivalent to a Hilbert scheme of points on a K3 surface.

Such moduli spaces deform together with a polarized deformation of the K3 surface S and they yield 19-dimensional families of hyperkähler manifolds. Here the square of $\mathbf{v}$ is taken with respect to the so called *Mukai pairing* which is defined on the total cohomology $H^*(S, \mathbb{Z})$ of S. The above statement is the result of many different contributions, starting with the foundational work of Mukai [125, 126], and it is due to Huybrechts [67], O'Grady [131], and Yoshioka. The statement in its final form is [164, Theorems 0.1 and 8.1]). It also gives a precise non-emptyness statement: if the Mukai vector is *positive*, then the moduli space is non-empty if and only if it has non-negative expected dimension.

More recently, after the works of Mukai [126] and Orlov [135], the bounded derived categories of coherent sheaves of K3 surfaces and their autoequivalence groups have been extensively studied. The Torelli theorem (K3.1) has a homological counterpart:

(K3.3) **Derived Torelli Theorem:** Given two K3 surfaces S_1 and S_2, then $D^b(S_1) \cong D^b(S_2)$ are isomorphic if and only if there is an isometry between the total cohomology lattices of S_1 and S_2 preserving the Mukai weight-2 Hodge structure.

The lattice and Hodge structure mentioned above will be explained in Sect. 3.4. In the geometric setting considered in (K3.3), it might be useful to keep in mind that the pairing coincides with the usual Euler form on the algebraic part of the total cohomologies of S_1 and S_2. On the other hand, the $(2, 0)$-part in the Hodge decomposition mentioned above is nothing but $H^{2,0}(S_i)$.

Thus the existence of an equivalence can be detected again by looking at the Hodge and lattice structure of the total cohomology (and not just of H^2). Pushing this further, one can observe that $D^b(S)$ carries additional structures: *Bridgeland stability conditions.* Indeed, after the works by Bridgeland [37, 38] we have the following fact:

(K3.4) If S is a K3 surface, then the manifold parameterizing stability conditions on $D^b(S)$ is non-empty and a connected component is a covering of a generalized period domain.

Following the pattern for stability of sheaves, one can take a stability condition σ and a primitive vector $\mathbf{v}$ as in (K3.2) such that σ is generic with respect to $\mathbf{v}$. If one considers the moduli space $M_\sigma(D^b(S), \mathbf{v})$ of σ-stable objects in $D^b(S)$ with Mukai vector $\mathbf{v}$, then by Toda [154], Bayer and Macrì [19, 20], Minamide et al. [123], we have the analogue of (K3.4):

(K3.5) The moduli space $M_\sigma(D^b(S), \mathbf{v})$ is a smooth projective hyperkähler manifold of dimension $\mathbf{v}^2+2$ which is deformation equivalent to a Hilbert scheme of points on a K3 surface.

Quite surprisingly, a similar picture appears in a completely different setting when the canonical bundle is far from being trivial: Fano varieties. The first key example is the case of smooth *cubic fourfolds*.

(C.1) **Torelli Theorem:** Two cubic fourfolds W_1 and W_2 defined over $\mathbb{C}$ are isomorphic if and only if there is a Hodge isometry $H^4(W_1, \mathbb{Z}) \cong H^4(W_2, \mathbb{Z})$ that preserves the square of a hyperplane class.

The result is due to Voisin [157] (other proofs were given in [48, 73, 113]). The striking similarity with K3 surfaces is confirmed by the fact that $H^4(W, \mathbb{Z})$ carries actually a weight-2 Hodge structure of K3 type. Moreover, it was observed by Hassett [66] that the 20-dimensional moduli space $\mathscr{C}$ of cubic fourfolds contain divisors of Noether-Lefschetz type and some of them parametrize cubic fourfolds W with a *Hodge-theoretically associated K3 surface*. Roughly, this simply means that the orthogonal in $H^4(W, \mathbb{Z})$ of a rank 2 sublattice, generated by the self-intersection H^2 of a hyperplane class and by some special surface which is not homologous to H^2, looks like the primitive cohomology of a polarized K3 surface.

The hyperkähler geometry that one can associate to a cubic fourfold is actually quite rich. Indeed, the Fano variety of lines in a cubic fourfold is a four-dimensional projective hyperkähler manifold which is deformation equivalent to a Hilbert scheme of points on a K3 surface (see [27]). More recently, it was proved by Lehn et al. [108] that the moduli space of generalized twisted cubics inside a cubic fourfold also gives rise to a eight-dimensional projective hyperkähler manifold which is again deformation equivalent to a Hilbert scheme of points on a K3 surface. All these constructions work in families and thus provide 20-dimensional locally complete families of four or eight dimensional polarized smooth projective hyperkähler manifolds.

From the homological point of view, it was observed by Kuznetsov [95] that the derived category $D^b(W)$ of a cubic fourfold W contains an admissible subcategory $\mathscr{K}u(W)$ which is the right orthogonal of the category generated by the three line bundles $\mathscr{O}_W$, $\mathscr{O}_W(H)$ and $\mathscr{O}_W(2H)$. We will refer to $\mathscr{K}u(W)$ as the *Kuznetsov component* of W. The category $\mathcal{K}u(W)$ has the same homological properties of $D^b(S)$, for S a K3 surface: it is an indecomposable category with Serre functor which is the shift by 2 and the same Hochschild homology as $D^b(S)$. But, for W very general, there cannot be a K3 surface S with an equivalence $\mathcal{K}u(W) \cong D^b(S)$. This is the reason why we should think of Kuznetsov components as *non-commutative K3 surfaces*.

The study of non-commutative varieties was started more than 30 years ago. Artin and Zhang [11] investigated the case of non-commutative projective spaces (see also the book in preparation [163]). In these notes we will follow closely the approach developed by Kuznetsov [95] and Huybrechts [71]. One important feature is that the Kuznetzov component comes with a naturally associated lattice $\widetilde{H}(\mathcal{K}u(W), \mathbb{Z})$ with a weight-2 Hodge structure. The lattice is actually isometric to the extended K3 lattice mentioned in (K3.3). Hence, as for K3 surfaces, it is natural to expect that (C.1) has a homological counterpart in the same spirit as (K3.3):

(C.2) **(Conjectural) Derived Torelli Theorem for Cubic Fourfolds:** We expect that, if W_1 and W_2 are cubic fourfolds, then $\mathcal{K}u(W_1) \cong \mathcal{K}u(W_2)$ if and only if there is a Hodge isometry $\widetilde{H}(\mathcal{K}u(W_1), \mathbb{Z}) \cong \widetilde{H}(\mathcal{K}u(W_2), \mathbb{Z})$ which preserves the orientation of 4-positive directions.

Such a conjecture will be explained later in this paper but it is worth pointing out that it has been proved generically by Huybrechts [71].

One of the aims of these notes is to show how, following [23, 24], properties (K3.4) and (K3.5) generalize to this setting. Namely, let W be a cubic fourfold and let **v** be a primitive vector in the algebraic part of the total cohomology $\widetilde{H}(\mathcal{K}u(W), \mathbb{Z})$ of $\mathcal{K}u(W)$.

(C.3) The manifold parameterizing stability conditions on $\mathcal{K}u(W)$ is non-empty and a connected component is a covering of a generalized period domain.
(C.4) The moduli space $M_\sigma(\mathcal{K}u(W), \mathbf{v})$ is a smooth projective hyperkähler manifold of dimension $\mathbf{v}^2 + 2$ which is deformation equivalent to a Hilbert scheme of points on a K3 surface, if σ is a stability condition as in (C.3) which is generic with respect to **v**.

Again, both constructions work in families (in an appropriate sense) and thus we get 20-dimensional locally complete families of smooth projective hyperkähler manifolds. Also, we recover the Fano variety of lines and the manifold related to twisted cubics as instances of (C.4). The geometric applications of (C.3) and (C.4) are even more and include a reproof of the Torelli Theorem (C.1), a new simple proof of the integral Hodge conjecture for cubic fourfolds and the extension of a result by Addington and Thomas [4] saying that for a cubic W having a Hodge-theoretically associated K3 surface is the same as having an equivalence $\mathcal{K}u(W) \cong D^b(S)$, for a smooth projective K3 surface S. All these results will be discussed in this paper.

The interesting point is that we expect (C.3) and (C.4) to hold for other interesting classes of Fano manifolds with naturally associated non-commutative K3 surfaces.

This is the case of Gushel-Mukai and Debarre-Voisin manifolds. We will discuss this later on in the paper.

Although we do our best to make this paper as much self-contained as possible, there are various Hodge-theoretical aspects of the theory of cubic fourfolds that are not discussed here and that can be found in the lecture notes of Huybrechts [72] that appear in the same volume as the present ones. Moreover, we only touch very briefly in our treatment the fundamental theory of Homological Projective Duality: we refer to the original articles [92, 100, 140]. In [99], the relation between rationality questions and geometric properties of Kuznetsov components is widely discussed. Finally, an excellent survey of several aspects of the theory of semiorthogonal decompositions is [97].

Plan of the Paper Let us briefly sketch how the paper is organized. In Sect. 2 we introduce the general material concerning non-commutative varieties with an emphasis on those that are of Calabi-Yau type. After presenting semiorthogonal decompositions and exceptional objects (see Sect. 2.1), we discuss in Sect. 2.2 the notion of non-commutative variety in general. In our framework this just means an admissible subcategory $\mathscr{D}$ of the bounded derived category $D^b(X)$, for X a smooth projective variety. As such, it comes with a Serre functor that we compute and with a well-defined notion of product, which we use to define the Hochschild homology and cohomology of $\mathscr{D}$. A non-commutative Calabi-Yau variety is a non-commutative variety whose Serre functor is the shift by an integer n. In Sect. 2.3 we discuss a general framework to construct examples of such non-commutative varieties.

Section 3 has a more geometric flavor. It deals with the constructions of non-commutative K3 categories out of the derived categories of Fano manifolds. The first case we analyze is the one of cubic fourfolds W (see Sect. 3.1) where the non-commutative K3 surface is the K3 category $\mathscr{K}u(W)$ mentioned above. As a result, we state a (generalized version of a) result of Addington and Thomas (see Theorem 3.7) which characterizes completely the loci in the moduli space of cubic fourfolds parameterizing the cubics whose Kuznetsov component is actually equivalent to the bounded derived category of an actual (twisted) K3 surface. Our proof is provided in Sect. 5.4 and it is based on the use of stability conditions and of moduli spaces of stable objects. In Sects. 3.2 and 3.3 we study two other classes of Fano manifolds: Gushel-Mukai and Debarre-Voisin manifolds. Finally, in Sect. 3.5, by using techniques developed in Sect. 3.4, we present derived variants of the Torelli theorem for cubic fourfolds and discuss conjectural relations to the birational type of these fourfolds.

Section 4 is about Bridgeland stability conditions on Kuznetsov components. After recalling the basic definitions and properties in Sect. 4.1, we outline a few techniques: the tilting procedure (see Sect. 4.3) and the way this induces stability conditions on semiorthogonal components (see Sect. 4.4).

We state and prove our first main results in Sect. 5, where we deal with stability conditions and moduli spaces of stable objects in the Kuznetsov component of cubic fourfolds. Theorems 5.5 and 5.7 are the main results. They prove essentially what we claim in (C.3). The proof requires the techniques in Sect. 4 and a non-commutative Bogomolov inequality proved in Theorem 5.3. In Sect. 5.3 we study moduli spaces and prove Theorem 5.11 which yields (C.4). In the rest of the paper (see Sect. 5.4) we discuss several applications of these theorems. These include the complete proof of Theorem 3.7, the proof of the integral Hodge Conjecture for cubic fourfolds (see Proposition 5.17) originally due to Voisin, and various constructions of moduli spaces associated to low degree curves in cubic fourfolds (see Theorems 5.19 and 5.23). We also briefly discuss a homological approach to the Torelli theorem (C.1) (see Theorem 5.21).

Notation and Conventions We work over an algebraically closed field $\mathbb{K}$; often, when the characteristic of the field is not zero, we will assume it to be sufficiently large. When $\mathbb{K} = \mathbb{C}$, the field of complex numbers, we set $i = \sqrt{-1}$. All categories will be $\mathbb{K}$-linear, namely the morphism sets have the structure of $\mathbb{K}$-vector space and the compositions of morphisms are $\mathbb{K}$-bilinear maps, and all varieties will be over $\mathbb{K}$.

We assume a basic knowledge of abelian, derived, and triangulated categories. Basic references are, for example, [62, 68]. Given an algebraic variety X, we will denote by $D^b(X) := D^b(\mathrm{coh}(X))$ the bounded derived category of coherent sheaves on X. All derived functors will be denoted as if they were underived. If X and Y are smooth projective varieties, a functor $F \colon D^b(X) \to D^b(Y)$ is called a *Fourier-Mukai functor* if it is isomorphic to $\Phi_P(\_) := p_{Y*}(P \otimes p_X^*(\_))$, for some object $P \in D^b(X \times Y)$.

We also expect the reader to have some familiarity with K3 surfaces [70] and projective hyperkähler manifolds [50, 63].

2   Non-commutative Calabi-Yau Varieties

In this section, we follow closely the presentation and main results in [98]; foundational references are also [35, 36, 92, 97, 140]. We start with a very short review on semiorthogonal decompositions and exceptional collections in Sect. 2.1. We then introduce the notion of non-commutative variety in Sect. 2.2 and study basic facts about Serre functors and Hochschild (co)homology. Finally, in Sect. 2.3 we sketch the proof of a result by Kuznetsov (Theorem 2.39), which provides a general method to construct non-commutative Calabi-Yau varieties.

2.1   Semiorthogonal Decompositions

Let $\mathscr{D}$ be a triangulated category.

Definition 2.1 A *semiorthogonal decomposition*

$$\mathscr{D} = \langle \mathscr{D}_1, \ldots, \mathscr{D}_m \rangle$$

is a sequence of full triangulated subcategories $\mathscr{D}_1, \ldots, \mathscr{D}_m$ of $\mathscr{D}$ — called the *components* of the decomposition — such that:

(1) $\mathrm{Hom}(F, G) = 0$ for all $F \in \mathscr{D}_i$, $G \in \mathscr{D}_j$ and $i > j$.
(2) For any $F \in \mathscr{D}$, there is a sequence of morphisms

$$0 = F_m \to F_{m-1} \to \cdots \to F_1 \to F_0 = F,$$

such that $\mathrm{cone}(F_i \to F_{i-1}) \in \mathscr{D}_i$ for $1 \leq i \leq m$.

Remark 2.2 Condition (1) of the definition implies that the "filtration" in (2) and its "factors" are unique and functorial. The functor $\delta_i : \mathscr{D} \to \mathscr{D}_i$ given by the i-th "factor", i.e.,

$$\delta_i(F) = \mathrm{cone}(F_i \to F_{i-1}),$$

is called the *projection functor* onto $\mathscr{D}_i$. In the special case $\mathscr{D} = \langle \mathscr{D}_1, \mathscr{D}_2 \rangle$, the functor δ_1 is the left adjoint of the inclusion $\mathscr{D}_1 \hookrightarrow \mathscr{D}$, while the functor δ_2 is the right adjoint of $\mathscr{D}_2 \hookrightarrow \mathscr{D}$.

Examples of semiorthogonal decompositions generally arise from exceptional collections, together with the concept of admissible subcategory.

Definition 2.3 A full triangulated subcategory $\mathscr{C} \subset \mathscr{D}$ is called *admissible* if the inclusion functor admits left and right adjoints.

For a subcategory $\mathscr{C} \subset \mathscr{D}$, we define its *left* and *right orthogonals* as

$$^{\perp}\mathscr{C} := \{ G \in \mathscr{D} \mid \mathrm{Hom}(G, F) = 0 \text{ for all } F \in \mathscr{C} \},$$

$$\mathscr{C}^{\perp} := \{ G \in \mathscr{D} \mid \mathrm{Hom}(F, G) = 0 \text{ for all } F \in \mathscr{C} \}.$$

The following is well-known (see [99, Lemma 2.3] for a more general statement).

Proposition 2.4 *Let $\mathscr{C} \subset \mathscr{D}$ be an admissible subcategory. Then there are semiorthogonal decompositions*

$$\mathscr{D} = \langle \mathscr{C}, {}^{\perp}\mathscr{C} \rangle \qquad and \qquad \mathscr{D} = \langle \mathscr{C}^{\perp}, \mathscr{C} \rangle. \tag{1}$$

Proof Let us denote by $\Psi_L : \mathscr{D} \to \mathscr{C}$ the left adjoint to the inclusion functor. Then, for all $G \in \mathscr{D}$, we have a morphism $G \to \Psi_L(G)$ in $\mathscr{D}$. The cone of this morphism is in $\mathscr{C}^{\perp}$, thus giving the first semiorthogonal decomposition. The second one is analogous, by using the right adjoint. $\square$

Proposition 2.4 also has a converse statement. Namely, if a category $\mathscr{C}$ arises as semiorthogonal component in both decompositions as in (1), then it must be admissible. This follows immediately from Remark 2.2.

Examples of admissible categories are given by exceptional collections.

Definition 2.5

(i) An object $E \in \mathcal{D}$ is *exceptional* if $\mathrm{Hom}(E, E[p]) = 0$, for all integers $p \neq 0$, and $\mathrm{Hom}(E, E) \cong \mathbb{K}$.

(ii) A set of objects $\{E_1, \ldots, E_m\}$ in $\mathcal{D}$ is an *exceptional collection* if E_i is an exceptional object, for all i, and $\mathrm{Hom}(E_i, E_j[p]) = 0$, for all p and all $i > j$.

(iii) An exceptional collection $\{E_1, \ldots, E_m\}$ in $\mathcal{D}$ is *full* if the smallest full triangulated subcategory of $\mathcal{D}$ containing the exceptional collection is equivalent to $\mathcal{D}$.

(iv) An exceptional collection $\{E_1, \ldots, E_m\}$ in $\mathcal{D}$ is *strong* if $\mathrm{Hom}(E_i, E_j[p]) = 0$, for all $p \neq 0$ and all $i < j$.

Proposition 2.6 *Let $\mathscr{C} := \langle E \rangle$ be the smallest full triangulated subcategory of a proper[1] triangulated category $\mathcal{D}$ and containing the exceptional object E. Then $\mathscr{C}$ is admissible.*

Proof This can be seen directly by constructing explicitly the left and right adjoints to the functor $\mathrm{D}^b(\mathbb{K}) \to \mathcal{D}$, $V \mapsto E \otimes_{\mathbb{K}} V$ (inducing an equivalence $\mathrm{D}^b(\mathbb{K}) \cong \mathscr{C}$). More explicitly, since E is exceptional, given an object C in $\mathcal{D}$, consider the (canonical) evaluation morphism

$$\bigoplus_k \mathrm{Hom}(E, C[k]) \otimes E[-k] \to C,$$

where the first object is in $\mathscr{C}$. Complete it to a distinguished triangle

$$\bigoplus_k \mathrm{Hom}(E, C[k]) \otimes E[-k] \to C \to D.$$

Since E is exceptional, we get $\mathrm{Hom}(E, D[p]) = 0$, for all integers p. Hence $D \in \langle E \rangle^{\perp}$.

Therefore, we constructed a semiorthogonal decomposition

$$\mathcal{D} = \langle \langle E \rangle^{\perp}, \langle E \rangle \rangle.$$

Similarly, we can construct a decomposition $\mathcal{D} = \langle \langle E \rangle, {}^{\perp}\langle E \rangle \rangle$, and so the category $\mathscr{C}$ is admissible. $\qquad\square$

---

[1] A triangulated category $\mathcal{D}$ is *proper* over $\mathbb{K}$ if, for all $F, G \in \mathcal{D}$, $\dim_{\mathbb{K}}(\oplus_p \mathrm{Hom}_{\mathcal{D}}(F, G[p])) < +\infty$.

By using Proposition 2.4, if $\{E_1, \ldots, E_m\}$ is an exceptional collection in a proper triangulated category $\mathscr{D}$ and $\mathscr{C}$ denotes the smallest full triangulated subcategory of $\mathscr{D}$ containing the E_i's, then we get two semiorthogonal decompositions

$$\mathscr{D} = \langle \mathscr{C}^{\perp}, E_1, \ldots, E_m \rangle = \langle E_1, \ldots, E_m, {}^{\perp}\mathscr{C} \rangle.$$

Here, for sake of simplicity, we write E_i for the category $\langle E_i \rangle$.

Example 2.7 (Beilinson) A beautiful example is provided by the n-dimensional projective space $\mathbb{P}^n$. In this case, by a classical result of Beilinson [28], the line bundles

$$\{\mathscr{O}_{\mathbb{P}^n}(-n), \mathscr{O}_{\mathbb{P}^n}(-n+1), \ldots, \mathscr{O}_{\mathbb{P}^n}\}$$

form a full strong exceptional collection and so they yield a semiorthogonal decomposition

$$\mathrm{D}^{\mathrm{b}}(\mathbb{P}^n) = \langle \mathscr{O}_{\mathbb{P}^n}(-n), \mathscr{O}_{\mathbb{P}^n}(-n+1), \ldots, \mathscr{O}_{\mathbb{P}^n} \rangle.$$

The exceptionality and strongness of the collection follow from Bott's theorem. The basic idea for the proof of fullness is to use the resolution of the diagonal $\Delta \subset \mathbb{P}^n \times \mathbb{P}^n$ given by the Koszul resolution

$$0 \to \mathscr{O}_{\mathbb{P}^n}(-n) \boxtimes \Omega^n(n) \to \ldots \to \mathscr{O}_{\mathbb{P}^n}(-1) \boxtimes \Omega^1(1) \to \mathscr{O}_{\mathbb{P}^n \times \mathbb{P}^n} \to \mathscr{O}_{\Delta} \to 0$$

associated to a natural section $\mathscr{O}_{\mathbb{P}^n}(-1) \boxtimes \Omega^1(1) \to \mathscr{O}_{\mathbb{P}^n \times \mathbb{P}^n}$ (see, for example, [68, Section 8.3] for the details).

Actually, any set of line bundles $\{\mathscr{O}_{\mathbb{P}^n}(k), \mathscr{O}_{\mathbb{P}^n}(k+1), \ldots, \mathscr{O}_{\mathbb{P}^n}(k+n)\}$, for k any integer, is a full strong exceptional collection in $\mathrm{D}^{\mathrm{b}}(\mathbb{P}^n)$.

Example 2.8 (Kapranov et al.) Consider now the Grassmannian $\mathrm{Gr}(k, n)$ of k-dimensional subspaces in an n-dimensional $\mathbb{K}$-vector space. Let $\mathscr{U}$ be the tautological subbundle and $\mathscr{Q}$ the tautological quotient. If $\mathrm{char}(\mathbb{K}) = 0$ (or sufficiently large), Kapranov [83] has constructed a full strong exceptional collection, and so a semiorthogonal decomposition

$$\mathrm{D}^{\mathrm{b}}(\mathrm{Gr}(k, n)) = \langle \Sigma^{\alpha} \mathscr{U}^{\vee} \rangle_{\alpha \in R(k, n-k)},$$

where $R(k, n-k)$ is the $k \times (n-k)$ rectangle, α is a Young diagram, and Σ^{α} is the associated Schur functor. The basic idea of the proof is similar to the projective space case, by using the Borel-Bott-Weil Theorem for proving that the above collection is exceptional and a Koszul resolution of the diagonal associated to a canonical section of $\mathscr{U}^{\vee} \boxtimes \mathscr{Q}$. When $\mathrm{char}(\mathbb{K}) > 0$, the situation is more complicated and described in [40]. In mixed characteristic, it is worth mentioning [58] where semiorthogonal decompositions for the derived categories of Grassmannians over the integers are studied.

For later use, we need a different exceptional collection, described more recently in [59] (see also [105] for related results). We assume that $\gcd(k, n) = 1$. Then we have a semiorthogonal decomposition in a form which is more similar to the case of the projective space

$$\mathrm{D}^b(\mathrm{Gr}(k, n)) = \langle \mathscr{B}, \mathscr{B}(1), \ldots, \mathscr{B}(n-1) \rangle,$$

with the category $\mathscr{B}$ generated by the exceptional collection formed by the vector bundles $\Sigma^\alpha \mathscr{U}^\vee$ associated to the corresponding Schur functors, where the Young diagram α has at most $k - 1$ rows and whose p-th row is of length at most $(n - k)(k - p)/k$, for $p = 1, \ldots, k - 1$. Explicitly, in the case $\mathrm{Gr}(2, 5)$, the category $\mathscr{B}$ is generated by 2 exceptional objects:

$$\mathscr{B} = \langle \mathscr{O}_{\mathrm{Gr}(2,5)}, \mathscr{U}^\vee \rangle.$$

In the case of $\mathrm{Gr}(3, 10)$, the category $\mathscr{B}$ is generated by 12 exceptional objects.

Example 2.9 (Kapranov) Let Q be an n-dimensional quadric in $\mathbb{P}^{n+1}$ defined by an equation $\{q = 0\}$; we assume that $\mathrm{char}(\mathbb{K}) \neq 2$. By [83], the category $\mathrm{D}^b(Q)$ has one of the following semiorthogonal decompositions, given by full strong exceptional collections, according to the parity of n. If $n = 2m + 1$ is odd, we have

$$\mathrm{D}^b(Q) = \langle S, \mathscr{O}_Q, \mathscr{O}_Q(1), \ldots, \mathscr{O}_Q(n-1) \rangle,$$

where S is the spinor bundle on Q defined as $\mathrm{Coker}(\varphi|_Q)(-1)$ and $\varphi \colon \mathscr{O}_{\mathbb{P}^{n+1}}(-1)^{2^{m+1}} \to \mathscr{O}_{\mathbb{P}^{n+1}}^{2^{m+1}}$ is such that $\varphi \circ (\varphi(-1)) = q \cdot \mathrm{Id} \colon \mathscr{O}_{\mathbb{P}^{n+1}}(-2)^{2^{m+1}} \to \mathscr{O}_{\mathbb{P}^{n+1}}^{2^{m+1}}$.

If $n = 2m$ is even, we have

$$\mathrm{D}^b(Q) = \langle S^-, S^+, \mathscr{O}_Q, \mathscr{O}_Q(1), \ldots, \mathscr{O}_Q(n-1) \rangle,$$

where $S^- := \mathrm{Coker}(\varphi|_Q)(-1)$, $S^+ := \mathrm{Coker}(\psi|_Q)(-1)$, and $\varphi, \psi \colon \mathscr{O}_{\mathbb{P}^{n+1}}(-1)^{2^m} \to \mathscr{O}_{\mathbb{P}^{n+1}}^{2^m}$ are such that $\varphi \circ (\psi(-1)) = \psi \circ (\varphi(-1)) = q \cdot \mathrm{Id}$. The reader can have a look at [137] for more results about spinor bundles.

It is a very interesting question to determine which varieties admit a full exceptional collection. A classical result of Bondal [33] shows that a smooth projective variety has a full strong exceptional collection if and only if its bounded derived category is equivalent to the bounded derived category of finitely generated right modules over an associative finite-dimensional algebra. For example, there is a vast literature on homogeneous spaces (which conjecturally should have a full strong exceptional collection; see [105], and the references therein) or in relation to Dubrovin's Conjecture [57]. Another example is the moduli space of stable rational curves with n punctures: by using results of Kapranov [84] and Orlov ([134]; see also Example 2.13 below), it is easy to show that a full exceptional collection exists, but conjecturally there exists a full strong exceptional collection which is S_n-invariant (see [47, 120]). We also mention the following question by Orlov relating the existence of a full exceptional collection to rationality; we will study further conjectural relations between non-commutative K3 surfaces and rationality in Sect. 3.

Question 2.10 (Orlov) Let X be a smooth projective variety over $\mathbb{K}$. If $\mathrm{D}^b(X)$ admits a full exceptional collection, then X is rational.

For later use, we want to show that the semiorthogonal decompositions in Examples 2.7 and 2.9 can be made relative to a positive dimensional base.

Example 2.11 (Orlov) Let F be a vector bundle of rank $r + 1$ over a smooth projective variety X. Consider the projective bundle $\pi: \mathbb{P}_X(F) \to X$. Then, by [134], the pull-back functor π^* is fully faithful and we have a semiorthogonal decomposition

$$\mathrm{D}^b(\mathbb{P}_X(F)) = \langle \pi^*(\mathrm{D}^b(X)) \otimes \mathcal{O}_{\mathbb{P}_X(F)/X}(-r), \ldots, \pi^*(\mathrm{D}^b(X)) \rangle.$$

Example 2.12 (Kuznetsov) Let X, S be smooth projective varieties, and let $f: X \to S$ be a (flat) quadric fibration. In other words, there is a vector bundle E on S of rank $r + 2$ such that X is a divisor in $\mathbb{P}_S(E)$ of relative degree 2 corresponding to a line bundle $L \subseteq S^2 E^\vee$. We assume that $\mathrm{char}(\mathbb{K}) \neq 2$. Such a quadric fibration comes with a sheaf $\mathcal{B}$ of Clifford algebras on S. The corresponding sheaf $\mathcal{B}_0$ of even parts of Clifford algebras can be described as an $\mathcal{O}_S$-module in the following terms:

$$\mathcal{B}_0 \cong \mathcal{O}_S \oplus (\wedge^2 E \otimes L) \oplus (\wedge^4 E \otimes L^2) \oplus \ldots$$

We can then take the abelian category $\mathrm{coh}(S, \mathcal{B}_0)$ of coherent $\mathcal{B}_0$-modules on S and the corresponding derived category $\mathrm{D}^b(S, \mathcal{B}_0) := \mathrm{D}^b(\mathrm{coh}(S, \mathcal{B}_0))$. One can also consider the sheaf $\mathcal{B}_1$ of odd parts of the Clifford algebras, which is a coherent $\mathcal{B}_0$-module. Actually, for all $i \in \mathbb{Z}$, we have the sheaves

$$\mathcal{B}_{2i} := \mathcal{B}_0 \otimes_{\mathcal{B}_0} L^{-i} \qquad \mathcal{B}_{2i+1} := \mathcal{B}_1 \otimes_{\mathcal{B}_0} L^{-i}.$$

The pull-back functor f^* is fully-faithful. Moreover, there is a fully-faithful functor $\Phi: \mathrm{D}^b(S, \mathcal{B}_0) \hookrightarrow \mathrm{D}^b(X)$, and a semiorthogonal decomposition

$$\mathrm{D}^b(X) = \langle \Phi(\mathrm{D}^b(S, \mathcal{B}_0)), f^*(\mathrm{D}^b(S)), \ldots, f^*(\mathrm{D}^b(S)) \otimes \mathcal{O}_{X/S}(r - 1) \rangle.$$

The functor Φ has a left adjoint that we denote by Ψ. The case when $\mathrm{char}(\mathbb{K}) = 0$ was treated in [93, Theorem 4.2]. This was generalized in [14, Theorem 2.2.1] to fields of arbitrary odd characteristic. The case $\mathrm{char}(\mathbb{K}) = 2$ is discussed in the same paper and it can also be described in a similar way, but the definition of $\mathscr{B}_0$ is different.

Finally, the last case we review is the blow-up along smooth subvarieties.

Example 2.13 (Orlov) Let $Y \subseteq X$ be a smooth subvariety of codimension c. Consider the blow-up $f: \widetilde{X} \to X$ of X along Y. Let $i: E \hookrightarrow \widetilde{X}$ be the exceptional divisor. The restriction $\pi := f|_E: E \to Y$ is a projective bundle, as $E = \mathbb{P}_Y(\mathscr{N}_{Y/X})$. By [134], the pullback functor f^* is fully faithful and, for any integer j, the functors

$$\Psi_j: D^b(Y) \to D^b(\widetilde{X}) \qquad G \mapsto i_*(\pi^*(G) \otimes \mathscr{O}_{E/Y}(j)),$$

are fully-faithful as well, giving the semiorthogonal decomposition

$$D^b(\widetilde{X}) = \langle f^* D^b(X), \Psi_0(D^b(Y)), \dots, \Psi_{c-2}(D^b(Y)) \rangle.$$

2.2 Non-commutative Smooth Projective Varieties

We will work with the following definition of non-commutative smooth projective variety; while this is not the most general notion (see [100, 136, 140]), it will suffice for these notes.[2]

Definition 2.14 Let $\mathscr{D}$ be a triangulated category linear over $\mathbb{K}$. We say that $\mathscr{D}$ is a *non-commutative smooth projective variety* if there exists a smooth projective variety X over $\mathbb{K}$ and a fully faithful $\mathbb{K}$-linear exact functor $\mathscr{D} \hookrightarrow D^b(X)$ having left and right adjoints.

By identifying $\mathscr{D}$ with its essential image in $D^b(X)$, then the definition is only asking that $\mathscr{D}$ is an admissible subcategory. Note also that, as a consequence of the main result in [36], being admissible is a notion which is intrinsic to the category $\mathscr{D}$, namely every fully faithful functor from a non-commutative smooth projective variety into the bounded derived category of a smooth projective variety will have both adjoints.

Products Non-commutative smooth projective varieties are closed under products (or more generally gluing of categories; see [102, 136]). More precisely, if $\mathscr{D}_1 \subset D^b(X_1)$ and $\mathscr{D}_2 \subset D^b(X_2)$, we can define $\mathscr{D}_1 \boxtimes \mathscr{D}_2$ as the smallest triangulated

---

[2]Note that in [136, Definition 4.3] it is used the terminology *geometric noncommutative scheme* for what we call non-commutative smooth projective variety.

subcategory of $D^b(X_1 \times X_2)$ which is closed under taking direct summands and contains all objects of the form $F_1 \boxtimes F_2$, with $F_1 \in \mathscr{D}_1$ and $F_2 \in \mathscr{D}_2$.[3]

Proposition 2.15 *The subcategory $\mathscr{D}_1 \boxtimes \mathscr{D}_2 \subset D^b(X_1 \times X_2)$ is admissible.*

Proof This is a special case of [96, Theorem 5.8]. In our context of smooth projective varieties, it becomes quite simple.

In fact, we claim that there is a semiorthogonal decomposition

$$D^b(X_1 \times X_2) = \langle \mathscr{D}_1 \boxtimes \mathscr{D}_2, {}^\perp\mathscr{D}_1 \boxtimes \mathscr{D}_2, \mathscr{D}_1 \boxtimes {}^\perp\mathscr{D}_2, {}^\perp\mathscr{D}_1 \boxtimes {}^\perp\mathscr{D}_2 \rangle.$$

As remarked in the previous section, this immediately implies that $\mathscr{D}_1 \boxtimes \mathscr{D}_2$ is admissible, and thus a non-commutative smooth projective variety.

To prove the claim, we need to check conditions (2.1) and (2.1) of Definition 2.1. Condition (2.1) follows immediately from the Künneth formula (see, for example, [68, Section 3.3]):

$$\mathrm{Hom}(F_1 \boxtimes G_1, F_2 \boxtimes G_2) = \bigoplus_{i \in \mathbb{Z}} \mathrm{Hom}(F_1, F_2[i]) \otimes \mathrm{Hom}(G_1, G_2[-i]).$$

Condition (2.1) follows directly from the following elementary but very useful fact. Let $F \in D^b(X_1 \times X_2)$. Then since $X_1 \times X_2$ is projective, we can find a bounded above locally-free resolution $P^\bullet$ of F, such that, for all i, $P^i = P_1^i \boxtimes P_2^i$. Since $X_1 \times X_2$ is smooth, by truncating this resolution at sufficiently large n (with respect to the stupid truncation), we obtain a split exact triangle

$$G \to \sigma^{\geq -n} P^\bullet \to F,$$

for some $G \in D^b(X_1 \times X_2)$, namely F is a direct factor of $\sigma^{\geq -n} P^\bullet$. Since $\mathscr{D}_1$ and $\mathscr{D}_2$ are part of semiorthogonal decompositions of respectively $D^b(X_1)$ and $D^b(X_2)$, this concludes the proof. □

Having a notion of product, we can define the slightly technical notion of Fourier-Mukai functor between non-commutative varieties (see [73, 92], and [46] for a survey on Fourier-Mukai functors).

Definition 2.16 Let X_1, X_2 be algebraic varieties. Let $\mathscr{D}_1 \hookrightarrow D^b(X_1)$ and $\mathscr{D}_2 \hookrightarrow D^b(X_2)$ be admissible categories. A functor $F \colon \mathscr{D}_1 \to \mathscr{D}_2$ is called a *Fourier-Mukai functor* if the composite functor

$$D^b(X_1) \xrightarrow{\delta_1} \mathscr{D}_1 \to \mathscr{D}_2 \hookrightarrow D^b(X_2)$$

is of Fourier-Mukai type.

---

[3]This is different from the definition which appears in [96, Equation (10)], but in our context of smooth projective varieties it is equivalent (see also the beginning of the proof of [96, Theorem 5.8]).

In the previous definition, δ_1 denotes the projection functor of Remark 2.2 with respect to the semiorthogonal decomposition $D^b(X) = \langle \mathscr{D}_1, \mathscr{D}_2 \rangle$; it coincides with the left adjoint to the inclusion functor. By using Proposition 2.15, it is not hard to see that if a functor between non-commutative varieties is of Fourier-Mukai type, then the kernel $P_F \in D^b(X_1 \times X_2)$ actually lives in $\mathscr{D}_1^{\perp\perp} \boxtimes \mathscr{D}_2$; for example, the proof in [73, Lemma 1.5] works in general.

Remark 2.17 By [96, Theorem 6.4], the definition of $\mathscr{D}_1 \boxtimes \mathscr{D}_2$ does not depend on the embedding. More precisely, given a Fourier-Mukai equivalence $\mathscr{D}_1 \xrightarrow{\sim} \mathscr{D}_1'$, then there is a Fourier-Mukai equivalence of triangulated categories $\mathscr{D}_1 \boxtimes \mathscr{D}_2 \cong \mathscr{D}_1' \boxtimes \mathscr{D}_2$.

An immediate corollary of Proposition 2.15 is that the projection functor δ is of Fourier-Mukai type. We will use this to define Hochschild (co)homology for a non-commutative variety.

Lemma 2.18 *Let X be a smooth projective variety, and let $\mathscr{D} \subset D^b(X)$ be an admissible category. Then the projection functor $\delta \colon D^b(X) \to \mathscr{D}$ is of Fourier-Mukai type.*

Proof This is a special case of [96, Theorem 7.1]. Again, in our smooth projective context, the proof is very simple. Indeed, by Proposition 2.15, the subcategory $D^b(X) \boxtimes \mathscr{D} \subset D^b(X \times X)$ is admissible. The kernel of the projection functor is simply the projection of the structure sheaf of the diagonal $\mathscr{O}_\Delta \in D^b(X \times X)$ onto the category $D^b(X) \boxtimes \mathscr{D}$. □

Remark 2.19 Not all functors $\mathscr{D}_1 \to \mathscr{D}_2$ are of Fourier-Mukai type (see [147, 160]). Nevertheless, fully faithful functors are expected to be of Fourier-Mukai type (as they are in the commutative case, by Orlov's Representability Theorem [135]). This goes under the name of Splitting Conjecture (see [92, Conjecture 3.7]).

The Numerical Grothendieck Group Given a triangulated category $\mathscr{D}$, we denote by $K(\mathscr{D})$ its Grothendieck group. If $\mathscr{D}$ is a non-commutative smooth projective variety, then the Euler characteristic

$$\chi(F, G) := \sum_i (-1)^i \dim_k \mathrm{Hom}_{\mathscr{D}}(F, G[i])$$

is well-defined for all $F, G \in \mathscr{D}$ and it factors through $K(\mathscr{D})$.

Definition 2.20 Let $\mathscr{D}$ be a non-commutative smooth projective variety. We define the *numerical Grothendieck group* as $K_{\mathrm{num}}(\mathscr{D}) := K(\mathscr{D})/\ker \chi$.

The numerical Grothendieck group is a free abelian group of finite rank. Indeed, since $\mathscr{D}$ is an admissible subcategory of $D^b(X)$, for X a smooth projective variety, then $K_{\mathrm{num}}(\mathscr{D})$ is a subgroup of the numerical Grothendieck group $N(X)$ of X which is a free abelian group of finite rank.

The Serre Functor Serre duality for non-commutative varieties is studied via the notion of Serre functor, introduced and studied originally in [34].

Definition 2.21 Let $\mathscr{D}$ be a triangulated category. A *Serre functor* in $\mathscr{D}$ is an autoequivalence $S_{\mathscr{D}} \colon \mathscr{D} \to \mathscr{D}$ with a bi-functorial isomorphism

$$\mathrm{Hom}(F, G)^{\vee} = \mathrm{Hom}(G, S_{\mathscr{D}}(F))$$

for all $F, G \in \mathscr{D}$.

If a Serre functor exists then it is unique up to a canonical isomorphism. If X is a smooth projective variety, the Serre functor is given by $S_{\mathrm{D^b}(X)}(\_) = \_ \otimes \omega_X[\dim X]$. In general, Serre functors exist for non-commutative smooth projective varieties as well. If $\mathscr{D}$ has a Serre functor and $\Phi \colon \mathscr{C} \hookrightarrow \mathscr{D}$ is an admissible subcategory, then the Serre functor of $\mathscr{C}$ is given by the following formula

$$S_{\mathscr{C}} = \Psi_R \circ S_{\mathscr{D}} \circ \Phi \qquad \text{and} \qquad S_{\mathscr{C}}^{-1} = \Psi_L \circ S_{\mathscr{D}}^{-1} \circ \Phi,$$

where Ψ_R and Ψ_L denote respectively the right and left adjoint to the inclusion functor. Notice also that the Serre functor is of Fourier-Mukai type.

Serre functors behave nicely with respect to products. Given two non-commutative smooth projective varieties $\mathscr{D}_1 \subset \mathrm{D^b}(X_1)$ and $\mathscr{D}_2 \subset \mathrm{D^b}(X_2)$, let us denote by $P_{S_{\mathscr{D}_1}} \in \mathrm{D^b}(X_1 \times X_1)$, respectively $P_{S_{\mathscr{D}_2}} \in \mathrm{D^b}(X_2 \times X_2)$, kernels representing the Serre functors. Then the Serre functor of the product $\mathscr{D}_1 \boxtimes \mathscr{D}_2 \subset \mathrm{D^b}(X_1 \times X_2)$ is representable by $P_{S_{\mathscr{D}_1}} \boxtimes P_{S_{\mathscr{D}_2}}$. This can be proved directly (for example, the argument in [73, Corollary 1.4] works in general), since it is true for products of varieties.

We can now define non-commutative Calabi-Yau smooth projective varieties.

Definition 2.22 Let $\mathscr{D}$ be a non-commutative smooth projective variety. We say that $\mathscr{D}$ is a non-commutative *Calabi-Yau* variety of dimension n if $S_{\mathscr{D}} = [n]$.

By what we observed before, the product of two non-commutative Calabi-Yau varieties of dimension n and m is again Calabi-Yau of dimension $n + m$.

Hochschild (Co)homology The analogue of Hodge cohomology for non-commutative varieties is Hochschild homology. This can be defined naturally in the context of dg-categories. In our smooth and projective case, this can also be done directly at the level of triangulated categories, in a similar way to [44, 45, 121] for bounded derived categories of smooth projective varieties.[4] The main reference is [94], and we are content here to briefly sketch the basic properties (see also [141]).

Let $\mathscr{D}$ be a non-commutative smooth projective variety. We also fix an embedding $\mathscr{D} \hookrightarrow \mathrm{D^b}(X)$, for X a smooth projective variety; the definition will of course be independent of this choice. We let $P_\delta \in \mathrm{D^b}(X \times X)$ be a kernel of the projection

---

[4] Hochschild (co)homology for schemes was originally defined and studied in [112, 153, 161].

functor δ which is of Fourier-Mukai type, by Lemma 2.18. We also let $P_{\delta!} \in \mathrm{D}^b(X \times X)$ be the kernel of $S_{\mathscr{D}}^{-1}$, the inverse of the Serre functor of $\mathscr{D}$. For $i \in \mathbb{Z}$, we define

$$\mathrm{HH}^i(\mathscr{D}) := \mathrm{Hom}_{\mathrm{D}^b(X \times X)}(P_\delta, P_\delta[i]) \qquad \text{(Hochschild cohomology)}$$

$$\mathrm{HH}_i(\mathscr{D}) := \mathrm{Hom}_{\mathrm{D}^b(X \times X)}(P_{\delta!}[i], P_\delta) \qquad \text{(Hochschild homology)}$$

Our choice of degree for Hochschild homology is different from [94]. It is coherent, though, with the definition of Hochschild homology for varieties. Indeed, when $\mathscr{D} = \mathrm{D}^b(X)$, this gives the usual definitions

$$\mathrm{HH}^i(\mathrm{D}^b(X)) = \mathrm{Hom}_{\mathrm{D}^b(X \times X)}(\mathscr{O}_\Delta, \mathscr{O}_\Delta[i]),$$

$$\mathrm{HH}_i(\mathrm{D}^b(X)) = \mathrm{Hom}_{\mathrm{D}^b(X \times X)}(S_\Delta^{-1}[i], \mathscr{O}_\Delta),$$

where $S_\Delta^{-1} := \Delta_* \omega_X^{-1}[-\dim(X)]$.

As in the commutative case, Hochschild cohomology has the structure of a graded algebra, and Hochschild homology is a right module over it. We can also define a *Mukai pairing*

$$(\_,\_) : \mathrm{HH}_i(\mathscr{D}) \otimes \mathrm{HH}_{-i}(\mathscr{D}) \to \mathbb{K}$$

which is a non-degenerate pairing, induced by Serre duality. Roughly, by reasoning as in [43, Section 4.9], one first considers the isomorphism $\tau : \mathrm{Hom}(P_{\delta!}, P_\delta[-i]) \to \mathrm{Hom}(P_\delta, P_\delta \otimes p_2^* \omega_X[n-i])$. The latter vector space is Serre dual to $\mathrm{HH}_{-i}(\mathscr{D})$. Thus (v, w) is defined as the trace of the composition $\tau(v) \circ w$.

The triple consisting of Hochschild (co)homology and the Mukai pairing is called the *Hochschild structure* associated to $\mathscr{D}$.

Example 2.23 Let E be an exceptional object and let $\mathscr{D} = \langle E \rangle$. Then $\mathrm{HH}_\bullet(\mathscr{D}) = \mathrm{HH}^\bullet(\mathscr{D}) = \mathbb{K}$, both concentrated in degree 0.

Remark 2.24 The *Hochschild–Kostant–Rosenberg isomorphism* $\Delta_X^* \mathscr{O}_{\Delta_X} \to \bigoplus_i \Omega_X^i[i]$ yields the graded isomorphisms

$$I_{\mathrm{HKR}}^X : \mathrm{HH}_*(X) \to \mathrm{H\Omega}_*(X) := \bigoplus_i \mathrm{H\Omega}_i(X)$$
$$\text{and} \quad I_X^{\mathrm{HKR}} : \mathrm{HH}^*(X) \to \mathrm{HT}^*(X) := \bigoplus_i \mathrm{HT}^i(X)$$

where $\mathrm{H\Omega}_i(X) := \bigoplus_{q-p=i} H^p(X, \Omega_X^q)$ and $\mathrm{HT}^i(X) := \bigoplus_{p+q=i} H^p(X, \wedge^q T_X)$.

The existence of such isomorphisms is due to Swan [153] for fields of characteristic zero (see also [45]). In [162] it is proved that if X is a smooth scheme defined over a field of characteristic $p > \dim X$, then the same result holds. In the recent paper [9], this was extended (in a slightly weaker form) to smooth proper schemes over fields of characteristic $p \geq \dim X$.

A key property of Hochschild homology is that it behaves well with respect to semiorthogonal decompositions.

Proposition 2.25 *Let $\mathscr{D}$ be a non-commutative variety, and let*

$$\mathscr{D} = \langle \mathscr{D}_1, \ldots, \mathscr{D}_r \rangle$$

be a semiorthogonal decomposition. Then, for all $i \in \mathbb{Z}$,

$$\mathrm{HH}_i(\mathscr{D}) \cong \bigoplus_{j=1}^{r} \mathrm{HH}_i(\mathscr{D}_j).$$

Moreover, this decomposition is orthogonal with respect to the Mukai pairing.

Proof The first statement is [94, Corollary 7.5]. Here, for simplicity, we sketch the proof for the case in which $\mathscr{D} = \mathrm{D}^b(X)$ and we have a semiorthogonal decomposition $\mathrm{D}^b(X) = \langle \mathscr{C}, {}^{\perp}\mathscr{C} \rangle$.

We observed above that the projection functors δ_1 and δ_2 for the admissible subcategories $\mathscr{C}$ and ${}^{\perp}\mathscr{C}$ respectively are of Fourier-Mukai type with kernels P_1 and P_2. These objects sit in the following distinguished triangle

$$P_2 \to \mathscr{O}_{\Delta} \to P_1.$$

Similarly, we have a triangle

$$P_{2!} \to S_{\Delta}^{-1} \to P_{1!}.$$

Since, by semiorthogonality, $\mathrm{Hom}(P_{2!}, P_1[j]) = 0$, for all $j \in \mathbb{Z}$, we have an induced map

$$\mathrm{Hom}(S_{\Delta}^{-1}[i], \mathscr{O}_{\Delta}) \longrightarrow \mathrm{Hom}(P_{1!}[i], P_1) \oplus \mathrm{Hom}(P_{2!}[i], P_2).$$

By using Serre duality, it is not hard to see that $\mathrm{Hom}(P_{1!}, P_2[j]) = 0$, for all $j \in \mathbb{Z}$ as well; hence, the above map is an isomorphism. Finally, by definition, $\mathrm{HH}_i(\mathscr{D}_1) = \mathrm{Hom}(P_{1!}[i], P_1)$; regarding the second factor, it is a small argument (see [94, Corollary 3.12]) to see that $\mathrm{HH}_i(\mathscr{D}_2) \cong \mathrm{Hom}(P_{2!}[i], P_2)$.

For the second statement, one needs to use the previous construction of the morphism, together with the isomorphism $\mathrm{Hom}(P_{\delta!}, P_{\delta}[-i]) \cong \mathrm{Hom}(P_{\delta}, P_{\delta} \otimes p_2^* \omega_X[n - i])$ mentioned before. $\qquad\square$

Example 2.26 Let $\{E_1, \ldots, E_m\}$ be an exceptional collection, and let $\mathscr{D} = \langle E_1, \ldots, E_m \rangle$. Then $\mathrm{HH}_{\bullet}(\mathscr{D}) = \mathbb{K}^{\oplus m}$, concentrated in degree 0.

From Proposition 2.25, we can deduce all other properties of the Hochschild structure.

Theorem 2.27 *Let $\mathscr{C}$, $\mathscr{D}$, $\mathscr{E}$ be non-commutative smooth projective varieties.*

(1) *Any Fourier-Mukai functor $\Phi \colon \mathscr{C} \to \mathscr{D}$ induces a morphism of graded k-vector spaces $\Phi_{HH} \colon HH_\bullet(\mathscr{C}) \to HH_\bullet(\mathscr{D})$ such that $\mathrm{Id}_{HH} = \mathrm{Id}$ and, given another functor $\Psi \colon \mathscr{D} \to \mathscr{E}$, we have $(\Psi \circ \Phi)_{HH} = \Psi_{HH} \circ \Phi_{HH}$.*

(2) *If (Ψ, Φ) is a pair of adjoint Fourier-Mukai functors, then*

$$\left(\_, \Phi_{HH}(\_)\right) = \left(\Psi_{HH}(\_), \_\right).$$

(3) *There is a Chern character $\mathrm{ch} \colon K(\mathscr{D}) \to HH_0(\mathscr{D})$ such that, for all $F, G \in \mathscr{D}$,*

$$(\mathrm{ch}(F), \mathrm{ch}(G)) = -\chi(F, G).$$

(4) *The Hochschild structure is invariant under exact equivalences of Fourier-Mukai type.*

Proof As for (1), since Φ is a Fourier-Mukai functor, it is induced by a Fourier-Mukai functor $\Psi \colon D^b(X_1) \to D^b(X_2)$, where $\mathscr{C}$ and $\mathscr{D}$ are admissible subcategories of $D^b(X_1)$ and $D^b(X_2)$ respectively. Let E be the Fourier-Mukai kernel of Ψ. It is not difficult to see that Ψ induces a morphism $\Psi_{HH} \colon HH_\bullet(X_1) \to HH_\bullet(X_2)$. For a given i consider $\mu \in HH_i(X_1)$ and define $\Psi_{HH}(\mu) \in HH_i(X_2)$ in the following way. We consider the composition

$$S_{\Delta_{X_2}}^{-1}[i] \xrightarrow{\gamma} E \circ E^\vee[i] \xrightarrow{\mathrm{id} \circ \eta \circ \mathrm{id}} E \circ S_{\Delta_{X_1}}^{-1}[i] \circ S_{\Delta_{X_1}} \circ E^\vee \xrightarrow{\mathrm{id} \circ \mu \circ \mathrm{id} \circ \mathrm{id}} E \circ S_{\Delta_{X_1}} \circ E^\vee \xrightarrow{h} \mathscr{O}_{\Delta_{X_2}},$$

where $\eta \colon \mathscr{O}_{\Delta_{X_1}} \to S_{\Delta_{X_1}}^{-1} \circ S_{\Delta_{X_1}}$ is the isomorphism coming from the easy fact that $\Phi_{S_{\Delta_{X_1}}^{-1}} \circ \Phi_{S_{\Delta_{X_1}}} = \mathrm{id}$ and the morphisms γ and h are the natural ones.

By Proposition 2.25, $HH_\bullet(\mathscr{D})$ is an orthogonal factor of $HH_\bullet(X_2)$. So we set Φ_{HH} in (1) to be the composition of $\Psi_{HH}|_{HH_\bullet(\mathscr{C})}$ with the orthogonal projection onto $HH_\bullet(\mathscr{D})$.

With this definition, (2) follows from a straightforward computation. As for (3), following [43], let us recall that, given $E \in \mathscr{D}$, we can think of E as an object on $\mathrm{pt} \times X_2$ and just set $\mathrm{ch}(E) := (\Phi_E)_{HH}(1)$, where $(\Phi_E)_{HH} \colon HH_0(\mathrm{pt})(\cong \mathbb{K}) \to HH_0(X_2)$. Since $E \in \mathscr{D}$, we have $\mathrm{ch}(E) \in HH_0(\mathscr{D})$. The compatibility between the Mukai pairing and χ is proven as in [43, Theorem 7.6] (the change of sign is harmless here).

For the last statement and more details on the first three, the reader can consult [94, Section 7]. □

Notice, in particular, that by Theorem 2.27,(2.27), the Hochschild structure does not depend on the choice of the embedding $\mathscr{D} \hookrightarrow D^b(X)$.

For Calabi-Yau varieties, Hochschild homology and cohomology coincide, up to shift. This is the content of the following result (see [98, Section 5.2]). With our definition, the proof is immediate.

Proposition 2.28 *Let $\mathscr{D}$ be a non-commutative Calabi-Yau variety of dimension n. Then, for all $i \in \mathbb{Z}$, we have*

$$\mathrm{HH}^i(\mathscr{D}) \cong \mathrm{HH}_{n-i}(\mathscr{D}).$$

In particular, $\mathrm{HH}_n(\mathscr{D}) \neq 0.$

Proof We only need to show the last statement. But, by definition, $\mathrm{HH}^0(\mathscr{D}) \neq 0$, since $\mathrm{Id}_{P_\delta} \in \mathrm{Hom}(P_\delta, P_\delta)$. $\quad\square$

We can use Hochschild cohomology to define the notion of connectedness for non-commutative varieties.

Definition 2.29 Let $\mathscr{D}$ be a non-commutative smooth projective variety. We say that $\mathscr{D}$ is *connected* if $\mathrm{HH}^0(\mathscr{D}) = \mathbb{K}$.

Lemma 2.30 (Bridgeland's Trick) *Let $\mathscr{D}$ be a non-commutative Calabi-Yau variety of dimension n. If $\mathscr{D}$ is connected, then $\mathscr{D}$ is indecomposable, namely it does not admit any non-trivial semiorthogonal decomposition.*

Proof Let $\mathscr{D} = \langle \mathscr{D}_1, \mathscr{D}_2 \rangle$ be a non-trivial semiorthogonal decomposition. Since $\mathscr{D}$ is a non-commutative Calabi-Yau variety of dimension n, then both $\mathscr{D}_1$ and $\mathscr{D}_2$ are non-commutative Calabi-Yau varieties of dimension n as well. In particular, by Proposition 2.28, $\mathrm{HH}_n(\mathscr{D}_i) \neq 0$, for $i = 1, 2$. But then, by Proposition 2.25, we have

$$\mathbb{K} = \mathrm{HH}^0(\mathscr{D}) = \mathrm{HH}_n(\mathscr{D}) = \mathrm{HH}_n(\mathscr{D}_1) \oplus \mathrm{HH}_n(\mathscr{D}),$$

which is a contradiction. $\quad\square$

We use Hochschild (co)homology to finally define non-commutative K3 surfaces. Recall that for a K3 surface S, Hochschild (co)homology can be easily computed, by using the Hochschild-Kostant-Rosenberg Theorem:

$$\mathrm{HH}_\bullet(S) = \mathbb{K}[-2] \oplus \mathbb{K}^{\oplus 22} \oplus \mathbb{K}[2].$$

Definition 2.31 Let $\mathscr{D}$ be a non-commutative smooth projective variety. We say that $\mathscr{D}$ is a *non-commutative K3 surface* if $\mathscr{D}$ is a non-commutative connected Calabi-Yau variety of dimension 2 and its Hochschild (co)homology coincides with the Hochschild (co)homology of a K3 surface.

2.3 Constructing Non-commutative Calabi-Yau Varieties

The goal of this section is to present a result by Kuznetsov which covers essentially all currently known examples of non-commutative Calabi-Yau varieties. We will then study examples of non-commutative K3 surfaces in details in Sect. 3, and then concentrate on the case of those non-commutative K3 surfaces associated to cubic fourfolds in Sect. 5.

We will need first to introduce a bit more terminology. First of all, we recall the notion of spherical functor ([1, 7, 8, 122, 148]). We will follow the definition in [98, Section 2.5].

Given a functor $\Phi \colon \mathscr{C} \to \mathscr{D}$, we keep following the convention of denoting by Ψ_R and Ψ_L respectively its right and left adjoint functors (if they exist).

Definition 2.32 Let X, Y be smooth projective varieties and let $\Phi \colon D^b(X) \to D^b(Y)$ be a Fourier-Mukai functor. We say that Φ is *spherical* if

(1) the natural transformation $\Psi_L \oplus \Psi_R \to \Psi_R \circ \Phi \circ \Psi_L$, induced by the sum of the two units of the adjunction, is an isomorphism.
(2) the natural transformation $\Psi_L \circ \Phi \circ \Psi_R \to \Psi_L \oplus \Psi_R$, induced by the sum of the two counits of the adjunction, is an isomorphism.

Given a spherical functor, we can define the associated *spherical twist* functors:

$$T_X \colon D^b(X) \to D^b(X), \qquad T_X(F) := \operatorname{cone}(\Psi_L \circ \Phi(F) \to F)$$

$$T_Y \colon D^b(Y) \to D^b(Y), \qquad T_Y(G) := \operatorname{cone}(G \to \Phi \circ \Psi_L(G))[-1]$$

The key result about spherical functors is the following.

Proposition 2.33 *Let* $\Phi \colon D^b(X) \to D^b(Y)$ *be a spherical functor. Then the twist functors* T_X *and* T_Y *are autoequivalences. Moreover, we have*

$$\Phi \circ T_X = T_Y \circ \Phi \circ [2].$$

Proof With the above definition, this is proved in detail in [98, Proposition 2.13 & Corollary 2.17]. The idea of the proof is not hard: if we define

$$T'_X \colon D^b(X) \to D^b(X), \qquad T'_X(F) := \operatorname{cone}(F \to \Psi_R \circ \Phi(F))[-1]$$

$$T'_Y \colon D^b(Y) \to D^b(Y), \qquad T'_Y(G) := \operatorname{cone}(\Phi \circ \Psi_R(G) \to G),$$

we can show with a direct argument that T_X and T'_X (respectively, T_Y and T'_Y) are mutually inverse autoequivalences. The last formula follows then easily from the definitions.   □

Secondly, we recall the important notion of rectangular Lefschetz decomposition (which is fundamental for Homological Projective Duality; see [92, 140]).

Definition 2.34 Let X be a smooth projective variety, and let L be a line bundle on X. A *Lefschetz decomposition* of $D^b(X)$ with respect to L is a semiorthogonal decomposition of the form

$$D^b(X) = \langle \mathscr{B}_0, \mathscr{B}_1 \otimes L, \ldots, \mathscr{B}_{m-1} \otimes L^{\otimes(m-1)} \rangle, \qquad \mathscr{B}_0 \supseteq \mathscr{B}_1 \supseteq \ldots \supseteq \mathscr{B}_{m-1}.$$

A Lefschetz decomposition is called *rectangular* if $\mathscr{B}_0 = \mathscr{B}_1 = \ldots = \mathscr{B}_{m-1}$.

To simplify notation, given a line bundle L on a smooth projective variety X, we denote by $L \colon D^b(X) \xrightarrow{\sim} D^b(X)$ the autoequivalence given by tensoring by L.

We can now state the main result for this section. This is our setup.

Setup 2.35 Let M and X be smooth projective varieties and let L_M and L_X be line bundles on M and X respectively. Let $m, d \in \mathbb{Z}$ be such that $1 \leq d < m$. We assume:

(1) $D^b(M)$ has a rectangular Lefschetz decomposition with respect to L_M

$$D^b(M) = \langle \mathscr{B}_M, \mathscr{B}_M \otimes L_M, \dots, \mathscr{B}_M \otimes L_M^{\otimes(m-1)} \rangle,$$

(2) there is a spherical functor $\Phi \colon D^b(X) \to D^b(M)$

which satisfy the following compatibilities:

  (i) $L_M \circ \Phi = \Phi \circ L_X$;
  (ii) $L_X \circ T_X = T_X \circ L_X$;
  (iii) $T_M(\mathscr{B}_M \otimes L_M^{\otimes i}) = \mathscr{B}_M \otimes L_M^{\otimes i-d}$, for all $i \in \mathbb{Z}$.

Proposition 2.36 *The left adjoint functor Ψ_L induces a fully faithful functor* $\mathscr{B}_M \hookrightarrow D^b(X)$.

Proof This is a direct check; see [98, Lemma 3.10] for the details. $\qquad\square$

We set $\mathscr{B}_X := \Psi_L(\mathscr{B}_M)$. By Proposition 2.36, and by using properties (i) and (iii), we have a semiorthogonal decomposition

$$D^b(X) = \langle \mathscr{K}u(X), \mathscr{B}_X, \mathscr{B}_X \otimes L_X, \dots, \mathscr{B}_X \otimes L_X^{\otimes(m-d-1)} \rangle, \tag{2}$$

where $\mathscr{K}u(X)$ is defined as

$$\mathscr{K}u(X) := \langle \mathscr{B}_X, \mathscr{B}_X \otimes L_X, \dots, \mathscr{B}_X \otimes L_X^{\otimes(m-d-1)} \rangle^{\perp}.$$

Definition 2.37 We say that $\mathscr{K}u(X)$ is the *Kuznetsov component* of X associated to our data: M, Φ, and the rectangular Lefschetz decomposition of $D^b(M)$.

Let us define the two autoequivalences:

$$\rho \colon D^b(X) \xrightarrow{\sim} D^b(X) \qquad\qquad \rho := T_X \circ L_X^{\otimes d}$$

$$\sigma \colon D^b(X) \xrightarrow{\sim} D^b(X) \qquad\qquad \sigma := S_{D^b(X)} \circ T_X \circ L_X^{\otimes m}$$

Lemma 2.38 *We keep our assumptions as in Setup 2.35. We have:*

(1) $\sigma \circ \rho = \rho \circ \sigma$;
(2) $S_{D^b(X)}^{-1} = L_X^{\otimes m} \circ T_X \circ \sigma^{-1}$;
(3) σ and ρ respect the semiorthogonal decomposition (2).

Proof The first statement follows immediately from the property (ii) and the fact that the Serre functor commutes with any autoequivalence. The second statement follows then immediately. For the last one, one can check it directly; see [98, Lemma 3.11] for the details.   □

Theorem 2.39 (Kuznetsov) *Let* $c := \gcd(d, m)$. *The Serre functor of the Kuznetsov component can be expressed as*

$$S_{\mathscr{K}u(X)}^{d/c} = \rho^{-m/c} \circ \sigma^{d/c}.$$

To prove the theorem, we introduce a fundamental functor for the Kuznetsov component, the *degree shift* functor[5]:

$$O_{\mathscr{K}u(X)} \colon \mathscr{K}u(X) \to \mathscr{K}u(X) \qquad O_{\mathscr{K}u(X)} := \delta_{\mathscr{K}u(X)} \circ L_X,$$

where $\delta_{\mathscr{K}u(X)}$ denotes as usual the projection onto the Kuznetsov component (or equivalently, the left adjoint functor of the inclusion).

Lemma 2.40 *We keep our assumptions as in Setup 2.35. We have:*

(1) $O_{\mathscr{K}u(X)}$ *is an autoequivalence;*
(2) $O_{\mathscr{K}u(X)} \circ \rho = \rho \circ O_{\mathscr{K}u(X)}$ *and* $O_{\mathscr{K}u(X)} \circ \sigma = \sigma \circ O_{\mathscr{K}u(X)}$;
(3) $O_{\mathscr{K}u(X)}^i = \delta_{\mathscr{K}u(X)} \circ L_X^{\otimes i}$, *for all* $0 \le i \le m - d$;
(4) $S_{\mathscr{K}u(X)}^{-1} = O_{\mathscr{K}u(X)}^{m-d} \circ \rho \circ \sigma^{-1}$;
(5) $O_{\mathscr{K}u(X)}^d = \rho$.

Proof This is the summary of various results in [98, Section 3]. Property (1) follows by either (4) or (5). Property (2) follows by a direct check. Property (4) follows from (3) and by Lemma 2.38, (2). The key results are (3) and (5).

To prove (3), observe that the formula is true for $i = 0$. Let us assume the formula is true for $0 \le i < m - d$; we want to show it is true for $i + 1$ as well. Let $F \in \mathscr{K}u(X)$. We can consider the composition

$$F \otimes L_X^{\otimes(i+1)} \to \delta_{\mathscr{K}u(X)}(F \otimes L_X^{\otimes i}) \otimes L_X = O_{\mathscr{K}u(X)}^i(F) \otimes L_X \to$$

$$\to \delta_{\mathscr{K}u(X)}(O_{\mathscr{K}u(X)}^i(F) \otimes L_X) = O_{\mathscr{K}u(X)}^{i+1}(F).$$

We need to show that $\delta_{\mathscr{K}u(X)}(F \otimes L_X^{\otimes(i+1)}) = O_{\mathscr{K}u(X)}^{i+1}(F)$, or equivalently that the cone G of the above composition is in $\langle \mathscr{B}_X, \mathscr{B}_X \otimes L_X, \ldots, \mathscr{B}_X \otimes L_X^{\otimes(m-d-1)} \rangle$. But, by the octahedral axiom, the cone is an extension of two objects

$$G_1 \otimes L_X \to G \to G_2,$$

---

[5]In [98] the functor O is defined on the whole derived category $D^b(X)$ and it is called *rotation functor*.

where $G_1 \in \langle \mathscr{B}_X, \mathscr{B}_X \otimes L_X, \ldots, \mathscr{B}_X \otimes L_X^{\otimes(i-1)} \rangle$ and $G_2 \in \mathscr{B}_X$, which is what we wanted.

We will only show (5) under the assumption $d \le m - d$. Since we are interested in the Calabi-Yau case (where, in particular, we will have that $c = d$ divides m), this is enough for us. By (3), we have that $O^d_{\mathscr{K}u(X)} = \delta_{\mathscr{K}u(X)} \circ L_X^{\otimes d}$. By definition, we also have that $\rho = T_X \circ L_X^{\otimes d}$.

Let $F \in \mathscr{K}u(X)$. Then $\rho(F)$ is defined by the following triangle

$$\Psi_L(\Phi(F \otimes L_X^{\otimes d})) \to F \otimes L_X^{\otimes d} \to \rho(F).$$

By Lemma 2.38,(3), we know that $\rho(F) \in \mathscr{K}u(X)$. Hence, we only need to show that $\Psi_L(\Phi(F \otimes L_X^{\otimes d})) \in {}^{\perp}\mathscr{K}u(X)$.

By using adjointness, it is easy to see that $\Phi(F \otimes L_X^{\otimes d}) \in \langle \mathscr{B}_M, \ldots, \mathscr{B}_M \otimes L_M^{\otimes d-1} \rangle$. The adjoint of property (i) shows that, for all $i \in \mathbb{Z}$,

$$\Psi_L(\mathscr{B}_M \otimes L_M^{\otimes i}) = \Psi_L(\mathscr{B}_M) \otimes L_X^{\otimes i} = \mathscr{B}_X \otimes L_X^{\otimes i}.$$

Hence

$$\Psi_L(\Phi(F \otimes L_X^{\otimes d})) \in \langle \mathscr{B}_X, \ldots, \mathscr{B}_X \otimes L_X^{\otimes d-1} \rangle \subset {}^{\perp}\mathscr{K}u(X),$$

as we wanted. □

We can now prove Kuznetsov's theorem.

Proof of Theorem 2.39 By Lemma 2.40,(4), we can express the (inverse of the) Serre functor in terms of the functors $O_{\mathscr{K}u(X)}$, ρ, and σ. By Lemma 2.38,(1), all these functors commute. By raising everything to the power d/c, and by using Lemma 2.40,(5), the statement follows. □

3 Fano Varieties and Their Kuznetsov Components: Examples

In this section we present a few examples of non-commutative K3 surfaces. The basic references are [95, 99, 104]. There are very interesting examples of non-commutative Calabi-Yau varieties in higher dimension as well; we will not cover them in these notes and we refer to [79, 80, 98] and references therein.

The general goal could be stated as follows.

Question 3.1 How to construct examples of non-commutative K3 surfaces? Is there a generalized period map and a (derived) Torelli Theorem? What is the image of the period map?

Already the first part of Question 3.1 is not easy to answer. The main issue is that the only way we have to construct non-commutative K3 surfaces is by embedding them in some (commutative) Fano variety. Currently, there are a few families which have been studied. We will present three of them in this section: cubic fourfolds, Gushel-Mukai manifolds, and Debarre-Voisin manifolds. All of them arise indeed as Kuznetsov components in the derived category of a certain smooth Fano variety, as in Theorem 2.39. Other examples, not covered in these notes, are Küchle manifolds [101]; in such examples, though, a rectangular Lefschetz decomposition is not yet known and Theorem 2.39 not yet applicable directly.

There is a common theme and expectation that such non-commutative K3 surfaces should contain deep birational properties on the Fano variety itself. We can then formulate the following question.

Question 3.2 (Kuznetsov) Let W be a Fano fourfold. Assume that W has a Kuznetsov component $\mathcal{K}u(W)$ which is a non-commutative K3 surface. If W is rational, then is $\mathcal{K}u(W)$ equivalent to the derived category of a K3 surface?

The above question is not well-defined in general, since there is no invariant definition yet of what a Kuznetsov component is for a general fourfold W (see [99, Section 3]). On the other hand, in the examples we will see (cubic fourfolds and Gushel-Mukai fourfolds), there is an evident choice for it, and the above question therefore makes sense.

Question 3.2 also motivates the understanding of when such a Kuznetsov component is actually equivalent to the derived category of a K3 surface. We can then formulate the following question, which we are going to answer in the case of cubic fourfolds (see Theorem 3.7).

Question 3.3 (Addington-Thomas, Huybrechts) Let W be a Fano variety. Assume that W has a Kuznetsov component $\mathcal{K}u(W)$ which is a non-commutative K3 surface. Is it true that $\mathcal{K}u(W)$ is equivalent to the derived category of a K3 surface if and only if there is a primitive embedding of the hyperbolic lattice $U \hookrightarrow K_{\mathrm{num}}(\mathcal{K}u(W))$ in the numerical Grothendieck group of W?

The hyperbolic lattice U in Question 3.3 has rank 2, is even unimodular and is defined by the bilinear form

$$\begin{pmatrix} 0 & 1 \\ 1 & 0 \end{pmatrix}.$$

It has a very neat interpretation in terms of moduli spaces of objects in $\mathcal{K}u(W)$. Indeed, the two square-zero classes correspond to skyscraper sheaves and ideal sheaves of points on the K3 surface. Also, the fact that the two classes have intersection 1 corresponds to the fact that both are fine moduli spaces. This is the way we will approach this question in Sect. 5: we will recover the K3 surface as moduli space of stable objects in $\mathcal{K}u(W)$, and use a universal family to induce the derived equivalence.

In this section we will discuss the above questions in three examples. Cubic fourfolds will be in Sect. 3.1; Gushel-Mukai manifolds in Sect. 3.2; Debarre-Voisin manifolds in Sect. 3.3. In Sect. 3.4 we review the integral Mukai structure in these examples, by using topological K-theory, as in [4]. Finally, in Sect. 3.5 we discuss Torelli-type statements.

Finally, for sake of completeness, we mention that conjecturally all non-commutative smooth projective varieties are expected to be admissible subcategories in a Fano manifold (we refer to [31, 60, 85, 86, 128] for recent advances on this conjecture).

Question 3.4 (Bondal) Let $\mathscr{D}$ be a non-commutative smooth projective variety. Does there exists a Fano manifold W and a fully faithful functor $\mathscr{D} \hookrightarrow \mathrm{D}^b(W)$?

3.1 Cubic Fourfolds

Let $\iota\colon W \hookrightarrow \mathbb{P}^5$ be a cubic fourfold over $\mathbb{K}$; we assume $\mathrm{char}(\mathbb{K}) \neq 2, 3$. We denote by $\mathscr{O}_W(1)$ the hyperplane section $\mathscr{O}_{\mathbb{P}^5}(1)|_W$.

The Kuznetsov Component We start by using Theorem 2.39 to show that the Kuznetsov component of a cubic fourfold is a non-commutative K3 surface.

Lemma 3.5 *The functor* $\iota_*\colon \mathrm{D}^b(W) \to \mathrm{D}^b(\mathbb{P}^5)$ *is spherical. The associated spherical twists are* $T_W = \mathscr{O}_W(-3)[2]$ *and* $T_{\mathbb{P}^5} = \mathscr{O}_{\mathbb{P}^5}(-3)$.

Proof All the statements can be checked by using the exact sequence

$$0 \to \mathscr{O}_{\mathbb{P}^5}(-3) \to \mathscr{O}_{\mathbb{P}^5} \to \mathscr{O}_W \to 0$$

and a direct computation.   □

By using Lemma 3.5 and the semiorthogonal decomposition

$$\mathrm{D}^b(\mathbb{P}^5) = \langle \mathscr{O}_{\mathbb{P}^5}, \mathscr{O}_{\mathbb{P}^5}(1), \mathscr{O}_{\mathbb{P}^5}(2), \mathscr{O}_{\mathbb{P}^5}(3), \mathscr{O}_{\mathbb{P}^5}(4), \mathscr{O}_{\mathbb{P}^5}(5) \rangle,$$

it is immediate to check that the compatibilities in Setup 2.35 are met ($d = 3$ and $m = 6$). Hence, we have a semiorthogonal decomposition

$$\mathrm{D}^b(W) = \langle \mathscr{K}u(W), \mathscr{O}_W, \mathscr{O}_W(1), \mathscr{O}_W(2) \rangle$$

and, by Theorem 2.39, the Serre functor $S_{\mathscr{K}u(W)} = [2]$. Hence, $\mathscr{K}u(W)$ is a non-commutative 2-Calabi-Yau category.

We can also compute Hochschild homology of W directly, by using the Hochschild-Kostant-Rosenberg Theorem, and the Hodge diamond for W:

$$\mathrm{HH}_\bullet(W) = \mathbb{K}[-2] \oplus \mathbb{K}^{\oplus 25} \oplus \mathbb{K}[2].$$

By Proposition 2.25, the Hochschild homology of $\mathcal{K}u(W)$ is therefore isomorphic
to the one of a K3 surface. Therefore, $\mathcal{K}u(W)$ is an example of a non-commutative
K3 surface.

Pfaffian Cubic Fourfolds Toward understanding both Question 3.2 and Question 3.3 for cubic fourfolds, the first example to analyze in detail is the case of
Pfaffian cubic fourfolds. They are all contained in a special divisor in the moduli
space of cubic fourfolds.

Let V be a $\mathbb{K}$-vector space of dimension 6. For $i = 2, 4$, let $\mathrm{Pf}(i, V)$ be the closed
subset of $\mathbb{P}(\Lambda^2 V)$ consisting of those forms having rank $\leq i$. Let $L \subset \mathbb{P}(\Lambda^2 V)$ be
a linear subspace of dimension 8. We set

$$S := \mathrm{Pf}(2, V) \cap L \qquad \text{and} \qquad W := \mathrm{Pf}(4, V^{\vee}) \cap L^{\perp}.$$

For a general L, both S and W are smooth: S is a K3 surface of degree 14 and
W a cubic fourfold. We call all cubic fourfolds obtained in this way *Pfaffian cubic
fourfolds*, and the K3 surface the *associated K3 surface*.

We define the correspondence

$$\Gamma := \{(s, w) \in S \times W \; : \; s \cap \ker(w) \neq 0\},$$

with the natural projections $p_S \colon \Gamma \to S$ and $p_W \colon \Gamma \to W$. The above definition
makes sense in view of the observation that $\mathrm{Pf}(2, V) = \mathrm{Gr}(2, V)$ and thus the points
of S are actually 2-dimensional subspaces of V. Moreover, we think of w as a point
of $\mathbb{P}(\Lambda^2 V^{\vee})$ so that $\ker(w)$ is also a subspace of V. We remark here that even though
the expected codimension of Γ is 3 a direct computation shows that it is actually 2.

Proposition 3.6 (Kuznetsov) *Let W be a Pfaffian cubic fourfold, and let S be
the associated K3 surface. Then the ideal sheaf $\mathscr{I}_{\Gamma}$ induces a Fourier-Mukai
equivalence*

$$\Phi_{\mathscr{I}_{\Gamma} \otimes p_W^* \mathcal{O}_W(1)} \colon \mathrm{D}^b(S) \xrightarrow{\sim} \mathcal{K}u(W) \subset \mathrm{D}^b(W).$$

Proof This is the content of [90, Theorem 2]. We follow the presentation given in
[3, Proposition 3], and we refer there for all details.

The argument goes as follows, under the additional assumption that L is general
(which is enough for our future purposes). In this case, indeed, S does not contain a
line and W does not contain a plane.

Consider two distinct points $p_1, p_2 \in S$ and set $\Gamma_i := p_S^{-1}(p_i)$. Note that Γ_i is
a quartic scroll, for $i = 1, 2$. Since $p_1 \neq p_2$, we have that Γ_1 and Γ_2 are distinct.
Indeed, if we identify p_i with the subspace it parametrizes, we have $p_1 \cap p_2 = \{0\}$
because, otherwise, S would contain a line. This implies that if $\Gamma_1 = \Gamma_2$, then the
maps $\pi_i \colon \Gamma_i \to \mathbb{P}(p_i)$ mapping w to $p_i \cap \ker(w)$ would define two different rulings
on $\Gamma_1 = \Gamma_2$. This is not possible.

To show that $\Phi := \Phi_{\mathscr{I}_\Gamma \otimes p_W^* \mathscr{O}_W(1)}$ is fully faithful, one just applies the standard criterion due to Bondal and Orlov (see, for example, [68, Proposition 7.1]). In particular, for $p_1, p_2 \in S$, we have to prove that

$$\dim \operatorname{Hom}(\Phi(\mathscr{O}_{p_1}), \Phi(\mathscr{O}_{p_2})[i]) = \dim \operatorname{Hom}(\mathscr{O}_{p_1}, \mathscr{O}_{p_2}[i]).$$

A simple computation shows that $\Phi(\mathscr{O}_{p_i}) = \mathscr{I}_{\Gamma_i}(1)$. Thus, the equality above can be rewritten as

$$\dim \operatorname{Hom}(\mathscr{I}_{\Gamma_1}(1), \mathscr{I}_{\Gamma_2}(1)[i]) = \dim \operatorname{Hom}(\mathscr{O}_{p_1}, \mathscr{O}_{p_2}[i]).$$

The equality is clearly trivial when $i < 0$. On the other hand, Γ_i has codimension 2 and Γ_1 and Γ_2 are distinct if $p_1 \neq p_2$. Hence the equality holds for $i = 0$ as well. Since the Serre functor of $\mathscr{K}u(W)$ is the shift by 2, the same results holds true for $i = 2$. Since $\chi(\mathscr{I}_{\Gamma_1}, \mathscr{I}_{\Gamma_2}) = 0$, the case $i = 1$ follows as well.

This implies that Φ is an equivalence, since we observed that $\mathscr{K}u(W)$ is a connected Calabi-Yau category of dimension 2 and thus cannot have a proper admissible subcategory. □

These cubic fourfolds are rational, as proven in [27, Proposition 5 ii)]. The argument goes as follows. Take V' a general codimension 1 linear subspace in V. The assignment that sends $w \in W$, to $\ker(w) \cap V'$ defines a birational map

$$W \dashrightarrow \mathbb{P}(V'),$$

which gives the rationality of W.

Cubic Fourfolds and K3 Surfaces In Sect. 5, we will develop the theory of moduli spaces for the Kuznetsov component of a cubic fourfold. This will allow us to give a complete answer to Question 3.3 for cubic fourfolds:

Theorem 3.7 (Addington–Thomas, Bayer–Lahoz–Macrì–Nuer–Perry–Stellari)
Let W be a cubic fourfold. Then $\mathscr{K}u(W)$ is equivalent to the derived category of a K3 surface if and only if there is a primitive embedding of the hyperbolic lattice $U \hookrightarrow K_{\mathrm{num}}(W)$ in the numerical Grothendieck group of W.

At the lattice level, the condition $U \hookrightarrow K_{\mathrm{num}}(\mathscr{K}u(W))$ implies that W is *special*, in the sense of Hassett [66]. Roughly speaking, a cubic fourfold W is *special* if $H^4(W, \mathbb{Z}) \cap H^{2,2}(W)$ contains the class of a surface which is not homologous to the self-intersection H^2 of a hyperplane class in W. These special cubic fourfolds organize themselves in divisors of Noether-Lefschetz type.

Theorem 3.7 was first proved by Addington and Thomas in [4] generically on these divisors. The completion of their result is in [23]. Both results though rely on Proposition 3.6 (or a variant of it, for cubic fourfolds containing a plane; see Remark 3.8 below).

Remark 3.8 An analogous result can be proved, to characterize cubic fourfolds W for which $\mathcal{K}u(W)$ is equivalent to $D^b(S, \alpha)$, where S is a K3 surface and α is an element in the Brauer group $\mathrm{Br}(S) := H^2(S, \mathcal{O}_S^*)_{\mathrm{tor}}$ of S. Given a twisted K3 surface, i.e., a pair (S, α) as above, we can define the abelian category $\mathrm{coh}(S, \alpha)$ of α-twisted coherent sheaves on S (see [42, Chapter 1] for an extensive introduction). We set $D^b(S, \alpha) := D^b(\mathrm{coh}(S, \alpha))$.

Following [95], it is not difficult to construct examples where $\mathcal{K}u(W) \cong D^b(S, \alpha)$. Indeed, consider a generic cubic fourfold W containing a plane P. Let P' another plane in $\mathbb{P}^5$ which is skew with respect to P. Let $\pi_P : W \dashrightarrow P'$ be the natural projection from P. Given $p \in P'$, the preimage $\pi_P^{-1}(p)$ is the union of P and a quadric Q_p. By blowing-up $\widetilde{\pi}_P : \widetilde{W} \to P'$, we get a quadric fibration. Given the double nature of $\widetilde{W}$ as a blow-up and as a quadric fibration, one can combine Examples 2.12 and 2.13 and show that $\mathcal{K}u(W) \cong D^b(P', \mathcal{B}_0)$.

Back to the geometric setting and due to the genericity assumption on W, the quadric Q_p is singular if and only if p belongs to a smooth sextic $C \subseteq P'$. The double cover S of P' ramified along C is a smooth K3 surface and the quadric fibration provided by π_P yields a natural class $\alpha \in \mathrm{Br}(S)$. Moreover $D^b(P', \mathcal{B}_0) \cong D^b(S, \alpha)$.

The rephrasing of Theorem 3.7 in the twisted setting is the following. Let W be a cubic fourfold. Then $\mathcal{K}u(W)$ is equivalent to the derived category of a twisted K3 surface (S, α) if and only if there is a primitive vector $\mathbf{v} \in K_{\mathrm{num}}(\mathcal{K}u(W))$ such that $\mathbf{v}^2 = 0$. This was proved generically on Hassett divisors by Huybrechts in [71], and the completion is in [23].

It should be noted that the condition of having an isotropic vector in $K_{\mathrm{num}}(\mathcal{K}u(W))$ mentioned above is equivalent to the condition of having a primitive embedding $U(n) \hookrightarrow K_{\mathrm{num}}(\mathcal{K}u(W))$. This shows the analogy with the untwisted case considered in the theorem above. We conclude by observing that a partial result in the case of cubics containing a plane is in [124].

3.2 Gushel-Mukai Manifolds

In this section, we assume $\mathrm{char}(\mathbb{K}) = 0$. Gushel-Mukai manifolds were introduced and studied in a series of papers [51–53, 55, 78], based on earlier classification results in [64, 127].

Definition 3.9 A *Gushel-Mukai (GM) manifold* is a smooth n-dimensional intersection

$$X := \mathrm{Cone}(\mathrm{Gr}(2, 5)) \cap \mathbb{P}^{n+4} \cap Q, \qquad 2 \le n \le 6,$$

where $\mathrm{Cone}(\mathrm{Gr}(2, 5)) \subset \mathbb{P}^{10}$ is the cone over the Grassmannian $\mathrm{Gr}(2, 5) \subset \mathbb{P}^9$ in its Plücker embedding, $\mathbb{P}^{n+4} \subset \mathbb{P}^{10}$ is a linear subspace, and $Q \subset \mathbb{P}^{n+4}$ is a quadric hypersurface.

The geometry of GM manifolds is very rich and it is the subject of much interest recently, also due to their similarity (and connections) with cubic fourfolds. We will only shortly recall the definition of Kuznetsov component for $n = 4, 6$, and mention a few results towards Questions 3.2 and 3.3 in these cases, mostly without proofs. Our main reference is [104], and we refer there for all details.

The Kuznetsov Component We assume $n = 4, 6$.[6] In this case X is a Fano manifold. Since X is smooth, the intersection cone(Gr(2, 5)) $\cap Q$ does not contain the vertex of the cone. Hence, we can consider the projection from the vertex of the cone in the Grassmannian

$$f \colon X \to \mathrm{Gr}(2, 5),$$

which is called the *Gushel map*.

There are two possibilities for the Gushel map. Either f is an embedding and its image is a quadric section of a smooth linear section of Gr(2, 5) (in such a case, we say that the GM manifold is *ordinary*), or f is a double covering onto a smooth linear section of Gr(2, 5), ramified along a quadric section (in such a case, we say that the GM manifold is *special*, and we denote by τ the involution). In either case, we denote the smooth linear section by M_X.

Lemma 3.10 *Let $\iota \colon M \hookrightarrow \mathrm{Gr}(2, 5)$ be a smooth linear section of dimension $N \geq 3$. Then M has a rectangular Lefschetz decomposition with respect to $\mathscr{O}_M(1) := \mathscr{O}_{\mathrm{Gr}(2,5)}(1)|_M$:*

$$\mathrm{D}^b(M) = \langle \mathscr{B}_M, \mathscr{B}_M(1), \ldots, \mathscr{B}_M(N - 2) \rangle,$$

where $\mathscr{B}_M = \{\mathscr{O}_M, \mathscr{U}_M^\vee\}$, $\mathscr{U}_M := \mathscr{U}_{\mathrm{Gr}(2,5)}|_M$.

Proof This is [91, Theorem 1.2 & Section 6.1] (see also [104, Lemma 2.2]). It can also be obtained, in an indirect way, from Theorem 2.39; we briefly sketch the argument. Indeed, assume, for simplicity, that M has dimension 5. Then, the functor ι_* is spherical and compatible with the rectangular Lefschetz decomposition of Example 2.8. The corresponding Kuznetsov component is Calabi-Yau of dimension -3. By Propositions 2.25 and 2.28, this is impossible since the Hochschild homology of M can be computed and it is concentrated in degree 0. $\qquad\square$

Lemma 3.11 *The functor $f_* \colon \mathrm{D}^b(X) \to \mathrm{D}^b(M_X)$ is spherical. The associated spherical twists are $T_X = \mathscr{O}_X(-2)[2]$, $T_{M_X} = \mathscr{O}_{M_X}(-2)$, if X is ordinary, and $T_X = \tau \circ \mathscr{O}_X(-1)[1]$, $T_{M_X} = \mathscr{O}_{M_X}(-1)[-1]$, if X is special.*

Proof This is a direct check; see [98, Proposition 3.4] for the details. $\qquad\square$

---

[6]If $n = 2$, then X is a K3 surface. If n is odd, everything goes through in the same way, but the Kuznetsov component is an Enriques-type category, with $S^2 = [4]$.

By using Lemma 3.11 and the rectangular Lefschetz semiorthogonal decomposition of M_X in Lemma 3.10, we can check that the compatibilities of Setup 2.35 are met.[7] Hence, we have a semiorthogonal decomposition

$$D^b(X) = \langle \mathcal{K}u(X), \mathcal{B}_X, \mathcal{B}_X(1), \ldots, \mathcal{B}_X(n-3) \rangle,$$

where $\mathcal{B}_X := f^*\mathcal{B} = \langle \mathcal{O}_X, \mathcal{U}_X^\vee \rangle$. By Theorem 2.39, the Serre functor $S_{\mathcal{K}u(X)} =$ [2].

The Hochscild homology of $\mathcal{K}u(X)$ can been computed, again by using Proposition 2.25 and the Hochschild-Kostant-Rosenberg Theorem, since the Hodge diamond of X is known (see [51, 78]; in particular, [104, Proposition 2.9]). It coincides with the Hochschild homology of a K3 surface. Therefore, $\mathcal{K}u(X)$ is a non-commutative K3 surface.

Ordinary GM Fourfolds Containing a Quintic del Pezzo Surface The analogous result of Proposition 3.6 for GM fourfolds is the following (see [104, Theorem 1.2]).

Theorem 3.12 *Let X be an ordinary GM fourfold containing a quintic del Pezzo surface. Then there is a K3 surface S such that $\mathcal{K}u(X) \cong D^b(S)$.*

The geometric construction of the K3 surface S in Theorem 3.12 is rather concrete. In the language of [104], S is a *generalized dual* of the Gushel-Mukai fourfold X. We do not need to be explicit here about this. But it is worth mentioning that S is a Gushel-Mukai surface (i.e., $n = 2$ in Definition 3.9).

With a view toward Question 3.2, GM fourfolds as in Theorem 3.12 are actually rational (see [104, Lemma 4.7]): very roughly, by blowing up a quintic del Pezzo surface, we get a fibration over $\mathbb{P}^2$ whose general fiber is a smooth quintic del Pezzo surface. Since, by a theorem of Enriques, Manin, and Swinnerton-Dyer, a quintic Del Pezzo surface defined over an infinite field k is k-rational [150], this shows rationality over $\mathbb{P}^2$, and so rationality of X.

Gushel-Mukai Manifolds and K3 Surfaces In [104] there are many interesting conjectures on the Kuznetsov components of GM manifolds, in particular related to duality. In the very recent preprint [100], the generalized duality conjecture [104, Conjecture 3.7] has been completely solved (see [100, Corollary 9.21]). This gives an analogue of Theorem 3.12 for GM sixfolds. On the other hand, we still do not know even a generic answer to Question 3.3 for GM fourfolds.

3.3 Debarre-Voisin Manifolds

Also in this section, we assume char($\mathbb{K}$) $= 0$. Debarre-Voisin manifolds were studied in [54] with the aim of constructing new examples of locally complete

---

[7]If $n = 4$, we have $d = 2$ and $m = 4$, for X ordinary, and $d = 1$ and $m = 3$, for X special; if $n = 6$, we have $d = 1$ and $m = 5$.

families of polarized hyperkähler fourfolds. Their derived categories are less studied: indeed much less is known with respect to the two previous examples, and all basic questions are still open.

Let $\iota\colon X \hookrightarrow \text{Gr}(3,10)$ be a smooth linear section. It is a Fano manifold of dimension 20. We denote by $\mathscr{O}_X(1)$ the restriction to X of the Plücker line bundle $\mathscr{O}_{\text{Gr}(3,10)}(1)$. A computation similar to what we saw before, gives:

Lemma 3.13 *The functor $\iota_*\colon D^b(X) \to D^b(\text{Gr}(3,10))$ is spherical. The associated spherical twists are $T_X = \mathscr{O}_X(-1)[2]$ and $T_{\text{Gr}(3,10)} = \mathscr{O}_{\text{Gr}(3,10)}(-1)$.*

By using Fonarev's rectangular Lefschetz decomposition of Example 2.8, the compatibilities of Setup 2.35 are met ($d = 1$ and $m = 10$), and we obtain a semiorthogonal decomposition

$$D^b(X) = \langle \mathscr{K}u(X), \mathscr{B}_X, \ldots, \mathscr{B}_X(8) \rangle,$$

where $\mathscr{B}_X := \iota^* \mathscr{B}_{\text{Gr}(3,10)}$ has a strong full exceptional collection of length 12. By Theorem 2.39, the category $\mathscr{K}u(X)$ is 2-Calabi-Yau. The Hodge numbers of X have been computed in [54, Theorem 1.1]; by using the Hochschild-Kostant-Rosenberg Theorem again, we have

$$\text{HH}_\bullet(X) = \mathbb{K}[-2] \oplus \mathbb{K}^{130} \oplus \mathbb{K}[2],$$

and so $\mathscr{K}u(X)$ is an example of non-commutative K3 surface.

There is no Debarre-Voisin manifold where Question 3.3 have been answered yet.

3.4 The Mukai Lattice of the Kuznetsov Component

We assume throughout this section that the base field is the complex numbers, $\mathbb{K} = \mathbb{C}$. We introduce a lattice structure in Hochschild homology in the non-commutative K3 surface examples discussed in the previous sections. This, together with the Hodge structure, corresponds to the usual Mukai structure for (derived categories of) K3 surfaces; we refer to [68, Chapter 10] for a summary of results on K3 surfaces.

Topological K-Theory Let $\mathscr{D}$ be a non-commutative smooth projective variety. A general construction of the *topological K-theory* associated to $\mathscr{D}$ is in [32]. In our setting, this can be introduced in a way more closely related to the usual K-theory of a complex manifold as follows.

Setup 3.14 Let X be a smooth projective variety over $\mathbb{C}$. We assume that:

- $H^*(X, \mathbb{Z})$ is torsion-free and $H^{\text{odd}}(X, \mathbb{Z}) = 0$;
- there is a semiorthogonal decomposition

$$D^b(X) = \langle \mathscr{D}_X, E_1, \ldots, E_m \rangle$$

with $\{E_1, \ldots, E_m\}$ an exceptional collection.

We consider the topological K-theory $K_{\text{top}}(X)$ of X. In our setup, since cohomology is torsion-free and odd cohomology vanishes, $K_{\text{top}}(X) = K_{\text{top}}^0(X)$. This is defined as the Grothendieck group of topological $\mathbb{C}$-vector bundles on X.

The basic properties of topological K-theory in our setup are the following (see [12, 13]; see also [4, Section 2]):

(1) $K_{\text{top}}(\text{pt}) = \mathbb{Z}$.
(2) Any morphism $f\colon X \to Y$ induces a pull-back morphism $f^*\colon K_{\text{top}}(Y) \to K_{\text{top}}(X)$ and push-forward morphism $f_*\colon K_{\text{top}}(X) \to K_{\text{top}}(Y)$. There is a projection formula and a Grothendieck-Riemann-Roch formula. One can also take tensor products and duals of classes in topological K-theory.
(3) The *Mukai vector*

$$\mathbf{v}\colon K_{\text{top}}(X) \to H^*(X, \mathbb{Q}) \qquad \mathbf{v} := \text{ch} \,.\sqrt{\text{td}_X}$$

is injective and induces an isomorphism over $\mathbb{Q}$. In particular, $K_{\text{top}}(X)$ is torsion-free.
(4) The *Mukai pairing* $(\_, \_)$ on $K_{\text{top}}(X)$ is defined as follows. Pick a map $p\colon X \to \text{pt}$ to a point and define the topological Euler pairing as

$$\chi(v_1, v_2) := p_*(v_1^\vee \otimes v_2) \in K_{\text{top}}(\text{pt}),$$

for all $v_1, v_2 \in K_{\text{top}}(X)$. Notice that, by (1), $\chi(v_1, v_2)$ is an integer. We can now set $(\_, \_) := -\chi(\_, \_)$.
(5) Consider the following modification of the Hochschild-Konstant-Rosemberg isomorphism introduced in Remark 2.24:

$$I_K^X := (\text{td}(X)^{-1/2} \lrcorner (-)) \circ I_{\text{HKR}}^X,$$

where $\text{td}(X)^{-1/2} \lrcorner (-)$ denotes the contraction by $\text{td}(X)^{-1/2}$. We can then take the following sequence of morphisms

$$K_{\text{top}}(X) \hookrightarrow K_{\text{top}}(X) \otimes \mathbb{C} \xrightarrow{\mathbf{v}} H^*(X, \mathbb{C}) \xrightarrow{(I_K^X)^{-1}} \text{HH}_*(X),$$

where $\mathbf{v}$ denotes here the $\mathbb{C}$-linear extension of the Mukai vector. The composition above is compatible with the various Mukai pairings defined on topological K-theory, singular cohomology and Hochschild homology. Indeed $\mathbf{v}$ preserves the Mukai pairing by Grothendieck-Riemann-Roch for complex vector bundles while I_K^X does the same by [145].

Now let X_1 and X_2 be smooth projective varieties over $\mathbb{C}$ and let $P \in D^b(X_1 \times X_2)$. Consider the Fourier-Mukai functor $\Phi_P(\_) := (p_2)_*(P \otimes p_1^*(\_))\colon D^b(X_1) \to D^b(X_2)$. Since, by (2), pull-back, push-forward and tensorization induce compatible morphisms $(\Phi_P)_K$ and $(\Phi_P)_H$ at the level of (topological) K-theory and singular cohomology (with $\mathbb{Q}$ coefficients), we can consider the following diagram:

$$
\begin{array}{ccccccccc}
D^b(X_1) & \xrightarrow{\sqcup} & K(X_1) & \hookrightarrow & K_{\mathrm{top}}(X_1) & \xrightarrow{\mathrm{v}} & H^*(X_1,\mathbb{Q}) & \xrightarrow{(I_K^{X_1})^{-1}} & HH_*(X_1) \\
\downarrow{\scriptstyle\Phi_P} & & \downarrow{\scriptstyle(\Phi_P)_K} & & \downarrow{\scriptstyle(\Phi_P)_K} & & \downarrow{\scriptstyle(\Phi_P)_H} & & \downarrow{\scriptstyle(\Phi_P)_{HH}} \\
D^b(X_2) & \xrightarrow{\sqcup} & K(X_2) & \hookrightarrow & K_{\mathrm{top}}(X_2) & \xrightarrow{\mathrm{v}} & H^*(X_2,\mathbb{Q}) & \xrightarrow{(I_K^{X_2})^{-1}} & HH_*(X_2).
\end{array}
$$

$$(3)$$

Lemma 3.15 *All squares in* (3) *are commutative.*

Proof The commutativity of the first three squares on the left follows directly by the definition of the induced morphisms. The commutativity of the rightmost square is proved in [118, Theorem 1.2]. □

The induced morphism $(\Phi_P)_H$ is compatible with the Hodge structure on $H^*(X_i,\mathbb{Q})$ induced by the isomorphisms $I_K^{X_i}$ between the total cohomology groups and the Hochschild homologies $HH_*(X_i)$.

Definition 3.16 Assume we are in Setup 3.14. We define the *topological K-theory* of $\mathscr{D}_X$ as

$$
K_{\mathrm{top}}(\mathscr{D}_X) := \left\{ u \in K_{\mathrm{top}}(X) \ : \ ([E_i], u) = 0, \ \text{for all } i = 1, \ldots, m \right\}.
$$

Now let X_1 and X_2 be as in Setup 3.14 and let $\Phi \colon \mathscr{D}_{X_1} \to \mathscr{D}_{X_2}$ be a Fourier-Mukai functor. In this setting, it is not hard to show that we can rewrite (3) in the following way:

$$
\begin{array}{ccccccc}
\mathscr{D}_{X_1} & \longrightarrow & K(\mathscr{D}_{X_1}) & \hookrightarrow & K_{\mathrm{top}}(\mathscr{D}_{X_1}) & \longrightarrow & HH_*(\mathscr{D}_{X_1}) \\
\downarrow{\scriptstyle\Phi} & & \downarrow{\scriptstyle\Phi_K} & & \downarrow{\scriptstyle\Phi_K} & & \downarrow{\scriptstyle\Phi_{HH}} \\
\mathscr{D}_{X_2} & \longrightarrow & K(\mathscr{D}_{X_2}) & \hookrightarrow & K_{\mathrm{top}}(\mathscr{D}_{X_2}) & \longrightarrow & HH_*(\mathscr{D}_{X_2}).
\end{array}
$$

The Mukai and Hodge structures in the above diagram are compatible as well.

Finally, the Mukai structure is invariant under deformations. More precisely, let C be a smooth quasi-projective curve over $\mathbb{C}$, and let $g \colon \mathscr{X} \to C$ be a smooth projective morphism. We let $\mathscr{E}_1, \ldots, \mathscr{E}_m \in D^b(\mathscr{X})$ be families of exceptional objects and we assume we have a C-linear semiorthogonal decomposition[8]

$$
D^b(\mathscr{X}) = \langle \mathscr{D}_{\mathscr{X}}, \mathscr{E}_1 \otimes D^b(C), \ldots, \mathscr{E}_m \otimes D^b(C) \rangle.
$$

---

[8]In our smooth setting, C-linearity simply means that each semiorthogonal factor is closed under tensorization by pull-backs of objects from $D^b(C)$.

By [96], for each closed point $c \in C$, we have a semiorthogonal decomposition

$$\mathrm{D}^b(\mathscr{X}_c) = \langle \mathscr{D}_{\mathscr{X}_c}, \mathscr{E}_1|_c, \ldots, \mathscr{E}_m|_c \rangle.$$

We assume that each closed fiber $\mathscr{X}_c$ and the above semiorthogonal decomposition are as in Setup 3.14. Then, since topological K-theory is invariant by smooth deformations, we have the following result.

Lemma 3.17 *In the above notation and assumptions, we have that the topological K-theory $K_{\mathrm{top}}(\mathscr{D}_{\mathscr{X}_c})$ and its Mukai structure are invariant as $c \in C$ varies. In particular, if there exists $c_0 \in C$ and a smooth projective K3 surface S such that $\mathscr{D}_{\mathscr{X}_{c_0}} \cong \mathrm{D}^b(S)$, we have $K_{\mathrm{top}}(\mathscr{D}_{\mathscr{X}_c}) \cong \widetilde{\Lambda} := E_8(-1)^{\oplus 2} \oplus U^{\oplus 4}$ as lattice, for all $c \in C$.*

Examples Let W be a cubic fourfold defined over $\mathbb{C}$. Since W can be deformed to a Pfaffian cubic fourfold W' and we proved in Proposition 3.6 that $\mathscr{K}u(W') \cong \mathrm{D}^b(S)$, for S a smooth projective K3 surface, Lemma 3.17 implies that $K_{\mathrm{top}}(\mathscr{K}u(W))$, endowed with the Mukai pairing, is isometric to the K3 lattice $\widetilde{\Lambda}$.

The lattice $K_{\mathrm{top}}(\mathscr{K}u(W))$ has a weight-2 Hodge structure coming from Hochschild homology; explicitly, it can be defined in terms of the weight-4 Hodge structure on $H^4(W, \mathbb{Z})$:

$$\widetilde{H}^{2,0}(\mathscr{K}u(W)) := \mathbf{v}^{-1}(H^{3,1}(W))$$

$$\widetilde{H}^{1,1}(\mathscr{K}u(W)) := \mathbf{v}^{-1}\left(\bigoplus_{p=0}^{4} H^{p,p}(W) \right)$$

$$\widetilde{H}^{0,2}(\mathscr{K}u(W)) := \mathbf{v}^{-1}(H^{1,3}(W)).$$

The lattice together with the Hodge structure is called the *Mukai lattice* of $\mathscr{K}u(W)$ and denoted by $\widetilde{H}(\mathscr{K}u(W), \mathbb{Z})$. We set

$$\widetilde{H}_{\mathrm{Hodge}}(\mathscr{K}u(W), \mathbb{Z}) := \widetilde{H}(\mathscr{K}u(W), \mathbb{Z}) \cap \widetilde{H}^{1,1}(\mathscr{K}u(W))$$

$$\widetilde{H}_{\mathrm{alg}}(\mathscr{K}u(W), \mathbb{Z}) := K_{\mathrm{num}}(\mathscr{K}u(W)).$$

We will see later in Theorem 5.11 that we have $\widetilde{H}_{\mathrm{Hodge}}(\mathscr{K}u(W), \mathbb{Z}) = \widetilde{H}_{\mathrm{alg}}(\mathscr{K}u(W), \mathbb{Z})$. This will imply the integral Hodge conjecture holds for cubic fourfolds (see Proposition 5.17).

Example 3.18 Let (S, α) be a twisted K3 surface. Then the total cohomology $H^*(S, \mathbb{Z})$ is endowed with a Mukai pairing and a weight-2 Hodge structure which depends on a lift to $H^2(S, \mathbb{Q})$ of α (see, for example, [74]). This lattice with this Hodge structure is called the Mukai lattice and it is denoted by $\widetilde{H}(S, \alpha, \mathbb{Z})$. When α is trivial we simply write $\widetilde{H}(S, \mathbb{Z})$ and we have that $\widetilde{H}^{2,0}(S) = H^{2,0}(S)$, $\widetilde{H}^{0,2}(S) = H^{0,2}(S)$ while $\widetilde{H}^{1,1}(S) = H^0(S, \mathbb{C}) \oplus H^{1,1}(S) \oplus H^4(S, \mathbb{C})$. If $\mathscr{K}u(W) \cong \mathrm{D}^b(S, \alpha)$, then the two integral Hodge structures coincide.

Remark 3.19 Consider the projection functor $\delta \colon D^b(W) \to \mathscr{K}u(W)$ and fix any line $L \subset W$ (as we will see later, there is always a 4-dimensional family of lines in a cubic fourfold). Define

$$\lambda_1 := \mathbf{v}(\delta(\mathscr{O}_L(1))) \qquad \lambda_2 := \mathbf{v}(\delta(\mathscr{O}_L(2))).$$

This vectors generate a primitive positive definite sublattice

$$A_2 = \begin{pmatrix} 2 & -1 \\ -1 & 2 \end{pmatrix} \subset \tilde{H}_{\text{Hodge}}(\mathscr{K}u(W), \mathbb{Z})$$

This embedding of A_2 moves in families. We should think of this primitive sublattice as the choice of a lattice polarization on $\mathscr{K}u(W)$: we will return to this when defining Bridgeland stability conditions on $\mathscr{K}u(W)$. By [4, Proposition 2.3] we have a Hodge isometry

$$\langle \lambda_1, \lambda_2 \rangle^\perp \cong H^4_{\text{prim}}(W, \mathbb{Z})(-1), \tag{4}$$

where $H^4_{\text{prim}}(W, \mathbb{Z})$ is the orthogonal complement of the self-intersection of a hyperplane class. By the Local Torelli theorem (see [158, Section 6.3.2]), it follows that the very general cubic fourfold W has the property that $A_2 = \tilde{H}_{\text{Hodge}}(\mathscr{K}u(W), \mathbb{Z})$.

Remark 3.20 A direct computation shows that the discriminant group of A_2 is the cyclic group $\mathbb{Z}/3\mathbb{Z}$. Thus [130, Theorem 1.6.1 and Corollary 1.5.2] implies that any autoisometry of A_2 extends to an autoisometry of $\tilde{H}(\mathscr{K}u(W), \mathbb{Z})$. Viceversa, the same results from [130] yield that any autoisometry of the orthogonal $A_2^\perp$ in $\tilde{H}(\mathscr{K}u(W), \mathbb{Z})$ extends to an autoisometry of the Mukai lattice.

Example 3.21 For a cubic fourfold W, we can consider the autoequivalence $O_{\mathscr{K}u(W)}$. Its action on $\tilde{H}(\mathscr{K}u(W), \mathbb{Z})$ was investigated in [71, Proposition 3.12]. In particular, $(O_{\mathscr{K}u(W)})_H$ fixes the sublattice A_2 and cyclicly permutes the elements λ_1, λ_2 and $-\lambda_1 - \lambda_2$. On the other hand, $(O_{\mathscr{K}u(W)})_H$ acts as the identity on $H^4_{\text{prim}}(W, \mathbb{Z})$.

As we observed in Example 3.18, an equivalence $D^b(S) \cong \mathscr{K}u(W)$ induces a Hodge isometry $\tilde{H}(S, \mathbb{Z}) \cong \tilde{H}(\mathscr{K}u(W), \mathbb{Z})$. If such a Hodge isometry exists, we say that W *has a Hodge theoretically associated K3 surface*. Notice that if W has an Hodge theoretically associated K3 surface S, then the copy of the hyperbolic lattice generated by $H^0(S, \mathbb{Z})$ and $H^4(S, \mathbb{Z})$ embeds primitively in $K_{\text{num}}(\mathscr{K}u(W)) \subseteq \tilde{H}_{\text{Hodge}}(\mathscr{K}u(W), \mathbb{Z})$. A simple application of [130, Theorem 1.6.1 and Corollary 1.5.2] shows that the converse is also true. Namely, if there is a primitive embedding $U \hookrightarrow K_{\text{num}}(\mathscr{K}u(W))$, then W has a Hodge theoretically associated K3 surfaces. All in all, these two conditions are equivalent.

If X is a Gushel-Mukai manifold, Theorem 3.12 (for fourfolds; in the six-folds case, this is [100, Corollary 9.21]) implies that X can be deformed to a

Gushel-Mukai manifold X' such that $\mathscr{K}u(X') \cong D^b(S)$, for a S a K3 surface. Hence the discussion above can be repeated for Gushel-Mukai manifolds. For Debarre-Voisin manifolds, this is not yet known even though it is expected to hold true as well. Moreover, in the case of Gushel-Mukai fourfolds, the concept of Hodge theoretically associated K3 surfaces is slightly different though from the case of cubic fourfolds. Indeed, these are special in the sense of [55], and clearly $\widetilde{H}(S, \mathbb{Z}) \cong \widetilde{H}(\mathscr{K}u(X), \mathbb{Z})$ is still equivalent to the condition $U \hookrightarrow K_{\mathrm{num}}(\mathscr{K}u(X))$. But, contrary to the cubic fourfolds case, having a Hodge theoretically associated K3 surface may not be divisorial for GM fourfolds, as observed in [143, Section 3.3].

3.5 Derived Torelli Theorem

Let us go back to the case of a twisted K3 surface (S, α). The total cohomology $H^*(S, \mathbb{R})$ comes with an orientation provided by the four positive (with respect to the Muaki pairing) vectors

$$v_1 := (0, \omega, 0) \qquad v_2 := \left(1, 0, -\frac{\omega^2}{2}\right) \qquad v_3 := \mathrm{Re}\, \psi \qquad v_4 := \mathrm{Im}\, \psi,$$

where ω is a positive real multiple of an ample line bundle and ψ is a generator of $H^{2,0}(S)$.

Remark 3.22 In the language of stability conditions that will be introduced in Sect. 4, the choice of v_1 and v_2 correspond to the choice of a stability condition in a connected component of the space of stability conditions of $D^b(S)$. Note that this space is expected to be connected.

The following result was first proved by Orlov in his seminal paper [135] (with the addition of [77]) and in [74, 75, 146] for twisted K3 surfaces:

Theorem 3.23 (Derived Torelli Theorem for K3 Surfaces) *Let (S_1, α_1) and (S_2, α_2) be twisted K3 surfaces. Then the following are equivalent:*

(1) *There exists an equivalence $D^b(S_1, \alpha_1) \cong D^b(S_2, \alpha_2)$;*
(2) *There exists an orientation preserving Hodge isometry $\widetilde{H}(S_1, \alpha_1, \mathbb{Z}) \cong \widetilde{H}(S_2, \alpha_2, \mathbb{Z})$.*

To formulate a similar result in the context of non-commutative K3 surfaces arising in one of the three classes of examples discussed above, we need to specify an orientation on $\widetilde{H}(\mathscr{K}u(X), \mathbb{Z})$, where X is either a cubic fourfold or a Gushel-Mukai manifold or a Debarre-Voisin manifold.

Inspired by Remark 3.22 we could proceed as follows. Assume that $\mathscr{K}u(X)$ has a stability condition σ. Then $\widetilde{H}(\mathscr{K}u(X), \mathbb{Z})$ contains four positive directions spanned by the real and imaginary part of the central charge Z of σ and the real and imaginary part of a generator of $\widetilde{H}^{2,0}(\mathscr{K}u(X))$. We can then formulate the following natural question:

Question 3.24 (Huybrechts) Let X_1 and X_2 be either cubic fourfold or Gushel-Mukai manifold or Debarre-Voisin manifold. Is it true that there exists a Fourier-Mukai equivalence $\mathscr{K}u(X_1) \cong \mathscr{K}u(X_2)$ if and only if there is an orientation preserving Hodge isometry $\widetilde{H}(\mathscr{K}u(X_1), \mathbb{Z}) \cong \widetilde{H}(\mathscr{K}u(X_2), \mathbb{Z})$?

A positive answer would give a non-commutative version of Theorem 3.23. This would lead us to explore the relations between the existence of equivalences between the Kuznetsov components and the birational type of the fourfolds:

Question 3.25 (Huybrechts) Let X_1 and X_2 be fourfolds as above. Is it true that the existence of a Fourier-Mukai equivalence $\mathscr{K}u(X_1) \cong \mathscr{K}u(X_2)$ implies that X_1 and X_2 are birational?

In the case of a cubic fourfold W, the above discussion about orientation can be made very precise, since $\widetilde{H}_{\mathrm{alg}}(\mathscr{K}u(W), \mathbb{Z})$ always contains a copy of the positive definite lattice A_2. Together with the real and imaginary part of a generator of $\widetilde{H}^{2,0}(\mathscr{K}u(W))$ this lattice provides a natural orientation on the Mukai lattice of $\mathscr{K}u(W)$.

Remark 3.26 It was observed in [71, Lemma 2.3] that the Mukai lattice $\widetilde{H}(\mathscr{K}u(W), \mathbb{Z})$ is always endowed with an orientation reversing Hodge isometry. A way to construct this is by taking the isometry of A_2 such that $\lambda_1 \mapsto -\lambda_1$ while $\lambda_2 \mapsto \lambda_1 + \lambda_2$. By Remark 3.20, this extends to a Hodge isometry of $\widetilde{H}(\mathscr{K}u(W), \mathbb{Z})$ which changes the orientation. Another more geometric way of describing such an orientation reversing Hodge isometry is by taking the action induced on the Mukai lattice by the autoequivalence of $\mathscr{K}u(W)$ obtained by taking the dualizing functor $\mathbb{D}(\_) = \mathbf{R}\mathscr{H}om(\_, \mathscr{O}_W)$ (post)composed with the tensorization by $\mathscr{O}_W(1)$.

For cubic fourfolds, Question 3.24 has the following (partial) answer which is a slightly more precise version of items (i) and (ii) of [71, Theorem 1.5].

Theorem 3.27 (Non-commutative Derived Torelli) *Let W_1 and W_2 be cubic fourfolds such that either W_1 is very general or $\widetilde{H}_{\mathrm{Hodge}}(\mathscr{K}u(W_1), \mathbb{Z})$ contains a primitive vector $\mathbf{v}$ with $\mathbf{v}^2 = 0$. Then the following are equivalent:*

(1) *There exists a Fourier-Mukai equivalence $\mathscr{K}u(W_1) \cong \mathscr{K}u(W_2)$;*
(2) *There exists a Hodge isometry $\widetilde{H}(\mathscr{K}u(W_1), \mathbb{Z}) \cong \widetilde{H}(\mathscr{K}u(W_2), \mathbb{Z})$ which is orientation preserving.*

Proof Condition (1) implies (2) in full generality, without the assumptions mentioned in the statement. Indeed, we observed in the previous section that a Fourier-Mukai equivalence $\mathscr{K}u(W_1) \cong \mathscr{K}u(W_2)$ induces an Hodge isometry $\widetilde{H}(\mathscr{K}u(W_1), \mathbb{Z}) \cong \widetilde{H}(\mathscr{K}u(W_2), \mathbb{Z})$. In case it is not orientation preserving, we can apply Remark 3.26.

Assume (2). If W_1 is very general, then $\widetilde{H}_{\mathrm{Hodge}}(\mathscr{K}u(W_1), \mathbb{Z}) = A_2$, and so the same holds for W_2. By Eq. (4) the lattices $H^4_{\mathrm{prim}}(W_i, \mathbb{Z})$ are the orthogonal complements of A_2. Thus, we get a Hodge isometry $H^4_{\mathrm{prim}}(W_1, \mathbb{Z}) \cong H^4_{\mathrm{prim}}(W_2, \mathbb{Z})$. The Torelli theorem for cubic fourfolds mentioned in the introduction and reproved later in the last section (see Theorem 5.21), implies then that $W_1 \cong W_2$. Hence, in particular, there is a Fourier-Mukai equivalence $\mathscr{K}u(W_1) \cong \mathscr{K}u(W_2)$.

If $\widetilde{H}_{\mathrm{Hodge}}(\mathscr{K}u(W), \mathbb{Z})$ contains a primitive vector $\mathbf{v}$ with $\mathbf{v}^2 = 0$, then, by Theorem 3.7 and Remark 3.8, there is a Fourier-Mukai equivalence $\mathscr{K}u(W) \cong D^b(S, \alpha)$, for some twisted K3 surface (S, α). The fact that (2) implies (1) is then an easy application of Theorem 3.23, as the orientation preserving Hodge isometry $\widetilde{H}(\mathscr{K}u(W_1), \mathbb{Z}) \cong \widetilde{H}(\mathscr{K}u(W_2), \mathbb{Z})$ yields an orientation preserving Hodge isometry $\widetilde{H}(S_1, \alpha_1, \mathbb{Z}) \cong \widetilde{H}(S_2, \alpha_2, \mathbb{Z})$. $\square$

Huybrechts' result [71, Theorem 1.5] has an additional part proving that the equivalence between (1) and (2) in Theorem 3.27 holds also for general points in the divisors of the moduli space $\mathscr{C}$ of cubic fourfolds parameterizing special cubic fourfolds.

Remark 3.28 The tight analogy between Kuznetsov components of cubic fourfolds and K3 surfaces, suggests that the number of isomorphisms classes of cubic fourfolds with equivalent Kuznetsov components (with the equivalence given by a Fourier-Mukai functor) should be finite. Indeed, for K3 surfaces such a finiteness result is due to Bridgeland and Maciocia [39, Corollary 1.2] (and extended in [74, Corollary 4.6] to twisted K3 surfaces). For Kuznetsov components, the same statement is proved in [71, Theorem 1.1]. Notice that the number of isomorphism classes of cubic fourfolds with equivalent Kuznetsov components can be arbitrarily large [142]. The same holds for (twisted) K3 surfaces [74, 132, 152].

In the presence of well defined period maps, we could wonder if for two manifolds X_1 and X_2 which are either cubic or Gushel-Mukai or Debarre-Voisin, the following two conditions are equivalent:

(1) There exists a Fourier-Mukai equivalence $\mathscr{K}u(X_1) \cong \mathscr{K}u(X_2)$ commuting with $O_{\mathscr{K}u(X_1)}$ and $O_{\mathscr{K}u(X_2)}$;
(2) X_1 and X_2 are points of the same fibre of the period map.

If X_1 and X_2 are cubic fourfolds, this is true and reduces once more to the Torelli Theorem. This will be explained in Theorem 5.20.

4 Bridgeland Stability Conditions

In this section we give a short review on the theory of Bridgeland stability conditions, with a particular emphasis on the case of non-commutative K3 surfaces. The main references are still the original works [37, 38, 88]. There are also lecture notes on the subject; see, for example, [15, 69, 117].

4.1 Definition and Bridgeland's Deformation Theorem

Let $\mathscr{D}$ be a non-commutative smooth projective variety. The first ingredient in Bridgeland stability condition is the notion of bounded t-structures. We will actually only define what a heart of a bounded t-structure is; in view of [37, Lemma 3.2], this definition uniquely determines the bounded t-structure.

Definition 4.1 A *heart of a bounded t-structure* in $\mathscr{D}$ is a full subcategory $\mathscr{A} \subset \mathscr{D}$ such that

(a) for $E, F \in \mathscr{A}$ and $n < 0$ we have $\mathrm{Hom}(E, F[n]) = 0$, and
(b) for every $E \in \mathscr{D}$ there exists a sequence of morphisms

$$0 = E_0 \xrightarrow{\varphi_1} E_1 \to \dots \xrightarrow{\varphi_m} E_m = E$$

such that the cone of φ_i is of the form $A_i[k_i]$ for some sequence $k_1 > k_2 > \cdots > k_m$ of integers and objects $A_i \in \mathscr{A}$.

If $\mathscr{D} = \mathrm{D}^b(X)$, where X is a smooth projective variety, then $\mathrm{coh}(X) \subset \mathrm{D}^b(X)$ satisfies the axioms above and thus it is the heart of a bounded t-structure. Other more elaborate ways of constructing these subcategories are discussed in Sect. 4.3. The heart of a bounded t-structure is always an abelian category.

Definition 4.2 Let Λ be a finite rank free abelian group and let $v \colon K(\mathscr{D}) \twoheadrightarrow \Lambda$ be a surjective group homomorphism. A *Bridgeland stability condition* on $\mathscr{D}$ (with respect to the pair (Λ, v)) is a pair $\sigma = (\mathscr{A}, Z)$ consisting of the heart of a bounded t-structure $\mathscr{A} \subset \mathscr{D}$ and a group homomorphism $Z \colon \Lambda \to \mathbb{C}$ (called *central charge*) such that:

(a) For every $0 \neq A \in \mathscr{A}$, $Z(A)^9$ lies in the extended upper-half plane, i.e., $\mathrm{Im}(Z(A)) \geq 0$ and if $\mathrm{Im}(Z(A)) = 0$, then $\mathrm{Re}(Z(A)) < 0$ (we say that Z is a *stability function*).
(b) The function Z allows one to define a *slope* by setting $\mu_\sigma := -\frac{\mathrm{Re}\,Z}{\mathrm{Im}\,Z}$ and a notion of stability: An object $0 \neq E \in \mathscr{A}$ is σ-*semistable* if for every proper subobject F, we have $\mu_\sigma(F) \leq \mu_\sigma(E)$. We then require any object A of $\mathscr{A}$ to have a Harder-Narasimhan filtration (HN filtration, for short) in semistable ones. This means that there is a finite sequence of monomorphisms in $\mathscr{A}$

$$0 = E_0 \hookrightarrow E_1 \hookrightarrow \cdots \hookrightarrow E_{n-1} \hookrightarrow E_n = A$$

such that the factors $F_j = E_j/E_{j-1}$ are μ_σ-semistable and

$$\mu_\sigma(F_1) > \mu_\sigma(F_2) \cdots > \mu_\sigma(F_n).$$

---

[9]We abuse notation and denote $Z(v(A))$ by $Z(A)$. We use the identifications $K(\mathscr{A}) = K(\mathscr{D})$.

(c) Finally, σ satisfies the *support property*: There exists a quadratic form Q on $\Lambda \otimes \mathbb{R}$ such that $Q|_{\mathrm{Ker}\, Z}$ is negative definite, and $Q(E) \geq 0$, for all σ-semistable objects $E \in \mathscr{A}$.

For an object $E \in \mathscr{A}$, the semistable objects in the filtration in Definition 4.2,(b) are called Harder-Narasimhan factors (HN factors, for short). The definition of semistable object can be extended to objects in $\mathscr{D}$: we say that an object $F \in \mathscr{D}$ is σ-semistable if $F = E[n]$, for $E \in \mathscr{A}$ σ-semistable and $n \in \mathbb{Z}$. We can also define the *phase* of a semistable object as follows: for $E \in \mathscr{A}$ σ-semistable, we set $\varphi(E) := \frac{1}{\pi} \arg(Z(E)) \in (0, 1]$ and $\varphi(E[n]) = \varphi(E) + n$. The subcategory given by the union of σ-semistable objects in $\mathscr{D}$ gives a *slicing* of $\mathscr{D}$ (this is important, but we will not need this explicitly in these notes; see [37, Section 3]). Finally, Harder-Narasimhan filtrations can be defined for any non-zero object $E \in \mathscr{D}$, by combining the two filtrations in Definition 4.2,(b) and in Definition 4.1,(b). We set $\varphi_\sigma^+(E)$, respectively $\varphi_\sigma^-(E)$, as the largest, respectively smallest, phase of the Harder-Narasimhan factors of E.

We denote by $\mathrm{Stab}_{(\Lambda, v)}(\mathscr{D})$ the set of stability conditions on $\mathscr{D}$ as in the above definition. To simplify the notation, we often write $\mathrm{Stab}_\Lambda(\mathscr{D})$ or $\mathrm{Stab}(\mathscr{D})$ when Λ and/or v are clear. This set is actually a topological space: the topology is given by the coarsest one such that, for any $E \in \mathscr{D}$, the maps

$$\mathscr{Z} : (\mathscr{A}, Z) \mapsto Z, \quad (\mathscr{A}, Z) \mapsto \varphi^+(E), \quad (\mathscr{A}, Z) \mapsto \varphi^-(E)$$

are continuous. More explicitly this topology is induced by the generalized (i.e., with values in $[0, +\infty]$) metric

$$d(\sigma_1, \sigma_2) = \sup_{0 \neq E \in \mathscr{D}} \left\{ |\varphi_{\sigma_1}^+(E) - \varphi_{\sigma_2}^+(E)|, |\varphi_{\sigma_1}^-(E) - \varphi_{\sigma_2}^-(E)|, \|Z_1 - Z_2\| \right\},$$

for $\sigma_1, \sigma_2 \in \mathrm{Stab}_\Lambda(\mathscr{D})$. Here $\|\_\|$ denotes the induced operator norm on $\mathrm{Hom}(\Lambda, \mathbb{C})$, with respect to the choice of any norm in Λ.

The key result in the theory of stability conditions is Bridgeland Deformation Theorem. This is the main result of [37].

Theorem 4.3 (Bridgeland) *The continuous map $\mathscr{Z}$ is a local homeomorphism and thus the topological space $\mathrm{Stab}_\Lambda(\mathscr{D})$ has a natural structure of a complex manifold of dimension* $\mathrm{rk}(\Lambda)$.

By acting on the central charge by a linear transformation, we have an action of $\mathbb{C}$ (or more generally of the universal cover of $\mathrm{GL}_2^+(\mathbb{R})$) on $\mathrm{Stab}(\mathscr{D})$. By using this action, the proof of Theorem 4.3 reduces to study deformations of the central charge Z with $\mathrm{Im}(Z)$ constant; in this case, the heart is also constant. Then the result follows from an elementary convex geometry argument. This is explained in full detail in [16]. The role of the support property and an effective deformation statement is discussed also in [22, Appendix A].

If we fix $\mathbf{v} \in \Lambda$, then by using the support property there is a locally-finite set of *walls* (real codimension one submanifolds with boundary) in $\mathrm{Stab}_\Lambda(\mathscr{D})$ where the set of semistable objects with class $\mathbf{v}$ changes.

Definition 4.4 Let $\mathbf{v} \in \Lambda$. A stability condition σ is called *generic* with respect to $\mathbf{v}$ (or $\mathbf{v}$-*generic*) if it does not lie on a wall for $\mathbf{v}$.

The main open problem in the theory is the lack of examples. While the surface case is now well-understood [10, 18, 38, 155], starting from threefolds the theory becomes quite scarce. In fact, for a long time no example of a stability condition was known for a projective Calabi-Yau threefold. The first Calabi-Yau examples were finally produced in [22, 115, 116]; the case of quintic threefolds has been recently addressed in [109].

A conjectural approach to construct stability condition is via the notion of tilt-stability [21, 22]. This is a weak stability condition, and we recall here the definition and the main example: we will use this to construct Bridgeland stability conditions on non-commutative K3 surfaces.

Definition 4.5 A *weak stability condition* on $\mathscr{D}$ is a pair $\sigma = (\mathscr{A}, Z)$ consisting of the heart of a bounded t-structure $\mathscr{A} \subset \mathscr{D}$ and a group homomorphism $Z \colon \Lambda \to \mathbb{C}$ satisfying (b), (c) in Definition 4.2 and such that

(a') for $E \in \mathscr{A}$, we have $\mathrm{Im} Z(E) \geq 0$, with $\mathrm{Im} Z(E) = 0 \Rightarrow \mathrm{Re} Z(E) \leq 0$.

Example 4.6 Let X be a smooth projective variety of dimension n and with an ample class H. Consider the lattice Λ_H^1 generated by the vectors

$$(H^n \, \mathrm{rk}(E), H^{n-1} \cdot \mathrm{ch}_1(E)) \in \mathbb{Q}^{\oplus 2},$$

for all $E \in \mathrm{coh}(X)$. Set $Z_{\mathrm{slope}}(E) := -H^{n-1} \cdot \mathrm{ch}_1(E) + \mathrm{i} H^n \, \mathrm{rk}(E)$. It is easy to verify that the pair $\sigma_{\mathrm{slope}} := (\mathrm{coh}(X), Z_{\mathrm{slope}})$ is a weak stability condition. Note that since Λ_H^1 has rank 2, the support property trivially holds.

The slope associated to σ_{slope} is the classical slope stability for sheaves. Hence, by the Bogomolov-Gieseker inequality, we have

$$\Delta_H(E) = \left(H^{n-1} \, \mathrm{ch}_1(E)\right)^2 - 2 \left(H^n \, \mathrm{ch}_0(E)\right) \left(H^{n-2} \, \mathrm{ch}_2(E)\right) \geq 0, \qquad (5)$$

for all σ_{slope}-semistable sheaves E.

4.2 Bridgeland's Covering Theorem

Let $\mathscr{D}$ be a non-commutative K3 surface such that $K_{\mathrm{num}}(\mathscr{D})$ is finitely generated. Set $\Lambda = K_{\mathrm{num}}(\mathscr{D})$. Consider the natural surjection $\mathbf{v} \colon K(\mathscr{D}) \twoheadrightarrow \Lambda$ and the Mukai pairing $(\_, \_)$ given by

$$(\mathbf{v}(E), \mathbf{v}(F)) = -\chi(E, F).$$

Example 4.7 If S is a K3 surface, then we have an identification $\Lambda = \tilde{H}_{\mathrm{alg}}(S, \mathbb{Z})$, and $\mathbf{v}$ is the Mukai vector. Similarly, if $\mathscr{D} = \mathscr{K}u(W)$, for W a cubic fourfold, then $\Lambda = \tilde{H}_{\mathrm{alg}}(\mathscr{K}u(W), \mathbb{Z})$, and $\mathbf{v}$ is as well the Mukai vector.

Under our assumptions, the pairing $(\_, \_)$ yields a natural identification between the vector spaces $\mathrm{Hom}(\Lambda, \mathbb{C})$ and $\Lambda \otimes \mathbb{C}$. In other words, the continuous maps $\mathscr{Z} \colon \mathrm{Stab}_\Lambda(\mathscr{D}) \to \mathrm{Hom}(\Lambda, \mathbb{C})$ can be rewritten as a continuous map

$$\eta \colon \mathrm{Stab}_\Lambda(\mathscr{D}) \to \Lambda \otimes \mathbb{C},$$

such that, for all $\sigma = (\mathscr{A}, Z) \in \mathrm{Stab}_\Lambda(\mathscr{D})$, we have $Z(\_) = (\eta(\sigma), \_)$.

Following [38], we define $\mathscr{P} \subset \Lambda \otimes \mathbb{C}$ as the open subset consisting of those vectors whose real and imaginary parts span positive-definite two-planes in $\Lambda \otimes \mathbb{R}$. Set

$$\mathscr{P}_0 := \mathscr{P} \setminus \bigcup_{\delta \in \Delta} \delta^\perp,$$

where $\Delta := \{\delta \in \Lambda : (\delta, \delta) = -2\}$.

Theorem 4.8 (Bridgeland's Covering) *If $\eta^{-1}(\mathscr{P}_0)$ is non-empty, then the restriction*

$$\eta \colon \eta^{-1}(\mathscr{P}_0) \to \mathscr{P}_0$$

is a covering map.

The proof is an application of Theorem 4.3. The actual statement is [16, Corollary 1.3] (based on [38, Proposition 8.3]).

Notice that $\mathscr{P}_0$ has two connected components. It is expected that the image $\mathrm{im}(\eta)$ is contained in only one connected component of $\mathscr{P}_0$ and that $\eta^{-1}(\mathscr{P}_0)$ is connected and simply-connected as well. This is known only for generic analytic K3 surfaces (and some generic twisted K3 surfaces) [76]; in the algebraic case, the strongest evidence is [17]. This is related to the choice of an orientation, as discussed in Remark 3.22.

Example 4.9 Let S be a K3 surface. We let $\mathscr{P}_0^+$ be the connected component of $\mathscr{P}_0$ containing vectors of the form $(1, 0, -\frac{\omega^2}{2}) + i(0, \omega, 0)$, where ω is a positive real multiple of an ample line bundle. The main result of [38] shows that there exist stability conditions on $\mathrm{D}^b(S)$ for which skyscraper sheaves are all stable of the same phase. They are all contained in the same connected component, denoted by $\mathrm{Stab}^\dagger(\mathrm{D}^b(S))$. Moreover, $\mathrm{Stab}^\dagger(\mathrm{D}^b(S)) \subset \eta^{-1}(\mathscr{P}_0^+)$. The proof is by using tilting of coherent sheaves: we will recall this construction in the next section.

4.3   Tilting and Examples

The aim of this section is to describe a way to produce (weak) stability condition by an iteration process based on tilting.

Let $\mathscr{D}$ be a non-commutative smooth projective variety and assume that we are given a weak stability condition $\sigma = (\mathscr{A}, Z)$ on $\mathscr{D}$. Let $\mu \in \mathbb{R}$. Consider the subcategories of $\mathscr{A}$ defined as follows:

$$\mathscr{T}_\sigma^\mu = \{E : \text{All HN factors } F \text{ of } E \text{ have slope } \mu_\sigma(F) > \mu\}$$
$$= \langle E : E \text{ is } \sigma\text{-semistable with } \mu_\sigma(E) > \mu \rangle,$$
$$\mathscr{F}_\sigma^\mu = \{E : \text{All HN factors } F \text{ of } E \text{ have slope } \mu_\sigma(F) \le \mu\}$$
$$= \langle E : E \text{ is } \sigma\text{-semistable with } \mu_\sigma(E) \le \mu \rangle,$$

where $\langle \_ \rangle$ denotes the extension closure. This notation will be used only in this section where there is no risk to confuse this with semiorthogonal decompositions.

The general theory of torsion pairs and tilting allows us to produce a new abelian category in $\mathscr{D}$ which is the heart of a bounded t-structure:

Proposition 4.10 *Given a weak stability condition* $\sigma = (Z, \mathscr{A})$ *and a choice of slope* $\mu \in \mathbb{R}$, *the category*

$$\mathscr{A}_\sigma^\mu = \langle \mathscr{T}_\sigma^\mu, \mathscr{F}_\sigma^\mu[1] \rangle$$

is the heart of a bounded t-structure on $\mathscr{D}$.

The proof is a direct check; see [65]. We will refer to $\mathscr{A}_\sigma^\mu$ as the *tilting* of $\mathscr{A}$ with respect to the weak stability condition σ at the slope μ. When σ is clear, we will sometimes just write $\mathscr{A}^\mu$.

Let us now consider the case $\mathscr{D} = D^b(X)$, where X is a smooth projective variety, and fix a hyperplane section H on X. By Example 4.6, we have the weak stability condition σ_{slope} and, given $\beta \in \mathbb{R}$, we can consider the tilt

$$\text{coh}^\beta(X) := (\text{coh}_{\sigma_{\text{slope}}}(X))^\beta \subseteq D^b(X).$$

We want now to go further and define a new weak stability condition whose heart is $\text{coh}^\beta(X)$. To this extent, for $E \in \text{coh}(X)$, set

$$\text{ch}^\beta(E) = e^{-\beta H} \text{ch}(E) \in H^*(X, \mathbb{R}).$$

We take Λ_H^2 to be the lattice generated by the vectors

$$(H^n \, \text{rk}(E), H^{n-1} \cdot \text{ch}_1(E), H^{n-2} \cdot \text{ch}_2(E)) \in \mathbb{Q}^{\oplus 3},$$

for all $E \in \mathrm{coh}(X)$. The classical Bogomolov inequality (5) defines a quadratic inequality on Λ^2_H, and it is the key ingredient in the following result:

Proposition 4.11 *Given $\alpha > 0, \beta \in \mathbb{R}$, the pair $\sigma_{\alpha,\beta} = (\mathrm{coh}^\beta(X), Z_{\alpha,\beta})$ with $\mathrm{coh}^\beta(X)$ as constructed above, and*

$$Z_{\alpha,\beta}(E) := \frac{1}{2}\alpha^2 H^n \mathrm{ch}_0^\beta(E) - H^{n-2}\mathrm{ch}_2^\beta(E) + i\, H^{n-1}\mathrm{ch}_1^\beta(E)$$

defines a weak stability condition on $\mathrm{D}^b(X)$ with respect to Λ^2_H. The quadratic form Q can be given by Δ_H. Moreover, these stability conditions vary continuously as $(\alpha, \beta) \in \mathbb{R}_{>0} \times \mathbb{R}$ varies. Finally, if $\dim(X) = 2$, then $\sigma_{\alpha,\beta}$ is a Bridgeland stability condition on $\mathrm{D}^b(X)$.

Proposition 4.11 was observed in the case of surfaces in [10, 38]. The higher dimensional version is in [21, 22]. When the choice of α and β are irrelevant, we will refer to the weak stability condition $\sigma_{\alpha,\beta}$ as *tilt stability*.

In the applications to cubic fourfolds, we will need to tilt $\mathrm{coh}^\beta(X)$ once more, where X will be some non-commutative Fano threefold. To this end, the only thing we will need is to define the analogue of the action of $\mathbb{C}$ on the central charge for tilt-stability, as for Bridgeland stability.

Consider the weak stability condition $\sigma_{\alpha,\beta}$ on $\mathrm{D}^b(X)$ as in Proposition 4.11. Let $\mu \in \mathbb{R}$. By Proposition 4.10, we get that

$$\mathrm{coh}^\mu_{\alpha,\beta}(X) := (\mathrm{coh}^\beta(X))^\mu_{\sigma_{\alpha,\beta}}.$$

is the heart of a bounded t-structure. Let $u \in \mathbb{C}$ be the unit vector in the upper half plane with $\mu = -\frac{\Re(u)}{\Im(u)}$ and consider the function

$$Z^\mu_{\alpha,\beta} := \frac{1}{u} Z_{\alpha,\beta}.$$

Then we have:

Proposition 4.12 *The pair $(\mathrm{coh}^\mu_{\alpha,\beta}(X), Z^\mu_{\alpha,\beta})$ is a weak stability condition on $\mathrm{D}^b(X)$.*

Proposition 4.12 was observed implicitly in [21]; the above statement is [24, Proposition 2.14].

4.4 Inducing Stability Conditions

In this last section, we discuss how stability conditions combine with semiorthogonal decompositions. We follow [24, Sections 4 & 5].

Let $\mathscr{D}$ be a non-commutative smooth projective variety. Let $E_1, \ldots, E_m$ be an exceptional collection; set $\mathscr{D}_2 := \langle E_1, \ldots, E_m \rangle$ and $\mathscr{D}_1 = \mathscr{D}_2^{\perp}$, so that we have a semiorthogonal decomposition

$$\mathscr{D} = \langle \mathscr{D}_1, \mathscr{D}_2 \rangle.$$

Proposition 4.13 *Let $\sigma = (\mathscr{A}, Z)$ be a weak stability condition on $\mathscr{D}$ with the following properties:*

(1) *$E_i \in \mathscr{A}$,*
(2) *$S_{\mathscr{D}}(E_i) \in \mathscr{A}[1]$, and*
(3) *$Z(E_i) \neq 0$,*

for all $i = 1, \ldots, m$. Assume moreover that there are no objects $0 \neq F \in \mathscr{A}_1 := \mathscr{A} \cap \mathscr{D}_1$ with $Z(F) = 0$ (i.e., $Z_1 := Z|_{K(\mathscr{A}_1)}$ is a stability function on $\mathscr{A}_1$). Then the pair $\sigma_1 = (\mathscr{A}_1, Z_1)$ is a stability condition on $\mathscr{D}_1$.

This is [24, Proposition 5.1]. In what follows we sketch why $\mathscr{A}$, under the above assumptions, induces the heart $\mathscr{A}_1$ of a bounded t-structure on $\mathscr{D}_1$.

Lemma 4.14 *Let $\mathscr{A} \subset \mathscr{D}$ be the heart of a bounded t-structure. Assume that $E_1, \ldots, E_m \in \mathscr{A}$ and $\mathrm{Hom}(E_i, F[p]) = 0$, for all $F \in \mathscr{A}$, $i = 1, \ldots, m$, and $p > 1$. Then $\mathscr{A}_1 := \mathscr{D}_1 \cap \mathscr{A}$ is the heart of a bounded t-structure on $\mathscr{D}_1$.*

Proof The category $\mathscr{A}_1$ satisfies condition (a) in Definition 4.1, since it holds for $\mathscr{A}$; hence, we only need to verify (b).

Consider $F \in \mathscr{D}_1$. For every $i = 1, \ldots, m$, there is a spectral sequence ([29, (3.1.3.4)]; see also [133, Proposition 2.4])

$$E_2^{p,q} = \mathrm{Hom}(E_i, H_{\mathscr{A}}^q(F)[p]) \Rightarrow \mathrm{Hom}(E_i, F[p+q]).$$

By assumption, these terms vanish except for $p = 0, 1$, and thus the spectral sequence degenerates at E_2. On the other hand, since $F \in \mathscr{D}_1 = \mathscr{D}_2^{\perp}$ we have $\mathrm{Hom}(E_i, F[p+q]) = 0$. Therefore, $\mathrm{Hom}(E_i, H_{\mathscr{A}}^q(F)[p]) = 0$, for all $p \in \mathbb{Z}$. This gives that $H_{\mathscr{A}}^q(F) \in \mathscr{A} \cap \mathscr{D}_1 = \mathscr{A}_1$, and so it proves the claim. □

Sketch of the Proof of Proposition 4.13 To induce the heart $\mathscr{A}_1$ we only need to verify the assumptions of Lemma 4.14. For all $i = 1, \ldots, m$, $p > 1$, and $F \in \mathscr{A}$, we have

$$\mathrm{Hom}(E_i, F[p]) = \mathrm{Hom}(F[p], S_{\mathscr{D}}(E_i))^{\vee} = \mathrm{Hom}(F, S_{\mathscr{D}}(E_i)[-p])^{\vee} = 0,$$

since $S_{\mathscr{D}}(E_i) \in \mathscr{A}[1]$. □

5 Cubic Fourfolds

The goal of this section is to present the main results in [23, 24]. First of all, for a cubic fourfold W, we show the existence of Bridgeland stability conditions on $\mathcal{K}u(W)$ (Theorem 5.5) and describe a connected component of the space of stability conditions (Theorem 5.7). Then we study moduli spaces of stable objects and we generalize the Mukai theory to the non-commutative setting; the main result is Theorem 5.11. Finally, in Sect. 5.4, we explain how to give a uniform setting to study the various interesting hyperkähler manifolds associated to a cubic fourfolds (e.g., the Fano variety of lines or the hyperkähler manifolds constructed by using twisted cubics). As consequences, we also get new proofs of the Torelli theorem (C.1) and the integral Hodge conjecture for cubic fourfolds.

The basic idea to construct stability conditions is to induce stability conditions as in Sect. 4.4. The problem is that we are currently not able to do this directly from the derived category of the cubic fourfold. Instead, we use its structure of conic fibration (see Sect. 5.1), to reduce to a (non-commutative) projective space, where this inducing procedure works. The key technical result is a generalization of the Bogomolov-Gieseker inequality (Theorem 5.3) to this non-commutative setting.

5.1 Conic Fibrations

In this section we assume that char$(\mathbb{K}) \neq 2$. Let W be a cubic fourfold and let $L \subseteq W$ be a line not contained in a plane in W.[10] We consider the projection

$$\pi_L \colon W \dashrightarrow \mathbb{P}^3$$

from L to a skew three-dimensional projective space in $\mathbb{P}^5$. Let $\sigma \colon \widetilde{W} := \mathrm{Bl}_L(W) \to W$ be the blow-up of W along L. The rational map π_L yields a conic fibration

$$\widetilde{\pi}_L \colon \widetilde{W} \to \mathbb{P}^3.$$

We use the following notation for divisor classes in $\widetilde{W}$: h is the pull-back of a hyperplane section on $\mathbb{P}^3$, H is the pull-back of a hyperplane section on W, and the exceptional divisor D of the blow-up has the form $D = H - h$.

---

[10]Note that such a line always exists as the family of lines in a smooth cubic fourfold are four-dimensional by [27]. On the other hand, such an hypersurface can contain only a finite number of planes.

As explained in Example 2.12, the conic fibration structure produces a sheaf $\mathscr{B}_0$ of even parts of Clifford algebras on $\mathbb{P}^3$ and a semiorthogonal decomposition

$$D^b(\widetilde{W}) = \langle \Phi(D^b(\mathbb{P}^3, \mathscr{B}_0)), \underbrace{\mathscr{O}_{\widetilde{W}}(-h), \mathscr{O}_{\widetilde{W}}, \mathscr{O}_{\widetilde{W}}(h), \mathscr{O}_{\widetilde{W}}(2h)}_{\widetilde{\pi}_L^* D^b(\mathbb{P}^3)} \rangle. \tag{6}$$

The Serre functor $S_{\mathscr{B}_0}$ of the non-commutative smooth projective variety $D^b(\mathbb{P}^3, \mathscr{B}_0)$ has the form

$$S_{\mathscr{B}_0}(\_) = \_ \otimes_{\mathscr{B}_0} \mathscr{B}_{-3}[3]$$

(see [24, Section 7]).

On the other hand, by Example 2.13 (and after some simple mutations), the blow-up structure gives a semiorthogonal decomposition

$$D^b(\widetilde{W}) = \langle \sigma^* \mathscr{K}u(W), \mathscr{O}_{\widetilde{W}}(h - H), \mathscr{O}_{\widetilde{W}}, \mathscr{O}_{\widetilde{W}}(h), \mathscr{O}_{\widetilde{W}}(H), \mathscr{O}_D(h), \mathscr{O}_{\widetilde{W}}(2h), \mathscr{O}_{\widetilde{W}}(H + h) \rangle. \tag{7}$$

If one compares (7) and (6) and perform some elementary mutations and adjunctions, one sees that $\mathscr{K}u(W)$ embeds into $D^b(\mathbb{P}^3, \mathscr{B}_0)$ together with three more exceptional objects in (7). The precise statement is the following result (see [24, Proposition 7.7]):

Proposition 5.1 *Under the above assumptions,*

$$D^b(\mathbb{P}^3, \mathscr{B}_0) = \langle \Psi(\sigma^* \mathscr{K}u(W)), \mathscr{B}_1, \mathscr{B}_2, \mathscr{B}_3 \rangle,$$

where Ψ is the left adjoint of Φ.

As a conclusion, the composition $\Psi \circ \sigma^*$ yields a fully faithful embedding of the Kuznetsov component of W into the derived category of a non-commutative Fano manifold of dimension 3. This is crucial to reduce the complexity of the computations in the next section.

Remark 5.2 It was observed in [23] that the above construction works for families of cubic fourfolds over a suitable base. One needs this family to have a section for the relative Fano variety of lines (and the lines should not be contained in planes in the corresponding cubic fourfold).

To apply the techniques discussed in Sect. 4.4 and produce stability conditions on $\mathscr{K}u(W)$, we need to be able to talk about slope and tilt stability for the abelian category $\mathrm{coh}(\mathbb{P}^3, \mathscr{B}_0)$ and its tilts, respectively.

Consider the forgetful functor $\mathrm{Forg} \colon D^b(\mathbb{P}^3, \mathscr{B}_0) \to D^b(\mathbb{P}^3)$ and the *twisted Chern character* defined as

$$\mathrm{ch}_{\mathscr{B}_0}(E) := \mathrm{ch}(\mathrm{Forg}(E)) \left(1 - \frac{11}{32} \ell \right), \tag{8}$$

for all $E \in D^b(\mathbb{P}^3, \mathscr{B}_0)$, where ℓ denotes the class of a line in $\mathbb{P}^3$. We denote by $\text{ch}_{\mathscr{B}_0, i}$ the degree i component of $\text{ch}_{\mathscr{B}_0}$, and identify them with rational numbers.

Define on $\text{coh}(\mathbb{P}^3, \mathscr{B}_0)$ the slope function

$$\mu_h^{\mathscr{B}_0}(E) := \begin{cases} \frac{\text{ch}_{\mathscr{B}_0, 1}(E)}{\text{ch}_{\mathscr{B}_0, 0}(E)} & \text{if } \text{ch}_{\mathscr{B}_0, 0}(E) \neq 0 \\ +\infty & \text{otherwise}, \end{cases}$$

This function induces a weak stability condition on $\text{coh}(\mathbb{P}^3, \mathscr{B}_0)$. Thus it makes sense to talk about $\mu_h^{\mathscr{B}_0}$-(semi)stable (or, simply, slope-(semi)stable) objects in $\text{coh}(\mathbb{P}^3, \mathscr{B}_0)$.

The key result is the following generalization of the Bogomolov-Gieseker inequality for slope-semistable sheaves [24, Theorem 8.3]:

Theorem 5.3 *For any $\mu_h^{\mathscr{B}_0}$-semistable sheaf $E \in \text{coh}(\mathbb{P}^3, \mathscr{B}_0)$, we have*

$$\Delta_{\mathscr{B}_0}(E) := \text{ch}_{\mathscr{B}_0, 1}(E)^2 - 2\, \text{ch}_{\mathscr{B}_0, 0}(E)\, \text{ch}_{\mathscr{B}_0, 2}(E) \geq 0.$$

It is not difficult to see that $\mathscr{B}_i$ is slope-stable and, with the correction given by $\frac{11}{32}$ in (8), we have $\Delta_{\mathscr{B}_0}(\mathscr{B}_i) = 0$, for all $i \in \mathbb{Z}$. The proof of Theorem 5.3 follows a similar approach as in Langer's proof of the usual Bogomolov-Gieseker inequality [107].

5.2 Existence of Bridgeland Stability Conditions

We keep assuming that $\text{char}(\mathbb{K}) \neq 2$. In this section we apply the construction of the previous section to construct Bridgeland stability conditions on $\mathscr{K}u(W)$ and describe a connected component of $\text{Stab}(\mathscr{K}u(W))$.

The discussion in Sect. 4.3 works verbatim also for the non-commutative variety $(\mathbb{P}^3, \mathscr{B}_0)$. In particular, for a given $\beta \in \mathbb{R}$, we consider the modified twisted Chern character

$$\text{ch}_{\mathscr{B}_0}^{\beta} := e^{-\beta} \cdot \text{ch}_{\mathscr{B}_0}$$

and take the lattice $\Lambda_{\mathscr{B}_0}$ generated by the vectors

$$\left(\text{ch}_{\mathscr{B}_0, 0}(E), \text{ch}_{\mathscr{B}_0, 1}(E), \text{ch}_{\mathscr{B}_0, 2}(E)\right) \in \mathbb{Q}^{\oplus 3},$$

for all $E \in D^b(\mathbb{P}^3, \mathscr{B}_0)$. By Theorem 5.3 we have a quadratic form $\Delta_{\mathscr{B}_0}$ on $\Lambda_{\mathscr{B}_0}$.

Let $\Lambda_{\mathscr{B}_0, \mathscr{K}u(W)} \subseteq \Lambda_{\mathscr{B}_0}$ be the lattice which is the image of $K(\mathscr{K}u(W))$ under the natural composition $K(\mathscr{K}u(W)) \to K(\mathrm{D}^b(\mathbb{P}^3, \mathscr{B}^0)) \to \Lambda_{\mathscr{B}_0}$. Hence we have a surjection

$$v \colon K(\mathscr{K}u(W)) \twoheadrightarrow \Lambda_{\mathscr{B}_0, \mathscr{K}u(W)}.$$

Remark 5.4 It is not difficult to see that v has the following alternative description. The composition of Forg and of the fully faithful functor $\Psi \circ \sigma^* \colon \mathscr{K}u(W) \to \mathrm{D}^b(\mathbb{P}^3, \mathscr{B}_0)$ induces a morphism at the level of numerical Grothendieck groups which, composed with the Chern character $\mathrm{ch}_{\mathscr{B}_0}$ truncated at degree 2, yields a surjective morphism

$$u \colon \widetilde{H}_{\mathrm{alg}}(\mathscr{K}u(W), \mathbb{Z}) \twoheadrightarrow \Lambda_{\mathscr{B}_0, \mathscr{K}u(W)}$$

Given the Mukai vector $\mathbf{v} \colon K(\mathscr{K}u(W)) \to \widetilde{H}_{\mathrm{alg}}(\mathscr{K}u(W), \mathbb{Z})$, we have $v = u \circ \mathbf{v}$.

For $\beta \in \mathbb{R}$, we consider the abelian category $\mathrm{coh}^\beta(\mathbb{P}^3, \mathscr{B}_0)$, which is the heart of a bounded t-structure obtained by tilting $\mathrm{coh}(\mathbb{P}^3, \mathscr{B}_0)$ with respect to slope-stability at the slope $\mu_h^{\mathscr{B}_0} = \beta$. Moreover, for $\alpha \in \mathbb{R}_{>0}$ and all $\beta \in \mathbb{R}$, consider the function

$$Z_{\alpha, \beta}(E) := \frac{1}{2} \alpha^2 \mathrm{ch}_{\mathscr{B}_0, 0}^\beta(E) - \mathrm{ch}_{\mathscr{B}_0, 2}^\beta(E) + \mathrm{i}\, \mathrm{ch}_{\mathscr{B}_0, 1}^\beta(E)$$

defined on $\Lambda_{\mathscr{B}_0, \mathscr{K}u(W)}$ and taking values in $\mathbb{C}$. By Proposition 4.11, the pair

$$\sigma_{\alpha, \beta} := (\mathrm{coh}^\beta(\mathbb{P}^3, \mathscr{B}_0), Z_{\alpha, \beta})$$

is a weak stability condition on $\mathrm{D}^b(\mathbb{P}^3, \mathscr{B}_0)$ with respect to $\Lambda_{\mathscr{B}_0, \mathscr{K}u(W)}$ (see [24, Proposition 9.3]). The support property is provided by the quadratic form given by $\Delta_{\mathscr{B}_0}$.

We want to use this together with Proposition 4.13 to prove the following result.

Theorem 5.5 *If W is a cubic fourfold, then* $\mathrm{Stab}_{\Lambda_{\mathscr{B}_0, \mathscr{K}u(W)}}(\mathscr{K}u(W))$ *is non-empty.*

Proof This is [24, Theorem 1.2]; it is now easy to sketch a proof. Let us fix a line $L \subset W$ not contained in a plane in W. By Proposition 5.1, we have

$$\mathrm{D}^b(\mathbb{P}^3, \mathscr{B}_0) = \langle \mathscr{K}u(W), \mathscr{B}_1, \mathscr{B}_2, \mathscr{B}_3 \rangle.$$

Consider the weak stability condition $\sigma_{\alpha, \beta}$ mentioned above and set $\beta = -1$. Define the slope function $\mu_{\alpha, -1} := \mu_{\sigma_{\alpha, -1}}$ associated to $\sigma_{\alpha, -1}$. Let us tilt $\mathrm{coh}^{-1}(\mathbb{P}^3, \mathscr{B}_0)$ again with respect to $\mu_{\alpha, -1}$ at $\mu_{\alpha, -1} = 0$, getting the heart $\mathrm{coh}_{\alpha, -1}^0(\mathbb{P}^3, \mathscr{B}_0)$ of a bounded t-structure on $\mathrm{D}^b(\mathbb{P}^3, \mathscr{B}_0)$. Set $Z_{\alpha, -1}^0 := -\mathrm{i} Z_{\alpha, -1}$. By Proposition 4.12, the pair

$$\sigma_{\alpha, -1}^0 := (\mathrm{coh}_{\alpha, -1}^0(\mathbb{P}^3, \mathscr{B}_0), Z_{\alpha, -1}^0)$$

is a weak stability condition on $D^b(\mathbb{P}^3, \mathscr{B}_0)$. We want to check that the assumptions in Proposition 4.13 are satisfied for $\mathscr{D}_1 = \mathscr{K}u(W)$, $\mathscr{D}_2 = \langle \mathscr{B}_1, \mathscr{B}_2, \mathscr{B}_3 \rangle$ and α sufficiently small. This would then conclude the proof.

Let us start form (1). A direct computation shows that $\mathscr{B}_1, \mathscr{B}_2, \mathscr{B}_3$, $\mathscr{B}_{-2}[1]$, $\mathscr{B}_{-1}[1]$, $\mathscr{B}_0[1]$ belong to $\mathrm{coh}^{-1}(\mathbb{P}^3, \mathscr{B}_0)$, and they are $\sigma_{\alpha,-1}$-stable for all $\alpha > 0$. For α sufficiently small, one directly proves that

$$\mu_{\alpha,-1}(\mathscr{B}_{-2}[1]) < \mu_{\alpha,-1}(\mathscr{B}_{-1}[1]) < \mu_{\alpha,-1}(\mathscr{B}_0[1]) < 0 < \mu_{\alpha,-1}(\mathscr{B}_1) < \mu_{\alpha,-1}(\mathscr{B}_2) < \mu_{\alpha,-1}(\mathscr{B}_3).$$

Hence $\mathscr{B}_1$, $\mathscr{B}_2$ and $\mathscr{B}_3$ are contained in $\mathrm{coh}^0_{\alpha,-1}(\mathbb{P}^3, \mathscr{B}_0)$, for α sufficiently small.

As for (2), observe that, by [93, Corollary 3.9], $\mathscr{B}_i \otimes_{\mathscr{B}_0} \mathscr{B}_j \cong \mathscr{B}_{i+j}$. Thus $S_{\mathscr{B}_0}(\mathscr{B}_j) \cong \mathscr{B}_{j-3}[3]$ and $S_{\mathscr{B}_0}(\mathscr{B}_j) \in \mathrm{coh}^0_{\alpha,-1}(\mathbb{P}^3, \mathscr{B}_0)[1]$, for α sufficiently small and $j = 1, 2, 3$. A very simple check yields (3).

Finally, by [24, Lemma 2.15], if $E \in \mathrm{coh}^0_{\alpha,-1}(\mathbb{P}^3, \mathscr{B}_0)$ is such that $Z^0_{\alpha,-1}(E) = 0$, then $\mathrm{Forg}(E)$ is a torsion sheaf supported in dimension 0. But then $F \notin \mathscr{K}u(X)$ because

$$\mathrm{Hom}_{\mathscr{B}_0}(\mathscr{B}_j, F) \cong \mathrm{Hom}_{\mathscr{B}_0}(\mathscr{B}_0, F) \cong \mathrm{Hom}_{\mathbb{P}^3}(\mathcal{O}_{\mathbb{P}^3}, \mathrm{Forg}(F)),$$

where, in the last isomorphism, we used the adjunction between the functors $\_ \otimes_{\mathcal{O}_{\mathbb{P}^3}} \mathscr{B}_0$ and Forg (see again [24, Section 7]). □

To finish the section, we enlarge the lattice with respect to which the support property holds to get the analogue of Bridgeland's result for K3 surfaces [38] in Example 4.9.

Definition 5.6 A *full numerical stability condition* on $\mathscr{K}u(W)$ is a Bridgeland stability condition on $\mathscr{K}u(W)$ whose lattice Λ is given by the Mukai lattice $\widetilde{H}_{\mathrm{alg}}(\mathscr{K}u(W), \mathbb{Z})$ and the map v is given by the Mukai vector $\mathbf{v}$.

Let $\mathrm{Stab}(\mathscr{K}u(W))$ be the set of full stability conditions. As explained in Sect. 4.2, we have a map

$$\eta \colon \mathrm{Stab}(\mathscr{K}u(W)) \to \widetilde{H}_{\mathrm{alg}}(\mathscr{K}u(W), \mathbb{C})$$

together with a period domain $\mathscr{P}_0 \subseteq \widetilde{H}_{\mathrm{alg}}(\mathscr{K}u(W), \mathbb{C})$ such that $\eta|_{\eta^{-1}(\mathscr{P}_0)}$ is a covering map (see Theorem 4.8).

Let $\sigma = (\mathscr{A}, Z)$ be the stability condition constructed in the proof of Theorem 5.5. Consider the pair $\sigma' := (\mathscr{A}, Z')$, where $Z' := Z \circ u$ (see Remark 5.4). Then σ' is a full stability condition and $\eta(\sigma') \in \mathscr{P}_0$. This is proven in [24, Proposition 9.10] and it implies that the open subset $\eta^{-1}(\mathscr{P}_0)$ is non-empty. Let $\mathscr{P}_0^+$ denote the connected component of $\mathscr{P}_0$ containing $\eta(\sigma')$, and let $\mathrm{Stab}^\dagger(\mathscr{K}u(W))$ denote the connected component of $\mathrm{Stab}(\mathscr{K}u(W))$ containing σ'.

Theorem 5.7 *The connected component* $\mathrm{Stab}^\dagger(\mathscr{K}u(W))$ *is contained in* $\eta^{-1}(\mathscr{P}_0^+)$. *In particular, the restriction* $\eta \colon \mathrm{Stab}^\dagger(\mathscr{K}u(W)) \to \mathscr{P}_0^+$ *is a covering map.*

Theorem 5.7 is proved in [23]. The key ingredient is Theorem 5.11 below and the notion of family of stability conditions to reduce to the K3 surface case (Example 4.9).

Remark 5.8

(i) If $\mathcal{K}u(W) \cong D^b(S)$, for S a K3 surface, then by construction $\mathrm{Stab}^\dagger(\mathcal{K}u(W))$ coincides with the connected component $\mathrm{Stab}^\dagger(D^b(S))$ discussed in Example 4.9.
(ii) We expect Theorems 5.5 and 5.7 to hold also in the case of Gushel-Mukai fourfolds. Indeed, those varieties have a conic fibration that makes it very plausible that the approach in Sect. 5.1 would work also in that case. One of the difficulties consists in proving the analogue of Theorem 5.3 in this new geometric setting.

5.3 Moduli Spaces

The most important consequence of Theorem 5.5 is to be able to construct and study moduli spaces of stable objects on Kuznetsov components in an analogous way as the Mukai theory for K3 surfaces.

General Properties of Moduli Spaces of Complexes Let $\mathcal{D} \subset D^b(X)$ be a non-commutative smooth projective variety, where X is a smooth projective variety over $\mathbb{K}$. In a similar way as in Sect. 2.2, given a scheme B, locally of finite type over $\mathbb{K}$, we can define a quasi-coherent product category

$$\mathcal{D}_{\mathrm{Qcoh}} \boxtimes D_{\mathrm{Qcoh}}(B) \subset D_{\mathrm{Qcoh}}(X \times B)$$

in the unbounded derived category of quasi-coherent sheaves on $X \times B$ (this is the smallest triangulated subcategory closed under arbitrary direct sums and containing $\mathcal{D} \boxtimes D^b(B)$; see [96]).

Definition 5.9 An object $E \in D_{\mathrm{Qcoh}}(X \times B)$ is B-*perfect* if it is, locally over B, isomorphic to a bounded complex of quasi-coherent sheaves on B which are flat and of finite presentation.

Roughly, complexes which are B-perfect are those which can be restricted to fibers over B. We denote by $D_{B\text{-perf}}(X \times B)$ the full subcategory of $D(\mathrm{Qcoh}(X \times B))$ consisting of B-perfect complexes, and

$$\mathcal{D}_{B\text{-perf}} := (\mathcal{D}_{\mathrm{Qcoh}} \boxtimes D_{\mathrm{Qcoh}}(B)) \cap D_{B\text{-perf}}(X \times B).$$

Consider the 2-functor

$$\mathfrak{M} \colon \mathbf{Sch} \to \mathbf{Grp}$$

which maps a scheme B which is locally of finite type over $\mathbb{K}$ to the groupoid

$$\mathfrak{M}(B) := \left\{ E \in \mathcal{D}_{B\text{-perf}} : \begin{array}{l} \text{Ext}^i(E|_{X \times \{b\}}, E|_{X \times \{b\}}) = 0, \text{ for all } i < 0 \\ \text{and all geometric points } b \in B \end{array} \right\}.$$

The following is the main result in [111]:

Theorem 5.10 (Lieblich) *The functor $\mathfrak{M}$ is an Artin stack, locally of finite type, locally quasi-separated and with separated diagonal.*

To be precise, in [111] only the case $\mathcal{D} = D^b(X)$ is considered; for the extension to non-commutative smooth projective varieties, simply observe that the property of an object in $D^b(X)$ to be contained in $\mathcal{D}$ is open. Recall also that a stack is locally quasi-separated if it admits a Zariski covering with substacks which are quasi-separated.

Consider the open substack $\mathfrak{M}_{\text{Spl}}$ of $\mathfrak{M}$ parameterizing simple objects. Recall that an object E is *simple* if $\text{Hom}(E, E) \cong \mathbb{K}$. This is again an Artin stack, locally of finite type, locally quasi-separated and with separated diagonal. One can take another functor

$$\underline{M}_{\text{Spl}} : \textbf{Sch} \to \textbf{Set}$$

obtained from $\mathfrak{M}_{\text{Spl}}$ by forgetting the groupoid structure and quotienting by the equivalence relation obtained by tensoring by pull-backs of line bundles on B. A previous result by Inaba [81] ensures that $\underline{M}_{\text{Spl}}$ is represented by an algebraic space M_{Spl} which is locally of finite type over $\mathbb{K}$.

Bridgeland Moduli Spaces Assume now we have a Bridgeland stability condition $\sigma = (\mathcal{A}, Z) \in \text{Stab}_\Lambda(\mathcal{D})$, and let $\mathbf{v} \in \Lambda$. To be precise, we also need to choose a phase $\varphi \in \mathbb{R}$ such that $Z(\mathbf{v}) \in \mathbb{R}_{>0} \exp(i\pi\varphi)$. We denote by $\mathfrak{M}_\sigma(\mathcal{D}, \mathbf{v}, \varphi)$ the substack of $\mathfrak{M}$ parameterizing σ-semistable objects in $\mathcal{D}$ with class $\mathbf{v}$ and phase φ. Often we will use the simplified notation $\mathfrak{M}_\sigma(\mathcal{D}, \mathbf{v})$.

A priori it does not follow from the definition of Bridgeland stability condition that this is an open substack of finite type over $\mathbb{K}$.[11] Also, even if this is satisfied, a priori it is not clear that a *good moduli space* (in the sense of Alper [5]) exists; a positive result on this direction is due to Alper et al. [6]. Still in the above assumptions, if the class $\mathbf{v}$ is primitive in Λ and the stability condition σ generic with respect to $\mathbf{v}$, then there are no properly semistable objects and so a good moduli space $M_\sigma(\mathcal{D}, \mathbf{v})$ exists, as a subspace of M_{Spl} of finite type over $\mathbb{K}$.

It is a key result by Toda [154] that the stability conditions constructed by tilting in Proposition 4.11 do satisfy openness and boundedness.

---

[11]This condition of openness and boundedness of stability should probably be assumed in the definition of Bridgeland stability condition; see indeed [23, 88].

More generally, for a K3 surface S, still by [154], this is true for every stability condition in $\mathrm{Stab}^\dagger(\mathrm{D}^b(S))$. Moreover, by Bayer and Macrì [19], Minamide et al. [123], if $\mathbf{v} \in \tilde{H}_{\mathrm{alg}}(S, \mathbb{Z})$ is a non-zero vector and σ is a stability condition which is generic with respect to $\mathbf{v}$, then $M_\sigma(\mathrm{D}^b(S), \mathbf{v})$ exists as a projective variety. If $\mathbf{v}$ is primitive with $\mathbf{v}^2 + 2 \geq 0$, then $M_\sigma(\mathrm{D}^b(S), \mathbf{v})$ is a non-empty smooth projective hyperkähler manifold of dimension $\mathbf{v}^2 + 2$ which is deformation equivalent to a Hilbert scheme of points on a K3 surface (this is (K3.5) in the introduction; the condition $\mathbf{v}^2 + 2 \geq 0$ is also necessary for non-emptyness).

We will not state here the actual result [19, Theorem 1.3], since we will state the analogous version for non-commutative K3 surfaces arising from cubic fourfolds in Theorem 5.11 below. On the other hand, we do need the result on K3 surfaces to prove the result for cubic fourfolds. Hence we give a very short idea of the proof: there exists a Fourier-Mukai partner of S (which may be twisted) such that $M_\sigma(\mathrm{D}^b(S), \mathbf{v})$ becomes a moduli space of Gieseker semistable (twisted) vector bundles. And so the result follows directly from the sheaf case in [164].

The Kuznetsov Component of a Cubic Fourfold The main result for non-commutative K3 surfaces associated to cubic fourfolds is the following theorem from [23]. We assume in this section for simplicity that the base field is $\mathbb{C}$; while essentially all results hold true more generally, the fact that moduli spaces are projective relies on an analytic result.

Theorem 5.11 *Let W be a cubic fourfold. Then*

$$\tilde{H}_{\mathrm{Hodge}}(\mathscr{K}u(W), \mathbb{Z}) = \tilde{H}_{\mathrm{alg}}(\mathscr{K}u(W), \mathbb{Z}).$$

Moreover, assume that $\mathbf{v} \in \tilde{H}_{\mathrm{alg}}(\mathscr{K}u(W), \mathbb{Z})$ is a non-zero primitive vector and let $\sigma \in \mathrm{Stab}^\dagger(\mathscr{K}u(W))$ be a stability condition on $\mathscr{K}u(W)$ that is generic with respect to $\mathbf{v}$. Then

(1) *$M_\sigma(\mathscr{K}u(W), \mathbf{v})$ is non-empty if and only if $\mathbf{v}^2 + 2 \geq 0$. Moreover, in this case, it is a smooth projective irreducible holomorphic symplectic variety of dimension $\mathbf{v}^2 + 2$, deformation-equivalent to a Hilbert scheme of points on a K3 surface.*

(2) *If $\mathbf{v}^2 \geq 0$, then there exists a natural Hodge isometry*

$$\theta : H^2(M_\sigma(\mathscr{K}u(W), \mathbf{v}), \mathbb{Z}) \xrightarrow{\sim} \begin{cases} \mathbf{v}^\perp & \text{if } \mathbf{v}^2 > 0 \\ \mathbf{v}^\perp / \mathbb{Z}\mathbf{v} & \text{if } \mathbf{v}^2 = 0, \end{cases}$$

where the orthogonal is taken in $\tilde{H}(\mathscr{K}u(W), \mathbb{Z})$.

Remark 5.12 It would certainly be very interesting to consider the case when $\mathbf{v} = m\mathbf{v}_0$, for some $m > 1$ (i.e., $\mathbf{v}$ is not primitive). Then Theorem 5.11 shows that $M_\sigma(\mathscr{K}u(W), \mathbf{v})$ is non-empty (i.e., it contains the class of a semistable object) if and only if $\mathbf{v}_0^2 \geq -2$ (see [20, Theorem 2.6]), for σ a $\mathbf{v}$-generic stability condition. On the other hand, by [6], $M_\sigma(\mathscr{K}u(W), \mathbf{v})$ admits a good moduli space. If we can prove that $M_\sigma(\mathscr{K}u(X), \mathbf{v})$ is also normal, then we can deduce further that $M_\sigma(\mathscr{K}u(W), \mathbf{v})$

is an irreducible proper algebraic space and either dim $M_\sigma(\mathscr{K}u(W), \mathbf{v}) = \mathbf{v}^2 + 2$ and the stable locus is non-empty, or $m > 1$ and $\mathbf{v}^2 \leq 0$.

The proof of (1), which is the only part of the statement we focus on, is mainly based on a deformation argument which we can sketch as follows. Suppose that $\mathbf{v}$ is such that $\mathbf{v}^2 \geq -2$. It is not difficult to see that one can construct a family $\mathscr{W} \to C$ of smooth cubic fourfolds over a smooth curve C such that $W \cong \mathscr{W}_p$, for some point $p \in C$ while $\mathscr{K}u(\mathscr{W}_q) \cong D^b(S)$, for some other point $q \in C$. To find q, it is enough to deform W to one of the divisors mentioned in Sect. 3.1, corresponding to cubic fourfolds with homologically associated K3 surface. In view of Proposition 3.6, we can simply consider the Noether-Lefschetz divisor containing Pfaffian cubic fourfolds.

If we assume further that $\mathbf{v}$ is contained in $\widetilde{H}_{\mathrm{Hodge}}(\mathscr{K}u(\mathscr{W}_c), \mathbb{Z})$, for all $c \in C$, that we can consider the relative moduli space of Bridgeland (semi)stable objects $M(\mathbf{v}) \to C$ in the Kuznetsov components such that $M(\mathbf{v})_p \cong M_\sigma(\mathscr{K}u(W), \mathbf{v})$ while $M(\mathbf{v})_q$ is a moduli space of stable objects in $\mathscr{K}u(\mathscr{W}_q)$. This morphism turns out to be smooth and proper. Thus, to prove that $M(\mathbf{v})_p$ is non-empty, we just need to prove that $M(\mathbf{v})_q \neq \emptyset$. Since $\mathscr{K}u(\mathscr{W}_q) \cong D^b(S)$, this follows from the analogous statement for K3 surfaces mentioned before [19, Theorem 1.3]. Notice also that, since $\widetilde{H}_{\mathrm{Hodge}}(D^b(S), \mathbb{Z}) = \widetilde{H}_{\mathrm{alg}}(D^b(S), \mathbb{Z})$, the previous deformation argument implies the analogous statement holds for $\mathscr{K}u(W)$ as well.

The fact that $M_\sigma(\mathscr{K}u(W), \mathbf{v})$ is symplectic follows from the fact that $\mathscr{K}u(W)$ is a non-commutative K3 surface. Indeed, the tangent space of $M_\sigma(\mathscr{K}u(W), \mathbf{v})$ at E is identified to $\mathrm{Hom}(E, E[1])$ and Serre duality for $\mathscr{K}u(W)$ yields a non-degenerate skew-symmetric pairing

$$\mathrm{Hom}(E, E[1]) \times \mathrm{Hom}(E, E[1]) \to \mathrm{Hom}(E, E[2]) \cong \mathrm{Hom}(E, E) \cong \mathbb{C}.$$

The last isomorphism follows from the fact that E is stable. Grothendieck-Riemann-Roch allows us to compute the dimension of $\mathrm{Hom}(E, E[1])$ and thus the dimension of $M_\sigma(\mathscr{K}u(W), \mathbf{v})$. The closedness of this symplectic form follows as in the case for K3 surfaces (proved in [82, Theorem 3.3] and [103]).

To prove that this symplectic manifold is irreducible and projective (and that the resulting variety is deformation-equivalent to a Hilbert scheme of points), we use a general fact in [19]. The stability condition σ induces a nef line bundle on $M_\sigma(\mathscr{K}u(W), \mathbf{v})$ as follows. Let E be the (quasi-)universal family in $D^b(M_\sigma(\mathscr{K}u(W), \mathbf{v}) \times W)$. Since $M_\sigma(\mathscr{K}u(W), \mathbf{v})$ is a moduli space of stable objects in $\mathscr{K}u(W)$, E is a family of objects in $\mathscr{K}u(W)$. We then define the numerical Cartier divisor $\ell_\sigma \in \mathrm{NS}(M_\sigma(\mathbf{v}))_\mathbb{R}$ via the following assignment:

$$C \mapsto \ell_\sigma . C := \mathrm{Im}\left(-\frac{Z(\mathbf{v}(\Phi_E(\mathscr{O}_C)))}{Z(\mathbf{v})} \right),$$

for every curve $C \subseteq M_\sigma(\mathscr{K}u(W), \mathbf{v})$. The Positivity Lemma of [19] implies that ℓ_σ is nef. A careful application of the main result in [139] implies that a positive multiple of ℓ_σ is big. Thus the Base Point Free Theorem (see [87, Theorem 3.3]) implies that a positive multiple of ℓ_σ is ample.

To make this rigorous one needs to develop a new theory of stability conditions in families and this is what is carried out in [23]. In particular, this includes the crucial construction of the relative moduli spaces with respect to a family of stability conditions.

Remark 5.13 Assume we know that, for a Gushel-Mukai fourfold X, the Kuznetsov component $\mathcal{K}u(X)$ carries a stability condition which behaves nicely in family (see Remark 5.8). Then Theorem 5.11 holds for X as well. Indeed, the same proof applies using again a degeneration to divisors in the moduli space of Gushel-Mukai fourfolds parameterizing fourfolds whose Kuznetsov component is equivalent to the derived category of a K3 surface.

As a consequence, one can construct 20-dimensional locally complete families of hyperkähler manifolds. Indeed, take a family $\mathcal{W} \to S$ of cubic fourfolds. Let $\mathbf{v}$ be a primitive section of the local system given by the Mukai lattices $\widetilde{H}(\mathcal{K}u(\mathcal{W}_s), \mathbb{Z})$ of the fibers over $s \in S$, such that $\mathbf{v}$ stays algebraic on all fibers. Assume that for $s \in S$ very general, there exists a stability condition $\sigma_s \in \mathrm{Stab}^\dagger(\mathcal{K}u(\mathcal{W}_s))$ that is generic with respect to $\mathbf{v}$, and such that the associated central charge $Z \colon \widetilde{H}_{\mathrm{alg}}(\mathcal{K}u(\mathcal{W}_s), \mathbb{Z}) \to \mathbb{C}$ is monodromy-invariant.

Theorem 5.14 *There exists a non-empty open subset $S^0 \subset S$ and a variety $M^0(\mathbf{v})$ with a projective morphism $M^0(\mathbf{v}) \to S^0$ that makes $M^0(\mathbf{v})$ a relative moduli space over S^0: the fiber over $s \in S^0$ is a moduli space $M_{\sigma_s}(\mathcal{K}u(\mathcal{W}_s), \mathbf{v})$ of stable objects in the Kuznetsov category of the corresponding cubic fourfold.*

As we mentioned above, this has the following nice application.

Corollary 5.15 *For any pair (a, b) of coprime integers, there is a unirational locally complete 20-dimensional family, over an open subset of the moduli space of cubic fourfolds, of polarized smooth projective irreducible holomorphic symplectic manifolds of dimension $2n + 2$, where $n = a^2 - ab + b^2$. The polarization has divisibility 2 and degree either $6n$ if 3 does not divide n, or $\frac{2}{3}n$ otherwise.*

In this framework we recover some of the classical families of hyperkähler manifolds associated to cubic fourfolds. This is the content of some of the applications discussed below.

5.4 Applications

In the remaining part of these lecture notes, we want to focus on some geometric applications of Theorem 5.11 and examples.

End of the Proof of Theorem 3.7 The easy implication in Theorem 3.7 was proved at the end of Sect. 3.4. Let us now show that if there is a primitive embedding

$U \hookrightarrow \widetilde{H}_{\mathrm{alg}}(\mathscr{K}u(W), \mathbb{Z})$, then there is a K3 surface S and an equivalence $\mathrm{D}^{\mathrm{b}}(S) \cong \mathscr{K}u(W)$.

Under our assumption, there is $\mathbf{w} \in \widetilde{H}_{\mathrm{alg}}(\mathscr{K}u(W), \mathbb{Z})$ such that $\mathbf{w}^2 = 0$. Pick a stability condition $\sigma \in \mathrm{Stab}^\dagger(\mathscr{K}u(W))$ which is generic with respect to $\mathbf{w}$. By Theorem 5.11, the moduli space $M_\sigma(\mathscr{K}u(W), \mathbf{w})$ of σ-stable objects in $\mathscr{K}u(W)$ is non empty, and in fact a K3 surface S. Such a moduli space comes with a quasi-universal family E yielding a fully faithful functor $\Phi_E: \mathrm{D}^{\mathrm{b}}(S, \alpha) \to \mathrm{D}^{\mathrm{b}}(W)$, for some $\alpha \in \mathrm{Br}(S)$. Since S parametrizes stable objects in $\mathscr{K}u(W)$, the functor Φ_E factors through $\mathscr{K}u(W)$ and thus provides a fully faithful functor $\Phi_E: \mathrm{D}^{\mathrm{b}}(S, \alpha) \hookrightarrow \mathscr{K}u(W)$. As $\mathscr{K}u(W)$ is connected, this functor is actually an equivalence. Notice that this already proves the non-trivial implication in the generalization of Theorem 3.7 in Remark 3.8.

Since U embeds in $\widetilde{H}_{\mathrm{alg}}(\mathscr{K}u(W), \mathbb{Z})$, we have further a vector $\mathbf{v}$ such that $(\mathbf{v}, \mathbf{w}) = 1$. A standard argument shows that, under these assumptions, the quasi-universal family E is actually universal. Thus α is trivial and we get an equivalence $\mathrm{D}^{\mathrm{b}}(S) \cong \mathscr{K}u(W)$.

Remark 5.16 Theorem 3.7 also shows that Kuznetsov's categorical conjectural condition for rationality (i.e., $\mathscr{K}u(W) \cong \mathrm{D}^{\mathrm{b}}(S)$) matches the more classical Hodge theoretical one due to Harris and Hassett [66] (i.e., W has a Hodge theoretically associated K3 surface). At the moment, the list of divisors parameterizing rational cubic fourfolds and verifying Harris-Hassett-Kuznetsov prediction is short but interesting (see [149] for recent developments and [89, 129] for very recent and interesting results about specialization for stably rationality and rationality).

The Integral Hodge Conjecture for Cubic Fourfolds The rational Hodge conjecture for cubic fourfolds was proved by Zucker in [165] (see also [49]). On the other hand, on cubic fourfolds the Hodge conjecture holds for integral coefficients as well; this is due to Voisin [159, Theorem 18] and can be reproved directly as a corollary of Theorem 5.11 as follows (see [23]).[12]

Proposition 5.17 (Voisin) *The integral Hodge conjecture holds for any cubic fourfold* W.

Proof Consider a class $v \in H^4(W, \mathbb{Z}) \cap H^{2,2}(W)$. By [12, Section 2.5] (see also [4, Theorem 2.1 (3)]), there exists $w \in K_{\mathrm{top}}(W)$ such that $\mathbf{v}(w) = v + \widetilde{v}$, where $\widetilde{v} \in H^6(W, \mathbb{Q}) \oplus H^8(W, \mathbb{Q})$.

Take the projection w' of w to $\widetilde{H}(\mathscr{K}u(W), \mathbb{Z})$ (induced by the projection functor). Then w differs from w' by a linear combination with integral coefficients $w' = w + a_0[\mathscr{O}_W] + a_1[\mathscr{O}_W(1)] + a_2[\mathscr{O}_W(2)]$ in $K_{\mathrm{top}}(W)$. Since the projection preserves the Hodge structure, w' is actually in $\widetilde{H}_{\mathrm{Hodge}}(\mathscr{K}u(W), \mathbb{Z}) = \widetilde{H}_{\mathrm{alg}}(\mathscr{K}u(W), \mathbb{Z})$, by Theorem 5.11.

---

[12]The argument was also suggested to us by Claire Voisin.

Let $E \in \mathscr{K}u(W)$ be such that $\mathbf{v}(E) = w'$ and set

$$F := E \oplus \mathscr{O}_W^{\oplus |a_0|}[\epsilon(a_0)] \oplus \mathscr{O}_W(1)^{\oplus |a_1|}[\epsilon(a_1)] \oplus \mathscr{O}_W(2)^{\oplus |a_2|}[\epsilon(a_2)],$$

where for an integer $a \in \mathbb{Z}$, we define $\epsilon(a) = 0$ (resp., $= 1$) if $a \geq 0$ (resp., $a < 0$). Then $c_2(F) = v$, which is therefore algebraic.   □

Remark 5.18 By Remark 5.13, if one could prove that the Kuznetsov component of a GM fourfold carries Bridgeland stability conditions, then the theory of moduli spaces would allow us to repeat verbatim the same argument above and prove the integral Hodge conjecture for GM fourfolds.

The Fano Variety of Lines and the Torelli Theorem In the seminal paper [27], Beauville and Donagi showed that the Fano variety of lines $F(W)$ of a cubic fourfold W is a smooth projective hyperkähler manifold of dimension 4. Moreover, $F(W)$ is deformation equivalent to the Hilbert scheme of length-2 zero-dimensional subschemes of a K3 surface. The embedding $F(W)$ inside the Grassmannian of lines in $\mathbb{P}^5$ endowes $F(W)$ with a privileged ample polarization induced by the Plücker embedding.

The study of $F(W)$ as a moduli space of stable objects was initiated in [119] for cubic fourfolds containing a plane and satisfying an additional genericity condition. But the techniques discussed here allow us to prove complete results. Indeed, for any cubic fourfold W, in the notation of Theorem 5.7, fix a stability condition $\sigma \in$ Stab$^\dagger(\mathscr{K}u(W))$ such that $\eta(\sigma) \in (A_2)_{\mathbb{C}} \cap \mathscr{P} \subseteq \mathscr{P}_0^+$. We then get the following general result, which is [110, Theorem 1.1].

Theorem 5.19 (Li-Pertusi-Zhao) *In the assumptions above, the Fano variety of lines in W is isomorphic to the moduli space $M_\sigma(\mathscr{K}u(W), \lambda_1)$. Moreover, the ample line bundle ℓ_σ on $M_\sigma(\mathscr{K}u(W), \lambda_1)$ is identified with a multiple of the Plücker polarization by this isomorphism.*

Sketch of the proof Following [24, Appendix A], we outline the proof under the genericity assumption that $\widetilde{H}(\mathscr{K}u(W), \mathbb{Z})$ does not contain (-2)-classes. This means that there is no class $v \in \widetilde{H}(\mathscr{K}u(W), \mathbb{Z})$ such that $v^2 = -2$. This will be enough for the application to the Torelli Theorem for cubic fourfold.

Let L be a line in W. Following [103], consider the kernel of the evaluation map

$$F_L := \mathrm{Ker}\left(\mathscr{O}_W^{\oplus 4} \twoheadrightarrow I_L(1)\right),$$

which is a torsion-free Gieseker-stable sheaf. A direct computation shows that

$$\mathrm{Hom}(F_L, F_L[i]) = 0 \qquad \mathrm{Hom}(F_L, F_L) \cong \mathbb{C} \qquad \mathrm{Hom}(F_L, F_L[1]) \cong \mathbb{C}^4,$$

for $i < 0$. Moreover, $\mathbf{v}(F_L) = \lambda_1$ and $\mathbf{v}(F_L)^2 = 2$.

The point is that F_L needs to be σ-stable for any $\sigma \in \mathrm{Stab}^\dagger(\mathscr{K}u(W))$, under our genericity assumptions. Indeed, if this is not the case, then one can show that there must exist a distinguished triangle

$$A \to F_L \to B,$$

where A and B are σ-stable and $\dim \mathrm{Hom}(A, A[1]) = \dim \mathrm{Hom}(B, B[1])$. Moreover, a direct computation shows that such dimensions are equal to 2. This means that $\mathbf{v}(A)^2 = \mathbf{v}(B)^2 = 0$ and $(\mathbf{v}(A) + \mathbf{v}(B))^2 = 2$. But then $(\mathbf{v}(A) - \mathbf{v}(B))^2 = -2$. This is a contradiction

The mapping $L \mapsto F_L$ yields an embedding $F(W) \hookrightarrow M_\sigma(\mathscr{K}u(W), \lambda_1)$. By Theorem 5.11, the latter space is a smooth projective hyperkähler manifold of dimension 4. Hence $F(W) \cong M_\sigma(\mathscr{K}u(W), \lambda_1)$.

We omit here the discussion about the polarization ℓ_σ: it is a straightforward computation (which uses [19, Lemma 9.2] and [2, Equation (6)]).   □

We are now ready to answer Question 3.24 in the case of cubic fourfolds.

Theorem 5.20 (Huybrechts-Rennemo) *Let W_1 and W_2 be smooth cubic four-folds. Then $W_1 \cong W_2$ if and only if there is an equivalence $\Phi \colon \mathscr{K}u(W_1) \to \mathscr{K}u(W_2)$ such that $O_{\mathscr{K}u(W_2)} \circ \Phi = \Phi \circ O_{\mathscr{K}u(W_1)}$.*[13]

Sketch of the proof This is proved in [73, Corollary 2.10], by using the Jacobian ring. We present a sketch of a different proof, by using Theorem 5.19: the advantage being that this holds over arbitrary characteristics $\neq 2$. It follows from Example 3.21 that up to composing Φ with a suitable power of the autoequivalence $O_{\mathscr{K}u(W_2)}$ defined in Sect. 2.3 and, possibly, with the shift by 1, we can assume without loss of generality that $\Phi_H \colon \widetilde{H}(\mathscr{K}u(W_1), \mathbb{Z}) \to \widetilde{H}(\mathscr{K}u(W_2), \mathbb{Z})$ is such that $\Phi_H(\lambda_1) = \lambda_1$. Let σ_1 be any stability condition in $\mathrm{Stab}^\dagger(\mathscr{K}u(W_1))$ and let $\sigma_2 := \Phi(\sigma_1)$. Then Φ induces a bijection between the moduli spaces $M_{\sigma_1}(\mathscr{K}u(W_1), \lambda_1)$ and $M_{\sigma_2}(\mathscr{K}u(W_2), \lambda_1)$. A more careful analysis, based on the same circle of ideas as in [30, Section 5.2], shows that we can replace Φ with a Fourier-Mukai functor with the same properties. Hence, one can show that such a bijection can be replaced by an actual isomorphism of smooth projective varieties preserving the ample polarizations ℓ_{σ_1} and ℓ_{σ_2} (here we use the same approach as in [30, Section 5.3]).

If we pick $\sigma_1 \in \mathrm{Stab}^\dagger(\mathscr{K}u(W_1))$ and $\eta(\sigma_1) \in (A_2)_\mathbb{C} \cap \mathscr{P}$ then, by using the results in [23], it can be proved that we can choose the functor Φ in such a way that $\sigma_2 \in \mathrm{Stab}^\dagger(\mathscr{K}u(W_2))$ and $\eta(\sigma_2) \in (A_2)_\mathbb{C} \cap \mathscr{P}$. Theorem 5.19 implies that

$$M_{\sigma_i}(\mathscr{K}u(W_i), \lambda_1) \cong F(W_i)$$

---

[13]It is actually enough to assume that the action of Φ on $\widetilde{H}_{\mathrm{alg}}$ commutes with the action of the degree-shift functor.

with this isomorphism sending ℓ_{σ_i} to the Plücker polarization. Summarizing, we have an isomorphism $F(W_1) \cong F(W_2)$ preserving the Plücker polarization. A classical result by Chow (see [48, Proposition 4]), implies that $W_1 \cong W_2$.   □

As a consequence, we present the recent proof of the Torelli theorem for cubic fourfolds by Huybrechts and Rennemo [73]. This result was originally proved by Voisin in [157] and subsequently reproved by Loojienga in [113]. Based on the Torelli theorem for hyperkähler manifolds [156], Charles [48] gave an elementary proof relying on the Fano variety of lines. The new approach combines Theorem 5.20 and the Derived Torelli Theorem for twisted K3 surfaces (Theorem 3.23). Notice that here we assume the local injectivity of the period map for cubic fourfolds which is a classical result (see, for example, [158, Section 6.3.2]).

Theorem 5.21 (Voisin) *Two smooth complex cubic fourfolds W_1 and W_2 are isomorphic if and only if there exists a Hodge isometry $H^4_{\mathrm{prim}}(W_1, \mathbb{Z}) \cong H^4_{\mathrm{prim}}(W_2, \mathbb{Z})$.*

Sketch of proof Let $\varphi \colon H^4_{\mathrm{prim}}(W_1, \mathbb{Z}) \xrightarrow{\sim} H^4_{\mathrm{prim}}(W_2, \mathbb{Z})$ be a Hodge isometry. By [73, Proposition 3.2], it induces a Hodge isometry $\varphi' \colon \widetilde{H}(\mathscr{K}u(W_1), \mathbb{Z}) \xrightarrow{\sim} \widetilde{H}(\mathscr{K}u(W_2), \mathbb{Z})$ that preserves the natural orientation. This can be explained in the following way. By Eq. (4) in Remark 3.19, $H^4_{\mathrm{prim}}(W_i, \mathbb{Z}) \cong \langle \lambda_1, \lambda_2 \rangle^{\perp}$. But $A_2 = \langle \lambda_1, \lambda_2 \rangle$ has discriminant group $\mathbb{Z}/3\mathbb{Z}$ (see again Remark 3.19). Thus, by Remark 3.20, we see that φ extends to a Hodge isometry $\widetilde{H}(\mathscr{K}u(W_1), \mathbb{Z}) \cong \widetilde{H}(\mathscr{K}u(W_2), \mathbb{Z})$. If the latter isometry is orientation preserving, then we are done. Otherwise, we compose with an orientation reversing Hodge isometry (see Remark 3.26). So we may assume that φ' is orientation preserving since the very beginning.

A general deformation argument based on [71] shows that φ' extends to a local deformation $\mathrm{Def}(W_1) \cong \mathrm{Def}(W_2)$. The set $D \subset \mathrm{Def}(W_1)$ of points corresponding to cubic fourfolds W such that $\mathscr{K}u(W) \cong D^b(S, \alpha)$, for (S, α) a twisted K3 surface, and $\widetilde{H}_{\mathrm{alg}}(\mathscr{K}u(W), \mathbb{Z})$ does not contain (-2)-classes is dense in the moduli space (this follows from Remark 3.8 and [76, Lemma 3.22]). As explained in [73, Section 4.2], for any $t \in D$ there is an orientation preserving Hodge isometry $\varphi_t \colon \widetilde{H}(\mathscr{K}u(W_t), \mathbb{Z}) \to \widetilde{H}(\mathscr{K}u(W_t''), \mathbb{Z})$ which lifts to an equivalence $\Phi_t \colon \mathscr{K}u(W_t) \to \mathscr{K}u(W_t'')$ such that $O_{\mathscr{K}u(W_t'')} \circ \Phi = \Phi \circ O_{\mathscr{K}u(W_t)}$. Theorem 5.20 implies that $W_t \cong W_t''$, for any $t \in D$. Since the moduli space of cubic fourfolds is separated, this yields $W_1 \cong W_2$.   □

Remark 5.22 One should not expect that the same argument used in the proof of Theorem 5.21 might work for GM fourfolds. Indeed, in this case the period map is known to have positive dimensional fibers (see [55, Theorem 4.4]).

Twisted Cubics We can analyze further the relation between the geometry of moduli spaces of rational curves inside cubic fourfolds and the existence of interesting hyperkähler varieties associated to the cubic.

The next case would be to study moduli of conics inside a cubic W. But any such conic would be contained in a plane cutting out a residual line on W. On the other hand, for a line L in W, we can take all projective planes in $\mathbb{P}^5$ passing through L. The residual curve in each of these planes is then a conic. In conclusion, the moduli space of conics would be (at least birationally) a $\mathbb{P}^3$-bundle over $F(W)$. Thus there is no interesting new hyperkähler manifold associated to conics.

Contrary to this, the case of rational normal curves of degree 3 is extremely interesting and has been studied in [108]. We can summarize their work in the following way. Let us start with a cubic fourfold W which does not contain a plane. The irreducible component $M_3(W)$ of the Hilbert scheme containing twisted cubic curves is a smooth projective variety of dimension 10.

The curves in $M_3(W)$ always span a $\mathbb{P}^3$, so there is a natural morphism from $M_3(W)$ to the Grassmannian $Gr(3, \mathbb{P}^5)$ of three-dimensional projective subspaces in $\mathbb{P}^5$. This morphism induces a fibration $M_3(W) \to Z'(W)$, which turns out to be a $\mathbb{P}^2$-fiber bundle. With some further work, the authors of [108] prove that the variety $Z'(W)$ is also smooth and projective of dimension 8.

The geometric nature of $Z'(W)$ is difficult to describe but roughly speaking, $Z'(W)$ is constructed as a moduli space of determinantal representations of cubic surfaces in W (see [26, 56]). Finally, in $Z'(W)$ there is an effective divisor coming from non-CM twisted cubics on W. This divisor can be contracted and after this contraction we get a new variety denoted by $Z(W)$ and which is a smooth projective hyperkähler manifold of dimension 8. It contains the cubic fourfold W as a Lagrangian submanifold and $Z'(W)$ is actually the blow-up of $Z(W)$ in W.

An approach to the study of this hyperkähler manifold by homological methods was initiated in [106] (see also [151]). The main result is the following and shows that the whole picture in [108] has a neat modular interpretation.

Theorem 5.23 *Let W be a smooth cubic fourfold not containing a plane and with hyperplane class H.*

(1) *Let $\mathbf{v}_1 = \left(0, 0, H^2, 0, -\frac{1}{4}H^4\right)$. Then $Z'(W)$ is isomorphic to an irreducible component of the moduli space of Gieseker stable sheaves on W with Chern character $\mathbf{v}_1$.*

(2) *Let $\mathbf{v}_2 = \left(3, 0, -H^2, 0, \frac{1}{4}H^4\right)$. Then:*

   (2a) *$Z'(W)$ is isomorphic to an irreducible component of the moduli space of Gieseker stable torsion free sheaves on W with Chern character $\mathbf{v}_2$.*

   (2b) *If W is very general, both $Z(W)$ and $Z'(W)$ are isomorphic to an irreducible component of the moduli space of tilt-stable objects on $D^b(W)$ with Chern character $\mathbf{v}_2$. The contraction $Z'(W) \to Z(W)$ is realized as a wall-crossing contraction in tilt-stability.*

   (2c) *If W is very general, then $Z(W)$ is isomorphic to a moduli space of Bridgeland stable objects in $\mathscr{K}u(W)$ with Chern character $\mathbf{v}_2$.*

Part (2c) of the above result, together with Theorem 5.11, implies that $Z(W)$ is deformation equivalent to a Hilbert scheme of 4 points on a K3 surface. This result was first proved by Addington and Lehn in [3].

The last part of Theorem 5.23 has been recently improved in [110, Theorem 1.2].

Theorem 5.24 (Li-Pertusi-Zhao) *Let W be a smooth cubic fourfold not containing a plane and let* $\sigma \in \text{Stab}^\dagger(\mathscr{K}u(W))$ *such that* $\eta(\sigma) \in (A_2)_\mathbb{C} \cap \mathscr{P} \subseteq \mathscr{P}_0$. *The smooth projective hyperkähler eightfold* $Z(W)$ *is isomorphic to the moduli space* $M_\sigma(\mathscr{K}u(W), 2\lambda_1 + \lambda_2)$.

It is quite natural to ask what happens to the moduli space $M_\sigma(\mathscr{K}u(W), 2\lambda_1 + \lambda_2)$ when W degenerates to a cubic fourfold containing a plane. In this situation, any $\mathbb{P}^3$ containing the plane cuts a non-integral surface of degree 3. Hence some objects in the moduli space are properly semistable. In [138], Ouchi described a moduli space of stable objects in $\mathscr{K}u(W)$, for W a cubic fourfold containing a plane, such that W embeds into it as a Lagrangian submanifold.

Acknowledgements Our warmest thank goes to Alexander Kuznetsov: his many suggestions, corrections and observations helped us very much to improve the quality of this article. It is also our great pleasure to thank Arend Bayer, Andreas Hochenegger, Martí Lahoz, Howard Nuer, Alex Perry, Laura Pertusi, and Xiaolei Zhao for their insightful comments on the subject of these notes and for carefully reading a preliminary version of this paper. We are very grateful to Nick Addington, Enrico Arbarello and Daniel Huybrechts for many useful conversations and for patiently answering our questions, and to Amnon Yekutieli for pointing out the references [11] and [163]. We would also like to thank Andreas Hochenegger and Manfred Lehn for their collaboration in organizing the school these notes originated from, and the audience for many comments, critiques, and suggestions for improvements. Part of this paper was written while the second author was visiting Northeastern University. The warm hospitality is gratefully acknowledged.

The author "Emanuele Macrì" was partially supported by the NSF grant DMS-1700751. The author "Paolo Stellari" was partially supported by the ERC Consolidator Grant ERC-2017-CoG-771507-StabCondEn and by the research projects FIRB 2012 "Moduli Spaces and Their Applications" and PRIN 2015 "Geometria delle varietà proiettive".

References

1. N. Addington, New derived symmetries of some hyperkähler varieties. Algebr. Geom. **3**, 223–260 (2016)
2. N. Addington, On two rationality conjectures for cubic fourfolds. Math. Res. Lett. **23**, 1–13 (2016)
3. N. Addington, M. Lehn, On the symplectic eightfold associated to a Pfaffian cubic fourfold. J. Reine Angew. Math. **731**, 129–137 (2017)
4. N. Addington, R. Thomas, Hodge theory and derived categories of cubic fourfolds. Duke Math. J. **163**, 1885–1927 (2014)
5. J. Alper, Good moduli spaces for Artin stacks. Ann. Inst. Fourier **63**, 2349–2402 (2013)
6. J. Alper, D. Halpern-Leistner, J. Heinloth, Existence of moduli spaces for algebraic stacks, eprint arXiv:1812.01128v1
7. R. Anno, Spherical functors, eprint arXiv:0711.4409v5
8. R. Anno, T. Logvinenko, Spherical DG-functors. J. Eur. Math. Soc. (JEMS) **19**, 2577–2656 (2017)
9. B. Antieau, G. Vezzosi, A remark on the Hochschild-Kostant-Rosenberg theorem in characteristic p. Ann. Sc. Norm. Super. Pisa Cl. Sci (to appear), eprint arXiv:1710.06039v1
10. D. Arcara, A. Bertram, Bridgeland-stable moduli spaces for K-trivial surfaces, with an appendix by M. Lieblich. J. Eur. Math. Soc. (JEMS) **15**, 1–38 (2013)
11. M. Artin, J.J. Zhang, Noncommutative projective schemes. Adv. Math. **109**, 228–287 (1984)

12. M. Atiyah, F. Hirzebruch, in *Vector Bundles and Homogeneous Spaces*. Proceedings of Symposia in Pure Mathematics, vol. III (American Mathematical Society, Providence, 1961), pp. 7–38

13. M. Atiyah, F. Hirzebruch, The Riemann-Roch theorem for analytic embeddings. *Topology* **1**, 151–166 (1962)

14. A. Auel, M. Bernardara, M. Bolognesi, Fibrations in complete intersections of quadrics, Clifford algebras, derived categories, and rationality problems. J. Math. Pures Appl. **102**, 249–291 (2015)

15. A. Bayer, A tour to stability conditions on derived categories (2011)

16. A. Bayer, A short proof of the deformation property of Bridgeland stability conditions, e print arXiv:1606.02169v1

17. A. Bayer, T. Bridgeland, Derived automorphism groups of K3 surfaces of Picard rank 1. Duke Math. J. **166**, 75–124 (2017)

18. A. Bayer, E. Macrì, The space of stability conditions on the local projective plane. Duke Math. J. **160**, 263–322 (2011)

19. A. Bayer, E. Macrì, Projectivity and birational geometry of Bridgeland moduli spaces. J. Am. Math. Soc. **27**, 707–752 (2014)

20. A. Bayer, E. Macrì, MMP for moduli of sheaves on K3s via wall-crossing: nef and movable cones, Lagrangian fibrations. Invent. Math. **198**, 505–590 (2014)

21. A. Bayer, E. Macrì, T. Toda, Bridgeland stability conditions on threefolds I: Bogomolov-Gieseker type inequalities. J. Algebr. Geom. **23**, 117–163 (2014)

22. A. Bayer, E. Macrì, P. Stellari, The space of stability conditions on abelian threefolds, and on some Calabi-Yau threefolds. Invent. Math. **206**, 869–933 (2016)

23. A. Bayer, M. Lahoz, E. Macrì, H. Nuer, A. Perry, P. Stellari, Stability conditions in family, eprint arXiv:1902.08184v1

24. A. Bayer, M. Lahoz, E. Macrì, P. Stellari, Stability conditions on Kuznetsov components, eprint arXiv:1703.10839v1

25. A. Beauville, Variétés Kähleriennes dont la première classe de Chern est nulle. J. Differ. Geom. **18**, 755–782 (1983)

26. A. Beauville, Determinantal hypersurfaces. Mich. Math. J. **48**, 39–64 (2000)

27. A. Beauville, R. Donagi, La variété des droites d'une hypersurface cubique de dimension 4. C. R. Acad. Sci. Paris Sér. I Math. **301**, 703–706 (1985)

28. A. Beĭlinson, Coherent sheaves on $\mathbb{P}^n$ and problems in linear algebra. Funct. Anal. Appl. **12**, 214–216 (1979)

29. A. Beĭlinson, J. Bernstein, P. Deligne, Faisceaux pervers, in *Analysis and Topology on Singular Spaces, I* (Luminy, 1981), 5–171 (French). Astérisque, vol. 100 (Société Mathématique de France, Paris, 1982)

30. M. Bernardara, E. Macrì, S. Mehrotra, P. Stellari, A categorical invariant for cubic threefolds. Adv. Math. **229**, 770–803 (2012)

31. M. Bernardara, M. Bolognesi, D. Faenzi, Homological projective duality for determinantal varieties. Adv. Math. **296**, 181–209 (2016)

32. A. Blanc, Topological K-theory of complex noncommutative spaces. Compos. Math. **152**, 489–555 (2016)

33. A. Bondal, Representations of associative algebras and coherent sheaves. Izv. Akad. Nauk SSSR Ser. Mat. **53**, 25–44 (1989, in Russian); translation in *Math. USSR-Izv.* **34** (1990), 23–42

34. A. Bondal, M. Kapranov, Representable functors, Serre functors, and reconstructions. Izv. Akad. Nauk SSSR Ser. Mat. **53**, 1183–1205, 1337 (1989, in Russian); translation in *Math. USSR-Izv.* **35** (1990), 519–541

35. A. Bondal, D. Orlov, Semiorthogonal decomposition for algebraic varieties, eprint arXiv:alg-geom/9506012v1

36. A. Bondal, M. Van den Bergh, Generators and representability of functors in commutative and noncommutative geometry. Mosc. Math. J. **3**, 1–36, 258 (2003)

37. T. Bridgeland, Stability conditions on triangulated categories. Ann. Math. (2) **166**, 317–345 (2007)
38. T. Bridgeland, Stability conditions on K3 surfaces. Duke Math. J. **141**, 241–291 (2008)
39. T. Bridgeland, A. Maciocia, Complex surfaces with equivalent derived categories. Math. Z. **236**, 677–697 (2001)
40. R.O. Buchweitz, G. Leuschke, M. Van den Bergh, On the derived category of Grassmannians in arbitrary characteristic. Compos. Math. **151**, 1242–1264 (2015)
41. D. Burns, M. Rapoport, On the Torelli problem for kählerian K3 surfaces. Ann. Sci. École Norm. Sup. **8**, 235–273 (1975)
42. A. Căldăraru, Derived categories of twisted sheaves on Calabi-Yau manifolds, PhD-Thesis, Cornell University (2000)
43. A. Căldăraru, The Mukai pairing. I. The Hochschild structure, eprint arXiv:math/0308079v2
44. A. Căldăraru, The Mukai pairing. II. The Hochschild-Kostant-Rosenberg isomorphism. Adv. Math. **194**, 34–66 (2005)
45. A. Căldăraru, S. Willerton, The Mukai pairing. I. A categorical approach. N. Y. J. Math. **16**, 61–98 (2010)
46. A. Canonaco, P. Stellari, Fourier-Mukai functors: a survey, in *Derived Categories in Algebraic Geometry* (Tokyo, 2011). EMS Series of Congress Reports (European Mathematical Society, Zürich, 2013), pp. 27–60
47. A.-M. Castravet, J. Tevelev, Derived category of moduli of pointed curves - I, eprint arXiv:1708.06340v2
48. F. Charles, A remark on the Torelli theorem for cubic fourfolds, eprint arXiv:1209.4509v1
49. A. Conte, J.-P. Murre, The Hodge conjecture for fourfolds admitting a covering by rational curves. Math. Ann. **238**, 79–88 (1978)
50. O. Debarre, Hyperkähler manifolds, eprint arXiv:1810.02087v1
51. O. Debarre, A. Kuznetsov, Gushel-Mukai varieties: linear spaces and periods. Kyoto J. Math. (to appear), eprint arXiv:1605.05648v3
52. O. Debarre, A. Kuznetsov, On the cohomology of Gushel-Mukai sixfolds, eprint arXiv:1606.09384v1
53. O. Debarre, A. Kuznetsov, Gushel-Mukai varieties: classification and birationalities. Algebr. Geom. **5**, 15–76 (2018)
54. O. Debarre, C. Voisin, Hyper-Kähler fourfolds and Grassmann geometry. J. Reine Angew. Math. **649**, 63–87 (2010)
55. O. Debarre, A. Iliev, L. Manivel, Special prime Fano fourfolds of degree 10 and index 2, in *Recent Advances in Algebraic Geometry*, ed. by C. Hacon, M. Mustață, M. Popa. London Mathematical Society Lecture Notes Series, vol. 417 (Cambridge University Press, Cambridge, 2014), pp. 123–155
56. I. Dolgachev, *Classical Algebraic Geometry. A Modern View* (Cambridge University Press, Cambridge, 2012)
57. B. Dubrovin, Geometry and analytic theory of Frobenius manifolds, in *Proceedings of the International Congress of Mathematicians*, vol. II (Berlin, 1998). Documenta Mathematica (1998), pp. 315–326
58. A.I. Efimov, Derived categories of Grassmannians over integers and modular representation theory. Adv. Math. **304**, 179–226 (2017)
59. A. Fonarev, On the Kuznetsov-Polishchuk conjecture. Proc. Steklov Inst. Math. **290**, 11–25 (2015)
60. A. Fonarev, A. Kuznetsov, Derived categories of curves as components of Fano manifolds. J. Lond. Math. Soc. **97**, 24–46 (2018)
61. R. Friedman, A new proof of the global Torelli theorem for K3 surfaces, Ann. Math. (2) **120**, 237–269 (1984)
62. S. Gelfand, Y. Manin, *Methods of Homological Algebra*. Springer Monographs in Mathematics, 2nd edn. (Springer, Berlin, 2003)

63. M. Gross, D. Huybrechts, D. Joyce, *Calabi-Yau Manifolds and Related Geometries*. Lectures from the Summer School held in Nordfjordeid, June 2001, Universitext (Springer, Berlin, 2003)

64. N. Gushel', On Fano varieties of genus 6. Izv. Akad. Nauk SSSR Ser. Mat. **46**, 1159–1174, 1343 (1982, in Russian); translation in *Izv. Math.* **21** (1983), 445–459

65. D. Happel, I. Reiten, S. Smalø, Tilting in abelian categories and quasitilted algebras. Mem. Am. Math. Soc. **120**, viii+ 88pp (1996)

66. B. Hassett, Special cubic fourfolds. Compos. Math. **120**, 1–23 (2000)

67. D. Huybrechts, Birational symplectic manifolds and their deformations. J. Differ. Geom. **45**, 488–513 (1997)

68. D. Huybrechts, *Fourier-Mukai Transforms in Algebraic Geometry*. Oxford Mathematical Monographs (Oxford University Press, Oxford, 2006)

69. D. Huybrechts, Introduction to stability conditions, in *Moduli Spaces*. London Mathematical Society Lecture Note Series, vol. 411 (Cambridge University Press, Cambridge, 2014), pp. 179–229

70. D. Huybrechts, *Lectures on K3 Surfaces*. Cambridge Studies in Advanced Mathematics (Cambridge University Press, Cambridge, 2016)

71. D. Huybrechts, The K3 category of a cubic fourfold. Compos. Math. **153**, 586–620 (2017)

72. D. Huybrechts, Hodge theory of cubic fourfolds, their Fano varieties, and associated K3 categories, eprint arXiv:1811.02876v2

73. D. Huybrechts, J. Rennemo, Hochschild cohomology versus the Jacobian ring, and the Torelli theorem for cubic fourfolds. Algebr. Geom. **6**, 76–99 (2019)

74. D. Huybrechts, P. Stellari, Equivalences of twisted K3 surfaces. Math. Ann. **332**, 901–936 (2005)

75. D. Huybrechts, P. Stellari, Proof of Căldăraru's conjecture. An appendix to a paper by K. Yoshioka, in *The 13th MSJ International Research Institute - Moduli Spaces and Arithmetic Geometry*. Advanced Studies in Pure Mathematics, vol. 45 (Mathematical Society of Japan, Tokyo, 2006), pp. 31–42

76. D. Huybrechts, E. Macrì, P. Stellari, Stability conditions for generic K3 categories. Compos. Math. **144**, 134–162 (2008)

77. D. Huybrechts, E. Macrì, P. Stellari, Derived equivalences of K3 surfaces and orientation. Duke Math. J. **149**, 461–507 (2009)

78. A. Iliev, L. Manivel, Fano manifolds of degree ten and EPW sextics. Ann. Sci. Éc. Norm. Supér. (4) **44**, 393–426 (2011)

79. A. Iliev, L. Manivel, On cubic hypersurfaces of dimensions 7 and 8. Proc. Lond. Math. Soc. (3) **108**, 517–540 (2014)

80. A. Iliev, L. Manivel, Fano manifolds of Calabi-Yau Hodge type. J. Pure Appl. Algebra **219**, 2225–2244 (2015)

81. M. Inaba, Toward a definition of moduli of complexes of coherent sheaves on a projective scheme. J. Math. Kyoto Univ. **42**, 317–329 (2002)

82. M. Inaba, Smoothness of the moduli space of complexes of coherent sheaves on an abelian or a projective K3 surface. Adv. Math. **227**, 1399–1412 (2011)

83. M. Kapranov, On the derived categories of coherent sheaves on some homogeneous spaces. Invent. Math. **92**, 479–508 (1988)

84. M. Kapranov, Veronese curves and Grothendieck-Knudsen moduli space $\overline{M}_{0,n}$. J. Algebr. Geom. **2**, 239–262 (1993)

85. Y.-H. Kiem, I.-K. Kim, H. Lee, K.-S. Lee, All complete intersection varieties are Fano visitors. Adv. Math. **311**, 649–661 (2017)

86. Y.-H. Kiem, K.-S. Lee, Fano visitors, Fano dimension and quasi-phantom categories, eprint arXiv:1504.07810v4

87. J. Kollár, S. Mori, *Birational Geometry of Algebraic Varieties*. Cambridge Tracts in Mathematics, vol. 134 (Cambridge University Press, Cambridge, 1998)

88. M. Kontsevich, Y. Soibelman, Stability structures, motivic Donaldson-Thomas invariants and cluster transformations, eprint arXiv:0811.2435v1

89. M. Kontsevich, Y. Tschinkel, Specialization of birational types, eprint arXiv:1708.05699v1
90. A. Kuznetsov, Homological projective duality for Grassmannians of lines, eprint arXiv:math/0610957v1
91. A. Kuznetsov, Hyperplane sections and derived categories. Izv. Ross. Akad. Nauk Ser. Mat. **70**, 23–128 (2006, in Russian); translation in *Izv. Math.* **70** (2006), 447–547
92. A. Kuznetsov, Homological projective duality. Publ. Math. Inst. Hautes Études Sci. **105**, 157–220 (2007)
93. A. Kuznetsov, Derived categories of quadric fibrations and intersections of quadrics. Adv. Math. **218**, 1340–1369 (2008)
94. A. Kuznetsov, Hochschild homology and semiorthogonal decompositions, eprint arXiv:0904.4330v1
95. A. Kuznetsov, Derived categories of cubic fourfolds, in *Cohomological and Geometric Approaches to Rationality Problems*. Progress in Mathematics, vol. 282 (Birkhäuser, Boston, 2010), pp. 219–243
96. A. Kuznetsov, Base change for semiorthogonal decompositions. Compos. Math. **147**, 852–876 (2011)
97. A. Kuznetsov, Semiorthogonal decompositions in algebraic geometry, in *Proceedings of the International Congress of Mathematicians - Seoul 2014* (Kyung Moon Sa, Seoul, 2014), pp. 635–660
98. A. Kuznetsov, Calabi-Yau and fractional Calabi-Yau categories. J. Reine Angew. Math. (to appear), eprint arXiv:1509.07657v2
99. A. Kuznetsov, Derived categories view on rationality problems, in *Rationality Problems in Algebraic Geometry*. Lecture Notes in Mathematics, vol. 2172 (Springer, Cham, 2016), pp. 67–104
100. A. Kuznetsov, Categorical joins, eprint arXiv:1804.00144v2
101. A. Kuznetsov, Küchle fivefolds of type c5. Math. Z. **284**, 1245–1278 (2016)
102. A. Kuznetsov, V.A. Lunts, Categorical resolutions of irrational singularities. Int. Math. Res. Not. IMRN **13**, 4536–4625 (2015)
103. A. Kuznetsov, D. Markushevich, Symplectic structures on moduli spaces of sheaves via the Atiyah class. J. Geom. Phys. **59**, 843–860 (2009)
104. A. Kuznetsov, A. Perry, Derived categories of Gushel-Mukai varieties. Compos. Math. **154**, 1362–1406 (2018)
105. A. Kuznetsov, A. Polishchuk, Exceptional collections on isotropic Grassmannians. J. Eur. Math. Soc. (JEMS) **18**, 507–574 (2016)
106. M. Lahoz, M. Lehn, E. Macrì, P. Stellari, Generalized twisted cubics on a cubic fourfold as a moduli space of stable objects. J. Math. Pures Appl. **114**, 85–117 (2018)
107. A. Langer, Semistable sheaves in positive characteristic. Ann. Math. (2) **159**, 251–276 (2004), and Addendum: *Ann. of Math. (2)* **160** (2004), 1211–1213
108. C. Lehn, M. Lehn, C. Sorger, D. van Straten, Twisted cubics on cubic fourfolds. J. Reine Angew. Math. **731**, 87–128 (2017)
109. C. Li, On stability conditions for the quintic threefold, eprint arXiv:1810.03434v1
110. C. Li, L. Pertusi, X. Zhao, Twisted cubics on cubic fourfolds and stability conditions, eprint arXiv:1802.01134v1
111. M. Lieblich, Moduli of complexes on a proper morphism. J. Algebr. Geom. **15**, 175–206 (2006)
112. J.-L. Loday, Cyclic homology, a survey, in *Geometric and Algebraic Topology*. Banach Center Publications, vol. **18** (PWN, Warsaw, 1986), pp. 281–303
113. E. Looijenga, The period map for cubic fourfolds. Invent. Math. **177**, 213–233 (2009)
114. E. Looijenga, C. Peters, Torelli theorems for Kähler K3 surfaces. Compos. Math. **42**, 145–186 (1980/81)
115. A. Maciocia, D. Piyaratne, Fourier-Mukai transforms and Bridgeland stability conditions on abelian threefolds. Algebr. Geom. **2**, 270–297 (2015)

116. A. Maciocia, D. Piyaratne, Fourier-Mukai transforms and Bridgeland stability conditions on abelian threefolds II. Int. J. Math. **27**, 1650007, 27 pp. (2016)

117. E. Macrì, B. Schmidt, Lectures on Bridgeland stability, in *Moduli of Curves*. Lecture Notes of the Unione Matematica Italiana, vol. **21** (Springer, Cham, 2017), pp. 139–211

118. E. Macrì, P. Stellari, Infinitesimal derived Torelli theorem for K3 surfaces (Appendix by S. Mehrotra). Int. Math. Res. Not. IMRN **2009**, 3190–3220 (2009)

119. E. Macrì, P. Stellari, Fano varieties of cubic fourfolds containing a plane. Math. Ann. **354**, 1147–1176 (2012)

120. Y. Manin, M. Smirnov, On the derived category of $\overline{M}_{0,n}$. Izv. Ross. Akad. Nauk Ser. Mat. **77**, 93–108 (2013, in Russian); translation in *Izv. Math.* **77** (2013), 525–540

121. N. Markarian, The Atiyah class, Hochschild cohomology and the Riemann-Roch theorem. J. Lond. Math. Soc. (2) **79**, 129–143 (2009)

122. C. Meachan, A note on spherical functors, eprint `arXiv:1606.09377v2`

123. H. Minamide, S. Yanagida, K. Yoshioka, The wall-crossing behavior for Bridgeland's stability conditions on abelian and K3 surfaces. J. Reine Angew. Math. **735**, 1–107 (2018)

124. R. Moschetti, The derived category of a non generic cubic fourfold containing a plane. Math. Res. Lett. (to appear), eprint `arXiv:1607.06392v2`

125. S. Mukai, Symplectic structure of the moduli space of sheaves on an abelian or K3 surface, *Invent. Math.* **77** (1984), 101–116.

126. S. Mukai, On the moduli space of bundles on K3 surfaces. I, in *Vector Bundles on Algebraic Varieties* (Bombay, 1984). Tata Institute of Fundamental Research Studies in Mathematics, vol. 11 (Tata Institute of Fundamental Research, Bombay, 1987), pp. 341–413

127. S. Mukai, Biregular classification of Fano 3-folds and Fano manifolds of coindex 3. Proc. Natl. Acad. Sci. U. S. A. **86**, 3000–3002 (1989)

128. M. Narasimhan, Derived categories of moduli spaces of vector bundles on curves. J. Geom. Phys. **122**, 53–58 (2017)

129. J. Nicaise, E. Shinder, The motivic nearby fiber and degeneration of stable rationality, eprint `arXiv:1708.027901v3`

130. V.V. Nikulin, Integral symmetric bilinear forms and some of their applications. Izv. Akad. Nauk SSSR Ser. Mat. **43**, 111–177, 238 (1979, in Russian); translation in *Math USSR Izvestija* **14** (1980), 103–167

131. K. O'Grady, The weight-two Hodge structure of moduli spaces of sheaves on a K3 surface. J. Algebr. Geom. **6**, 599–644 (1997)

132. K. Oguiso, K3 surfaces via almost-prime. Math. Res. Lett. **9**, 47–63 (2002)

133. S. Okada, On stability manifolds of Calabi-Yau surfaces. Int. Math. Res. Not. **2006**, Art. ID 58743, 16 pp (2006)

134. D. Orlov, Projective bundles, monoidal transformations, and derived categories of coherent sheaves. Izv. Akad. Nauk SSSR Ser. Mat. **56**, 852–862 (1992, in Russian); translation in *Russian Acad. Sci. Izv. Math.* **41** (1993) 133–141

135. D. Orlov, Equivalences of derived categories and K3 surfaces. J. Math. Sci. **84**, 1361–1381 (1997)

136. D. Orlov, Smooth and proper noncommutative schemes and gluing of DG categories. Adv. Math. **302**, 59–105 (2016)

137. G. Ottaviani, Spinor bundles on quadrics. Trans. Am. Math. Soc. **307**, 301–316 (1988)

138. G. Ouchi, Lagrangian embeddings of cubic fourfolds containing a plane. Compos. Math. **153**, 947–972 (2017)

139. A. Perego, Kählerness of moduli spaces of stable sheaves over non-projective K3 surfaces, eprint `arXiv:1703.02001v1`

140. A. Perry, Noncommutative homological projective duality, eprint `arXiv:1804.00132v1`

141. A. Perry, Hochschild cohomology and group actions, eprint `arXiv:1807.09268v1`

142. L. Pertusi, Fourier-Mukai partners for general special cubic fourfolds, eprint `arXiv:1611.06687v2`

143. L. Pertusi, On the double EPW sextic associated to a Gushel-Mukai fourfold. J. Lond. Math. Soc. (to appear), eprint `arXiv:1709.02144v1`

144. I. Pijateckiĭ-Šapiro, I. Šafarevič, A Torelli theorem for algebraic surfaces of type K3. Izv. Akad. Nauk SSSR Ser. Mat. **35**, 530–572 (1971, in Russian); translation in *Math. USSR Izvestija* **5** (1971), 547–588

145. A. Ramadoss, The relative Riemann–Roch theorem from Hochschild homology. N. Y. J. Math. **14**, 643–717 (2008).

146. E. Reinecke, Autoequivalences of twisted K3 surfaces. Compos. Math. (to appear), eprint arXiv:1711.00846v1

147. A. Rizzardo, M. Van den Bergh, An example of a non-Fourier–Mukai functor between derived categories of coherent sheaves, with an appendix by A. Neeman. Invent. Math. (to appear), eprint arXiv:1410.4039v2

148. R. Rouquier, Categorification of $\mathfrak{sl}_2$ and braid groups, in *Trends in Representation Theory of Algebras and Related Topics*. Contemporary Mathematics, vol. 406 (American Mathematical Society, Providence, 2006), pp. 137–167

149. F. Russo, G. Staglianò, Congruences of 5-secant conics and the rationality of some admissible cubic fourfolds. Duke Math. J. (to appear), eprint arXiv:1707.00999v3

150. N. Shepherd-Barron, The rationality of quintic Del Pezzo surfaces-a short proof. Bull. Lond. Math. Soc. **24**, 249–250 (1992)

151. E. Shinder, A. Soldatenkov, On the geometry of the Lehn–Lehn–Sorger–van Straten eightfold. Kyoto J. Math. **57**, 789–806 (2017)

152. P. Stellari, Some remarks about the FM-partners of K3 surfaces with Picard numbers 1 and 2. Geom. Dedicata **108**, 1–13 (2004)

153. R. Swan, Hochschild cohomology of quasi-projective schemes. J. Pure Appl. Algebra **110**, 57–80 (1996)

154. Y. Toda, Moduli stacks and invariants of semistable objects on K3 surfaces. Adv. Math. **217**, 2736–2781 (2008)

155. Y. Toda, Stability conditions and extremal contractions. Math. Ann. **357**, 631–685 (2013)

156. M. Verbitsky, Mapping class group and a global Torelli theorem for hyperkähler manifolds, with an appendix by E. Markman. Duke Math. J. **162**, 2929–2986 (2013)

157. C. Voisin, Théorème de Torelli pour les cubiques de $\mathbb{P}^5$. Invent. Math. **86**, 577–601 (1986), and Erratum: Invent. Math. **172** (2008), 455–458

158. C. Voisin, *Hodge Theory and Complex Algebraic Geometry. II.* Cambridge Studies in Advanced Mathematics (Cambridge University Press, Cambridge, 2007)

159. C. Voisin, Some aspects of the Hodge conjecture. Jpn. J. Math. **2**, 261–296 (2007)

160. V. Vologodsky, Triangulated endofunctors of the derived category of coherent sheaves which do not admit DG liftings, eprint arXiv:1604.08662v1

161. C. Weibel, Cyclic homology for schemes. Proc. Am. Math. Soc. **124**, 1655–1662 (1996)

162. A. Yekutieli, The continuous Hochschild cochain complex of a scheme. Can. J. Math. **54**, 1319–1337 (2002)

163. A. Yekutieli, Derived categories, eprint arXiv:1610.09640v4

164. K. Yoshioka, Moduli spaces of stable sheaves on abelian surfaces. Math. Ann. **321**, 817–884 (2001)

165. S. Zucker, The Hodge conjecture for cubic fourfolds. Compos. Math. **34**, 199–209 (1977)

Appendix: Introduction to Derived Categories of Coherent Sheaves

Andreas Hochenegger

Abstract In these notes, an introduction to derived categories and derived functors is given. The main focus is the bounded derived category of coherent sheaves on a smooth projective variety.

1 Introduction

One way to get your hands on coherent sheaves is by short exact sequences. To name three important ones:

- $0 \to \mathcal{O}_{\mathbb{P}^n} \to \mathcal{O}_{\mathbb{P}^n}(1)^{\oplus(n+1)} \to \mathcal{T}_{\mathbb{P}^n} \to 0$ *Euler sequence*
- $0 \to \mathcal{I}_{Y|X} \to \mathcal{O}_X \to \mathcal{O}_Y \to 0$ *Ideal sheaf sequence*
- $0 \to \mathcal{T}_Y \to \mathcal{T}_X|_Y \to \mathcal{N}_{Y|X} \to 0$ *Normal sheaf sequence*

where $Y \subset X$ is a closed embedding.

Such sequences are usually the starting point for computations. But by applying any meaningful operation to such a sequence one will almost inevitably lose the exactness on the left or right end. Examples for such operations are

- $\mathcal{H}om(\mathcal{F}, -), \operatorname{Hom}(\mathcal{F}, -)$
- $\mathcal{F} \otimes -$
- $f_*, \Gamma(X, -)$
- f^*

where $\mathcal{F}$ is a coherent sheaf, and $f : X \to Y$ a morphism. Another issue is that the projection formula and flat base change work only for specific classes of coherent sheaves such as for locally free sheaves. That exactness gets lost, should not be seen as a failure but an indication that there is something more to say.

A. Hochenegger (✉)
Dipartimento di Matematica "Federigo Enriques", Università degli Studi di Milano, Milano, Italy
e-mail: andreas.hochenegger@unimi.it

© Springer Nature Switzerland AG 2019
A. Hochenegger et al. (eds.), *Birational Geometry of Hypersurfaces*,
Lecture Notes of the Unione Matematica Italiana 26,
https://doi.org/10.1007/978-3-030-18638-8_7

Example Let C and C' be two rational curves on a smooth projective surface X. Applying $\text{Hom}(-, \mathcal{O}_C)$ to the ideal sheaf sequence of C' yields

$$0 \to \text{Hom}(\mathcal{O}_{C'}, \mathcal{O}_C) \to H^0(\mathcal{O}_C) \to H^0(\mathcal{O}_C(C'))$$

This sequence is a short exact sequence if and only if

- C and C' are disjoint, then $\text{Hom}(\mathcal{O}_{C'}, \mathcal{O}_C) = 0$ and $\mathcal{O}_C \cong \mathcal{O}_C(C')$; or
- $C = C'$ and $H^0(\mathcal{O}_C(C))$ vanishes, as $\text{Hom}(\mathcal{O}_C, \mathcal{O}_C) \to H^0(\mathcal{O}_C)$ is an isomorphism.

These are quite special situations (note that the second case implies that $C^2 < 0$). In particular, if C and C' intersect, this sequence can to be continued with Ext-groups. The intersection number can be easily computed using the Euler characteristic:

$$C'.C = -\chi(\mathcal{O}_{C'}, \mathcal{O}_C) = -\dim \text{Hom}(\mathcal{O}_{C'}, \mathcal{O}_C) + \dim \text{Ext}^1(\mathcal{O}_{C'}, \mathcal{O}_C).$$

In order to deal with such examples, homological algebra proposes to replace sheaves by adapted resolutions and the derived category of sheaves will become the proper framework for such computations.

Aim These notes serve as a companion to the lecture notes [11] and give the necessary background on derived categories. The motivating question is how to change (or better: derive) a functor between abelian categories in order to keep exactness. We hope to convince the reader that this question leads quite inevitably to the notion of a derived category and derived functors.

In the first part, we give the general construction of derived categories of an abelian category and derived functors. The motivating question leads to the notion of adapted resolutions and quasi-isomorphisms in Sect. 2.1. As an intermediate step we arrive at the notion of a homotopy category in Sect. 2.2. In Sect. 2.3, the derived category is constructed and its triangulated structure discussed. Finally in Sect. 2.4, we will see that derived functors become exact on the derived level.

In the second part, we focus on the derived category of sheaves, especially on the construction of derived functors. There we deal with left-exact functors like Hom and push-forward in Sect. 3.1, and then with right-exact functors like $\otimes$ and pull-back in Sect. 3.2. Moreover, we give some compatibilities among these functors in Sect. 3.3. Finally, we discuss a bit the important notion of Fourier-Mukai transforms in Sect. 3.4.

The third section can be seen as an application of the theory of Fourier-Mukai transforms. Moreover, it should pave the way for [11]. There we present some comparatively recent results on the auto-equivalences of the derived category of a complex projective K3 surface.

For full details, we refer to the wonderful books [3] and [7] which these notes follow to quite some extent. But we also want to mention the books [4], [9] and [10] which were very helpful when compiling these notes. In this text, we do not give proper references, because all results are nowadays pretty standard and can be found in any of the above mentioned sources. The only exception is the last section were more recent results are presented and therefore some references given. We want to stress that most of the proofs below are just indications of the main ideas, and usually borrowed from one of the above mentioned books. We hope that these indications give the novice a good feeling about what is going on, and ideally leave such a reader well-prepared for a closer study using a textbook.

Prerequisites We assume that the reader has a background in algebraic geometry and is acquainted with basic notions from homological algebra.

Conventions With $\Bbbk$, we denote a field which is not necessarily algebraically closed or of characteristic zero. When we speak of categories, we implicitly assume that they are $\Bbbk$-linear (even though this is not strictly necessary for most of the abstract theory). By a variety we mean an integral separated scheme of finite type over $\Bbbk$.

2 From Abelian to Derived Categories

2.1 Adapted Resolutions

In this section, we will introduce the central notion of quasi-isomorphism and speak about adapted resolutions. Moreover, we will give a definition of the derived category by a universal property.

We fix some notation. Let $\mathcal{A}$ be an abelian category, i.e. we can speak of short exact sequences. We denote by $\mathrm{Com}(\mathcal{A})$ the *category of (cochain) complexes* in $\mathcal{A}$, i.e. its objects are sequences

$$C^\bullet: \quad \cdots \to C^{i-1} \xrightarrow{d^{i-1}} C^i \xrightarrow{d^i} C^{i+1} \to \cdots$$

with $C^i \in \mathcal{A}$ and $d^i \circ d^{i-1} = 0$ for all $i \in \mathbb{Z}$, and morphisms are *maps of complexes*, i.e.

$$
\begin{array}{ccccccc}
C^\bullet & \cdots \longrightarrow & C^i & \xrightarrow{d^i} & C^{i+1} & \longrightarrow & \cdots \\
f^\bullet \downarrow & & f^i \downarrow & \circlearrowleft & \downarrow f^{i+1} & & \\
D^\bullet & \cdots \longrightarrow & D^i & \xrightarrow{d^i} & D^{i+1} & \longrightarrow & \cdots
\end{array}
$$

With $\mathrm{Com}^+(\mathcal{A})$, $\mathrm{Com}^-(A)$ and $\mathrm{Com}^b(\mathcal{A})$ we denote the full subcategory of bounded below, bounded above, and bounded complexes, respectively. For example, $C^\bullet \in \mathrm{Com}^+(\mathcal{A})$ if $C^i = 0$ for $i \ll 0$.

By slight abuse of notation, given some class of objects $\mathcal{G}$ in $\mathcal{A}$, we will write $\mathrm{Com}(\mathcal{G})$ for the (full) subcategory consisting of those complexes in $\mathrm{Com}(\mathcal{A})$ which are sequences of objects in $\mathcal{G}$.

Due to $d^i \circ d^{i-1} = 0$ or, equivalently, im $d^{i-1} \subseteq \ker d^i$, we can take *cohomology* of any $C^{\bullet} \in \mathrm{Com}(\mathcal{A})$, i.e.

$$\mathcal{H}^i(C^{\bullet}) = \frac{\ker d^i}{\mathrm{im}\, d^{i-1}}.$$

Note that we can consider $\mathcal{H}^{\bullet}(C^{\bullet})$ when equipped with the zero-differential again as an element of $\mathrm{Com}(\mathcal{A})$. We say that $C^{\bullet}$ is *acyclic* (or *exact*) if it has no cohomology, i.e. $\mathcal{H}^{\bullet}(C^{\bullet}) = 0$.

Moreover, a map $f^{\bullet} \colon C^{\bullet} \to D^{\bullet}$ of complexes induces a map

$$\mathcal{H}^{\bullet}(f^{\bullet}) \colon \mathcal{H}^{\bullet}(C^{\bullet}) \to \mathcal{H}^{\bullet}(D^{\bullet}).$$

We say that $f^{\bullet}$ is a *quasi-isomorphism* if $\mathcal{H}^{\bullet}(f^{\bullet})$ is an isomorphism.

Definition 2.1 Let $F \colon \mathcal{A} \to \mathcal{B}$ be a left-exact functor between abelian categories. Let $\mathcal{I}_F$ be a class of objects in $\mathcal{A}$. We say that $\mathcal{I}_F$ is *F-adapted* if

- $F(I^{\bullet})$ is acyclic for any acyclic complex $I^{\bullet} \in \mathrm{Com}^+(\mathcal{I}_F)$;
- for any $A \in \mathcal{A}$ there is an injection $A \hookrightarrow I$ with $I \in \mathcal{I}_F$.

The first property says in particular that F preserves exactness of short exact sequences of objects in $\mathcal{I}_F$. The second property ensures that we can replace any A by an adapted resolution, as the following lemma shows.

Lemma 2.2 *Let $F \colon \mathcal{A} \to \mathcal{B}$ be a left-exact functor between abelian categories, and let $\mathcal{I}_F$ be an F-adapted class. Then for any $A \in \mathcal{A}$, there is a complex $I^{\bullet} \in \mathrm{Com}^+(\mathcal{I}_F)$ such that*

$$
\begin{array}{ccccccccc}
A & & \cdots \longrightarrow & 0 & \longrightarrow & A & \longrightarrow & 0 & \longrightarrow \cdots \\
{\scriptstyle f}\downarrow & & & & & \downarrow{\scriptstyle f} & & & \\
I^{\bullet} & & \cdots \longrightarrow & 0 & \longrightarrow & I^0 & \longrightarrow & I^1 & \longrightarrow \cdots
\end{array}
$$

is a quasi-isomorphism. We call $I^{\bullet}$ an F-adapted resolution of A.

Proof We only indicate how $I^{\bullet}$ can be constructed. By the second property of an F-adapted class, there is an injection $f \colon A \hookrightarrow I^0$ for some $I^0 \in \mathcal{I}_F$. Now continue inductively, by choosing I^{i+1} to contain the cokernel of the previous map, and setting $d^i \colon I^i \to I^{i+1}$ to be the composition $I^i \twoheadrightarrow \mathrm{coker} \hookrightarrow I^{i+1}$. $\square$

Remark 2.3 The above lemma can be generalised to complexes, i.e. for any $A^{\bullet} \in \mathrm{Com}^+(\mathcal{A})$ there is an adapted $I^{\bullet} \in \mathrm{Com}^+(\mathcal{I}_F)$ and a quasi-isomorphism $f^{\bullet} \colon A^{\bullet} \to I^{\bullet}$.

Proposition 2.4 *If $\mathcal{A}$ contains enough injective objects, i.e. for any $A \in \mathcal{A}$ there is an inclusion $A \hookrightarrow I$ with I injective, then the class $\mathcal{I}_{\mathcal{A}}$ of all injective objects in $\mathcal{A}$ is adapted for all left-exact functors starting in $\mathcal{A}$.*

Proof This question can be reduced to short exact sequences, by breaking up $I^\bullet$ into $0 \to \ker(d^i) \to I^i \to \operatorname{im}(d^i) \to 0$. Now the statement can be shown using two standard facts about injective objects:

- Any short exact sequence $0 \to I \to A \to B \to 0$ in $\mathcal{A}$ with I injective splits. In particular, its image under F is still exact.
- For a short exact sequence $0 \to I' \to I \to A \to 0$ in $\mathcal{A}$ with I, I' injective, also A is injective.   □

Remark 2.5 We have dealt here only with left-exact functors, but there is a dual story. For a right-exact functor $F: \mathcal{A} \to \mathcal{B}$, an F-adapted class $\mathcal{P}_F$ should satisfy

- $F(P^\bullet)$ is acyclic for any acyclic complex $P^\bullet \in \operatorname{Com}^-(\mathcal{P}_F)$;
- for any $A \in \mathcal{A}$ there is an surjection $P \twoheadrightarrow A$ with $P \in \mathcal{P}_F$.

Moreover, we get an F-adapted resolution $P^\bullet \to A$ in $\operatorname{Com}^-(\mathcal{P}_F)$. Finally, if there are enough projective objects, the class of projective objects is adapted for all right-exact functors.

The discussion of this section shows, that we want to identify quasi-isomorphic complexes, as such an identification allows us to pass from an object to an adapted resolution. This aim is summarised in the following definition.

Definition 2.6 Let $\mathcal{A}$ be an abelian category. A category $\mathcal{D}$ together with a functor $Q: \operatorname{Com}(\mathcal{A}) \to \mathcal{D}$ is called *derived category* of $\mathcal{A}$ if

- $Q(f^\bullet)$ is an isomorphism for any quasi-isomorphism $f^\bullet$;
- any other functor $F: \operatorname{Com}(\mathcal{A}) \to \mathcal{T}$ which maps quasi-isomorphisms to isomorphism factors uniquely through $\mathcal{D}$:

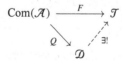

Analogously, we can define the bounded below, bounded above and bounded derived category of $\mathcal{A}$.

This definition by a universal property automatically yields the uniqueness up to equivalence, but we have yet to provide existence.

2.2 The Homotopy Category

In this section, we will introduce homotopies and show that they induce quasi-isomorphisms. In the case that there are enough injective objects (or dually, projective objects), these are all quasi-isomorphisms.

There is a cheap way to build a map of complexes:

Lemma 2.7 *Let $C^\bullet$ and $D^\bullet$ be two complexes and $\{h^i : C^i \to D^{i-1}\}_i$ be a sequence of morphisms in $\mathcal{A}$. Then $f^i = h^{i+1}d^i + d^{i-1}h^i : C^i \to D^i$ fit together to a map of complexes:*

$$
\begin{array}{ccc}
C^\bullet & & \cdots \longrightarrow C^i \xrightarrow{d^i} C^{i+1} \longrightarrow \cdots \\
f^\bullet \downarrow & & \quad{}^{h^i}\swarrow \;\; \downarrow f^i \;\; \swarrow_{h^{i+1}} \\
D^\bullet & & \cdots \longrightarrow D^{i-1} \xrightarrow{d^{i-1}} D^i \longrightarrow \cdots
\end{array}
$$

Proof We only have to check that $d^i(h^{i+1}d^i + d^{i-1}h^i) = (h^{i+2}d^{i+1} + d^i h^{i+1})d^i$, which holds as $C^\bullet$ and $D^\bullet$ are complexes. $\qquad\square$

Definition 2.8 Let $f^\bullet, g^\bullet : C^\bullet \to D^\bullet$ be two maps of complexes. We say that $f^\bullet$ and $g^\bullet$ are *homotopic*, if there is a sequence of morphisms $h^i : C^i \to D^{i-1}$, such that $f^i - g^i = h^{i+1}d^i + d^{i-1}h^i$ for all $i \in \mathbb{Z}$. We write $f^\bullet \sim g^\bullet$ in this case.

Lemma 2.9 *Let $f^\bullet, g^\bullet : C^\bullet \to D^\bullet$ be two maps of complexes. If $f^\bullet$ and $g^\bullet$ are homotopic, then the induced maps $\mathcal{H}^\bullet(f^\bullet)$ and $\mathcal{H}^\bullet(g^\bullet)$ are equal. Moreover, homotopy $\sim$ defines an equivalence relation for maps of complexes.*

As a corollary we get that homotopies are a source of quasi-isomorphisms:

Remark 2.10 Let $f^\bullet : C^\bullet \to D^\bullet$ and $g^\bullet : D^\bullet \to C^\bullet$ be two maps of complexes such that $f^\bullet \circ g^\bullet \sim \mathrm{id}_{D^\bullet}$ and $g^\bullet \circ f^\bullet \sim \mathrm{id}_{C^\bullet}$. Then both f and g are quasi-isomorphisms, as

$$
\mathcal{H}^\bullet(f^\bullet) \circ \mathcal{H}^\bullet(g^\bullet) = \mathcal{H}^\bullet(f^\bullet \circ g^\bullet) = \mathcal{H}^\bullet(\mathrm{id}_{D^\bullet}) = \mathrm{id}_{\mathcal{H}^\bullet(D^\bullet)}
$$

and similarly for $\mathcal{H}^\bullet(g^\bullet) \circ \mathcal{H}^\bullet(f^\bullet)$.

Definition 2.11 Let $\mathcal{A}$ be an abelian category. The *homotopy category* $\mathrm{Hot}(\mathcal{A})$ of $\mathcal{A}$ consists of

- objects: complexes of objects in $\mathcal{A}$;
- morphisms: maps of complexes modulo homotopy

$$
\mathrm{Hom}_{\mathrm{Hot}(\mathcal{A})}(C^\bullet, D^\bullet) := \mathrm{Hom}_{\mathrm{Com}(\mathcal{A})}(C^\bullet, D^\bullet)/\sim
$$

Moreover, we can define $\mathrm{Hot}^+(\mathcal{A})$, $\mathrm{Hot}^-(\mathcal{A})$ and $\mathrm{Hot}^b(\mathcal{A})$ as the full subcategories of $\mathrm{Hot}(\mathcal{A})$ consisting of bounded below, bounded above, and bounded complexes, respectively. Similarly for any full additive subcategory $\mathcal{C}$ of $\mathcal{A}$, we can define the homotopy category $\mathrm{Hot}(\mathcal{C})$ (and bounded analogues) by restricting to complexes of objects in $\mathcal{C}$.

Enough Injective Objects In the case that enough injectives are present, we can say even more about the homotopy category.

Proposition 2.12 *Let $f: C \to D$ be a morphism in an abelian category $\mathcal{A}$ with enough injectives. Then the following holds*

- *For any choice of injective resolutions $C \to I^\bullet$ and $D \to J^\bullet$, f can be lifted to a map of complexes $f^\bullet: I^\bullet \to J^\bullet$, in particular the following diagram commutes:*

$$
\begin{array}{ccc}
C & \longrightarrow & I^0 \\
f\downarrow & & \downarrow f^0 \\
D & \longrightarrow & J^0
\end{array}
$$

- *any two such lifts are homotopic.*

Proof We only show existence, because uniqueness up to homotopy can be shown similarly. By injectivity of J^0, there is the lift f^0 of the composition $C \to D \hookrightarrow J^0$:

$$
\begin{array}{ccc}
C & \longrightarrow & I^0 \\
f\downarrow & & \downarrow f^0 \\
D & \longrightarrow & J^0
\end{array}
$$

The statement can be shown by induction, continuing the argument like in that proof of Lemma 2.2. □

Remark 2.13 Actually, a similar proof which is (notationally) more involved shows that any $f^\bullet: C^\bullet \to D^\bullet$ in $\mathrm{Com}^+(\mathcal{A})$ for an abelian category $\mathcal{A}$ with enough injectives can be lifted to a map of complexes $\tilde{f}^\bullet: I^\bullet \to J^\bullet$ with $I^\bullet$ and $J^\bullet$ injective resolutions of $C^\bullet$ and $D^\bullet$. Again, any two such lifts are homotopic.

Remark 2.14 For $f^\bullet = \mathrm{id}_{C^\bullet}: C^\bullet \to C^\bullet$, we get as an important special case that any two injective resolutions of $C^\bullet$ are homotopic.

Finally, there is a converse to Remark 2.10, whose proof needs Remark 2.13.

Proposition 2.15 *Let $\mathcal{A}$ be an abelian category with enough injectives. Let $f^\bullet: I^\bullet \to J^\bullet$ be a quasi-isomorphism of injective complexes in $\mathrm{Com}^+(\mathcal{A})$. Then there is a quasi-isomorphism $g^\bullet: J^\bullet \to I^\bullet$ with homotopies $f^\bullet \circ g^\bullet \sim \mathrm{id}_{J^\bullet}$ and $g^\bullet \circ f^\bullet \sim \mathrm{id}_{I^\bullet}$.*

Given an abelian category $\mathcal{A}$ with enough injectives $\mathcal{I}$, the last proposition shows that quasi-isomorphisms become invertible in $\mathrm{Hot}^+(\mathcal{I})$, but even more is true.

Proposition 2.16 *Let $\mathcal{A}$ be an abelian category with enough injectives $\mathcal{I}$. Then $\mathrm{Hot}^+(\mathcal{I})$ is the bounded below derived category $\mathcal{D}^+(\mathcal{A})$ of $\mathcal{A}$.*

For an arbitrary F-adapted class, a quasi-isomorphism might not have a homotopy inverse like in Proposition 2.15. The crucial ingredient there is the lifting property of injective objects. As usual, we can enforce the existence of such homotopy inverses by formally introducing them. This will be done in the following section.

Remark 2.17 In the presence of enough projective objects $\mathcal{P}$ in an abelian category $\mathcal{A}$, we get statements dual to those in this subsection. Most notably, in this case $\text{Hot}^-(\mathcal{P})$ is the bounded above derived category $\mathcal{D}^-(\mathcal{A})$ of $\mathcal{A}$.

2.3 The Derived Category

In this section, we will finally give a construction of the derived category and speak about its triangulated structure.

For an abelian category $\mathcal{A}$ let *qis* denote the class of all quasi-isomorphisms. We finally state the existence of the derived category in general, which is due to Verdier.

Theorem 2.18 *Let $\mathcal{A}$ be an abelian category. The category $\mathcal{D}(\mathcal{A})$:= $\text{Hot}(\mathcal{A})[qis^{-1}]$ given by*

- *objects: complexes of objects in $\mathcal{A}$;*
- *morphisms: the same as in $\text{Hot}(\mathcal{A})$ but with quasi-isomorphisms formally inverted:*

$$\text{Hom}_{\mathcal{D}(\mathcal{A})}(C^{\bullet}, D^{\bullet}) := \text{Hom}_{\text{Hot}(\mathcal{A})}(C^{\bullet}, D^{\bullet})[qis^{-1}] =$$

$$= \left\{ \begin{array}{c} \tilde{C}^{\bullet} \\ {}^{s^{\bullet}}\!\!\nearrow \quad \searrow^{\tilde{f}^{\bullet}} \\ C^{\bullet} \dashrightarrow_{f} D^{\bullet} \end{array} \,\middle|\, \begin{array}{l} \tilde{C}^{\bullet} \text{ complex of objects in } \mathcal{A}, \\ s^{\bullet} \in \text{Hom}_{\text{Hot}(\mathcal{A})}(C^{\bullet}, \tilde{C}^{\bullet}) \text{ quasi-isomorphism,} \\ \tilde{f}^{\bullet} \in \text{Hom}_{\text{Hot}(\mathcal{A})}(\tilde{C}^{\bullet}, D^{\bullet}). \end{array} \right\}$$

is the derived category of $\mathcal{A}$.

Remark 2.19 The definition of the morphisms above as *roofs* is a bit informal. For example one needs to show that a zig-zag of two such roofs can be composed to a single roof. For this the key ingredient is that quasi-isomorphisms form a *localising class* of morphisms inside $\text{Hot}(\mathcal{A})$. To invert such a class is also called *Verdier localisation*.

One can construct the derived category of $\mathcal{A}$ also by formally inverting quasi-isomorphisms in $\text{Com}(\mathcal{A})$, see [3, §III.2.2]. But this causes several technical problems which can be avoided by passing first to $\text{Hot}(\mathcal{A})$.

Remark 2.20 There is a natural functor

$$\mathcal{A} \to \mathcal{D}(\mathcal{A}), \ C \mapsto [\cdots \to 0 \to C \to 0 \to \cdots]$$

mapping any $C \in \mathcal{A}$ to the complex with C at the zero position. By slight abuse of notation, we will denote this complex again by C.

This functor is fully faithful, i.e. for any two $C, D \in \mathcal{A}$ holds

$$\operatorname{Hom}_{\mathcal{D}(\mathcal{A})}(C, D) = \operatorname{Hom}_{\mathcal{A}}(C, D).$$

Moreover, the essential image of this functor consists of all complexes $C^\bullet$ such that $\mathcal{H}^i(C^\bullet) = 0$ for $i \neq 0$.

Triangulated Structure As the objects of $\mathcal{D}(\mathcal{A})$ are complexes, there is the *shift functor*:

$$[1]: \mathcal{D}(\mathcal{A}) \to \mathcal{D}(\mathcal{A}), \ C^\bullet \mapsto C^\bullet[1] := C^{\bullet+1}.$$

The usual convention here is that the sign of the differential changes under shift, i.e.

$$d^i_{C^\bullet[1]} = -d^{i+1}_{C^\bullet}.$$

With this shift functor, we can define the *(mapping) cone* of a map of complexes $f^\bullet: C^\bullet \to D^\bullet$ as the complex $\operatorname{Cone}(f^\bullet)$ with

$$\operatorname{Cone}^i(f^\bullet) = C^{i+1} \oplus D^i, \ d^i_{\operatorname{Cone}(f^\bullet)} = \begin{pmatrix} -d^{i+1}_{C^\bullet} & 0 \\ f^{i+1} & d^i_{D^\bullet} \end{pmatrix}$$

We will also write $\operatorname{Cone}(f^\bullet) = C^\bullet[1] \oplus D^\bullet$ as a semi-direct sum. With these definitions f induces a triangle of morphisms in $\mathcal{D}(\mathcal{A})$:

$$C^\bullet \xrightarrow{f^\bullet} D^\bullet \xrightarrow{j^\bullet} \operatorname{Cone}(f^\bullet) \xrightarrow{p^\bullet} C^\bullet[1].$$

where $j^\bullet$ is the inclusion of the semi-direct summand $D^\bullet$ and $p^\bullet$ the projection onto $C^\bullet[1]$.

Remark 2.21 We want to stress that only for honest maps of complexes we have an explicit construction of the mapping cone. Notationally, we will therefore mark a map of complexes $f^\bullet$ always with a dot, in order to distinguish them from (general) morphisms f in $\mathcal{D}(\mathcal{A})$ which are roofs.

Definition 2.22 We call a sequence of morphisms $C^\bullet \to D^\bullet \to E^\bullet \to C^\bullet[1]$ an *exact triangle* (or *distinguished triangle*)

We call a sequence of morphisms $C^\bullet \to D^\bullet \to E^\bullet \to C^\bullet[1]$ an *exact triangle* (or *distinguished triangle*) if there is a commutative diagram in $\mathcal{D}(\mathcal{A})$ of the form

$$
\begin{array}{ccccccc}
C^\bullet & \xrightarrow{f} & D^\bullet & \longrightarrow & E^\bullet & \longrightarrow & C^\bullet[1] \\
{\scriptstyle c}\downarrow{\scriptstyle \wr} & & {\scriptstyle d}\downarrow{\scriptstyle \wr} & & {\scriptstyle e}\downarrow{\scriptstyle \wr} & & {\scriptstyle c[1]}\downarrow{\scriptstyle \wr} \\
C'^\bullet & \xrightarrow{\tilde{f}^\bullet} & D'^\bullet & \longrightarrow & \operatorname{C}(\tilde{f}^\bullet) & \longrightarrow & C'^\bullet[1]
\end{array}
$$

with $\tilde{f}^\bullet$ a map of complexes.

The complex $E^\bullet$ is called the *cone* of the morphism $f \colon C^\bullet \to D^\bullet$ and denoted by $\mathrm{Cone}(f)$.

Remark 2.23 The triangle is often visualised in the following way:

$$
\begin{array}{ccc}
C^\bullet & \longrightarrow & D^\bullet \\
& \underset{[1]}{\nwarrow} \quad \swarrow & \\
& E^\bullet &
\end{array}
$$

where the lower left arrow involves a shift by one. Note that a triangle can also be extended to a long sequence

$$\cdots \to E^\bullet[-1] \to C^\bullet \to D^\bullet \to E^\bullet \to C^\bullet[1] \to D^\bullet[1] \to \cdots$$

which is actually a complex in $\mathcal{D}(\mathcal{A})$, see Remark 2.30 below.

Distinguished triangles generalise short exact sequences in a very precise way.

Proposition 2.24 *Let* $0 \to C \xrightarrow{f} D \xrightarrow{g} E \to 0$ *be a short exact sequence in the abelian category* $\mathcal{A}$*. Considering these objects in* $\mathcal{D}(\mathcal{A})$*, they form the exact triangle* $C \xrightarrow{f} D \xrightarrow{g} E \to C[1]$.

Proof We consider the exact triangle $C \xrightarrow{f} D \xrightarrow{j} \mathrm{Cone}(f) \xrightarrow{p} C[1]$. Note that $\mathrm{Cone}(f)$ is a two-term complex, quasi-isomorphic to E:

$$
\begin{array}{ccc}
C(f) & \qquad & 0 \longrightarrow C \xrightarrow{f} D \longrightarrow 0 \\
{\scriptstyle g}\big\downarrow & & \hspace{3.5em} \big\downarrow{\scriptstyle g} \\
E & & 0 \longrightarrow E \longrightarrow 0
\end{array}
$$

One can check that this quasi-isomorphism can be completed to a diagram of quasi-isomorphisms, which shows the claim:

$$
\begin{array}{ccccccc}
C & \xrightarrow{f} & D & \xrightarrow{j} & C(f) & \xrightarrow{p} & C[1] \\
\big\| & & \big\| & & {\scriptstyle g \in qis}\big\downarrow & & \big\| \\
C & \xrightarrow{f} & D & \xrightarrow{g} & E & \xrightarrow{h} & C[1]
\end{array}
$$

$\square$

Remark 2.25 The mindful reader may ask about the third morphism in the triangle, namely $h \colon E \to C[1]$. Note that $h \in \mathrm{Hom}_{\mathcal{D}(\mathcal{A})}(E, C[1]) = \mathrm{Ext}^1_{\mathcal{A}}(E, C)$, see Example 2.40. It is well-known that $\mathrm{Ext}^1(E, C)$ corresponds to extensions, so h encodes the middle term D; see [3, §III.6.2] for a discussion of this.

Theorem 2.26 *Let $\mathcal{A}$ be an abelian category. Then its derived category $\mathcal{D}(\mathcal{A})$ is a triangulated category, i.e. it satisfies the four axioms* **TR1–TR4**.

TR1 The triangle $C^\bullet \xrightarrow{\mathrm{id}} C^\bullet \to 0 \to C^\bullet[1]$ is exact.

Any triangle isomorphic to an exact one is again exact.

Any morphism $f: C^\bullet \to D^\bullet$ can be completed to an exact triangle.

Proof of **TR1** *for* $\mathcal{D}(\mathcal{A})$ For the derived category $\mathcal{D}(\mathcal{A})$ the second clause is satisfied by definition. For the first clause, one only needs to check that the cone Cone(id) is homotopic to the zero complex. To see the last, write a morphism $f: C^\bullet \to D^\bullet$ as a roof $f = \tilde{f}^\bullet \circ (s^\bullet)^{-1}$, which fits into the following commutative diagram of exact triangles:

$$
\begin{array}{ccccccc}
C^\bullet & \xrightarrow{f} & D^\bullet & \dashrightarrow{j^\bullet} & C(f) & \xrightarrow{s^\bullet[1]\circ p^\bullet} & C^\bullet[1] \\
{\scriptstyle (s^\bullet)^{-1}}\downarrow{\scriptstyle \wr} & & \| & & \| & & \wr\downarrow{\scriptstyle (s^\bullet)^{-1}[1]} \\
\tilde{C}^\bullet & \xrightarrow{\tilde{f}^\bullet} & D^\bullet & \xrightarrow{j^\bullet} & C(f^\bullet) & \xrightarrow{p^\bullet} & \tilde{C}^\bullet[1]
\end{array}
$$

$\square$

Remark 2.27 By the last clause of **TR1**, cones in the derived category $\mathcal{D}(\mathcal{A})$ unifying both kernel and cokernel of the abelian category $\mathcal{A}$. More precisely, considering a map $f: C \to D$ in $\mathcal{A}$ as a map of complexes in $\mathcal{D}(\mathcal{A})$, one can check that $\mathcal{H}^{-1}(\mathrm{Cone}(f)) = \ker(f)$ and $\mathcal{H}^0(\mathrm{Cone}(f)) = \mathrm{coker}(f)$.

TR2 The triangle $C^\bullet \xrightarrow{f} D^\bullet \xrightarrow{g} E^\bullet \xrightarrow{h} C^\bullet[1]$ is exact if and only if $D^\bullet \xrightarrow{g} E^\bullet \xrightarrow{h} C^\bullet[1] \xrightarrow{-f[1]} D^\bullet[1]$ is.

Proof of **TR2** *for* $\mathcal{D}(\mathcal{A})$ We only discuss " $\Longrightarrow$ " a bit (as the converse direction is analogous). By **TR1** we may assume that $f = f^\bullet$ is a map of complexes, $E^\bullet \cong \mathrm{Cone}(f^\bullet)$ and $g^\bullet: D^\bullet \to \mathrm{Cone}(f^\bullet)$ is the inclusion as a semi-direct summand. We have to show that $C^\bullet[1]$ is isomorphic to $\mathrm{Cone}(g^\bullet)$. Note that by our simplifications $\mathrm{Cone}(g^\bullet) = D^\bullet[1] \oplus C^\bullet[1] \oplus D^\bullet$. One can now check that

$$(-f^\bullet[1], \mathrm{id}, 0): D^\bullet[1] \oplus C^\bullet[1] \oplus D^\bullet \to \mathrm{Cone}(g^\bullet)$$

gives the desired isomorphism. $\square$

TR3 Given two exact triangles and two morphisms c and d as below:

$$
\begin{array}{ccccccc}
C^\bullet & \xrightarrow{f} & D^\bullet & \longrightarrow & E^\bullet & \longrightarrow & C^\bullet[1] \\
\downarrow{\scriptstyle c} & & \downarrow{\scriptstyle d} & & \downarrow{\scriptstyle e} & & \downarrow{\scriptstyle c[1]} \\
C'^\bullet & \xrightarrow{f'} & D'^\bullet & \longrightarrow & E'^\bullet & \longrightarrow & C'^\bullet[1]
\end{array}
$$

then there is a (not necessarily unique) morphism e making this diagram commutative.

Proof of **TR3** *for* $\mathcal{D}(\mathcal{A})$ After replacing f and f' by maps of complexes, and $E^\bullet$ and $E'^\bullet$ by the respective cones, one can check that $e = (c[1], d)$ fits into the diagram.   □

Remark 2.28 One might suppose by our reasoning about the existence of the dashed morphism in $\mathcal{D}(\mathcal{A})$, that cones are functorial in $\mathcal{D}(\mathcal{A})$, i.e. given a natural transformation $\eta \colon F \to G$ between functors preserving exact triangles, there exist the functor $\mathrm{Cone}(\eta)$ of cones.

In a naive way, such a statement is wrong. Take for example the exact triangle $C^\bullet \xrightarrow{\mathrm{id}} C^\bullet \to 0 \to C^\bullet[1]$ for any non-trivial $C^\bullet \in \mathcal{D}(\mathcal{A})$. After shifting, we can write down the following diagram

$$
\begin{array}{ccccccc}
C^\bullet & \longrightarrow & 0 & \longrightarrow & C^\bullet[1] & \xrightarrow{-\mathrm{id}[1]} & C^\bullet[1] \\
\downarrow & & \downarrow & & \downarrow & & \downarrow \\
0 & \longrightarrow & C^\bullet[1] & \xrightarrow{-\mathrm{id}[1]} & C^\bullet[1] & \longrightarrow & 0
\end{array}
$$

All the non-labelled solid arrows are just zero morphisms. For the dashed arrow, we can choose any morphism $C^\bullet[1] \to C^\bullet[1]$.

But in a more sophisticated way, such a statement is true for derived categories using dg-enhancements, a topic that we will not enter here.

Remark 2.29 By [12, Lem. 2.2], **TR3** is not necessary as an axiom, it follows from the other three axioms. But we prefer to keep it in this list, as it is an often used property of triangulated categories. Finally, this shows also that the non-functoriality of cones inside a triangulated category goes deeper than **TR3**.

Remark 2.30 From **TR1–TR3** follows that in exact triangles, the composition of two consecutive morphisms is zero. This follows from the following diagram (and shifted versions):

$$
\begin{array}{ccccccc}
C^\bullet & \xrightarrow{\mathrm{id}} & C^\bullet & \longrightarrow & 0 & \longrightarrow & C^\bullet[1] \\
\downarrow{\scriptstyle\mathrm{id}} & & \downarrow{\scriptstyle f} & \circlearrowleft & \downarrow & & \downarrow{\scriptstyle\mathrm{id}} \\
C^\bullet & \xrightarrow{f} & D^\bullet & \xrightarrow{g} & E^\bullet & \xrightarrow{h} & C^\bullet[1]
\end{array}
$$

TR4 Given two morphisms $f \colon C^\bullet \to D^\bullet$ and $g \colon D^\bullet \to E^\bullet$, there is a triangle of cones

$$\mathrm{Cone}(f) \to \mathrm{Cone}(g \circ f) \to \mathrm{Cone}(g) \to \mathrm{Cone}(f)[1]$$

which fits into the following commutative diagram (where we suppress for simplicity the last degree-increasing morphism in the exact triangles):

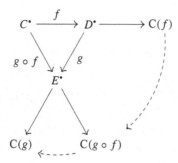

We omit the proof of **TR4** for $\mathcal{D}(\mathcal{A})$, as it is more technical.

Remark 2.31 The last axiom goes under the name *octahedral axiom* as it can be pictured by a diagram in the form of an octahedron, but we think that the above diagram is more helpful. It comes from the following lemma about abelian categories, which one might call *windmill lemma*:

Given $f: C \to D$ and $g: D \to E$ in an abelian category $\mathcal{A}$. Then there is an exact sequence of kernels and cokernels fitting into the commutative diagram of Fig. 1. The proof is an exercise in homological algebra, but may also be deduced from the octahedral axiom using Remark 2.27.

Remark 2.32 Just by restricting the class of exact triangles, one can also see that $\mathcal{D}^-(\mathcal{A})$, $\mathcal{D}^+(\mathcal{A})$ and $\mathcal{D}^b(\mathcal{A})$ are triangulated categories.

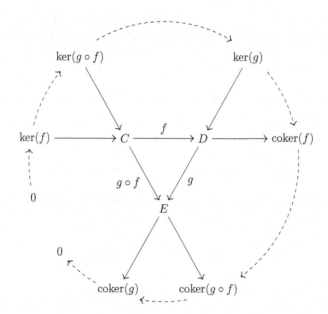

Fig. 1 The windmill lemma

2.4 Exact Functors

In this section, we introduce the notion of an exact functor between triangulated categories.

Definition 2.33 A functor $F: \mathcal{T} \to \mathcal{T}'$ between triangulated categories is called *exact*, if

- F commutes with shifts, i.e. there is a functor isomorphism $F \circ [1]_{\mathcal{T}} \cong [1]_{\mathcal{T}'} \circ F$;
- for any exact triangle $A \to B \to C \to A[1]$ in $\mathcal{T}$, its image $F(A) \to F(B) \to F(C) \to F(A)[1]$ is exact in $\mathcal{T}'$.

Proposition 2.34 *Let $\mathcal{A}$ be an abelian category and $\mathcal{D}(\mathcal{A})$ its derived category. Then there is the functor $\mathcal{H}^{\bullet}: \mathcal{D}(\mathcal{A}) \to \mathcal{D}(\mathcal{A})$ which sends each complex $C^{\bullet}$ to its cohomology $\mathcal{H}^{\bullet}(C^{\bullet})$ equipped with the zero differential. This functor is an exact functor.*

Remark 2.35 The statement of this proposition is usually formulated differently. For any exact triangle $C^{\bullet} \xrightarrow{c} D^{\bullet} \xrightarrow{d} E^{\bullet} \xrightarrow{e} C^{\bullet}[1]$ in $\mathcal{D}(\mathcal{A})$, its image under $\mathcal{H}^{\bullet}$ can be rolled out to a *long exact sequence in cohomology*:

$$\cdots \to \mathcal{H}^i(C^{\bullet}) \xrightarrow{\mathcal{H}^i(c)} \mathcal{H}^i(D^{\bullet}) \xrightarrow{\mathcal{H}^i(d)} \mathcal{H}^i(E^{\bullet}) \xrightarrow{\mathcal{H}^i(e)} \mathcal{H}^{i+1}(C^{\bullet}) \to \cdots$$

For defining derived functors in general, we first give an analogue of Proposition 2.16.

Proposition 2.36 *Let $F: \mathcal{A} \to \mathcal{B}$ be a left-exact functor, and let $\mathcal{I}_F$ be an F-adapted class. Then the inclusion $\mathrm{Com}^+(\mathcal{I}_F) \subset \mathrm{Com}^+(\mathcal{A})$ induces an equivalence*

$$\iota_F: \mathrm{Hot}^+(\mathcal{I}_F)[qis^{-1}] \to \mathcal{D}^+(\mathcal{A}).$$

Note that the inverse ι_F^{-1} replaces a complex $C^{\bullet}$ by an F-adapted resolution.

Definition 2.37 Let $F: \mathcal{A} \to \mathcal{B}$ be a left-exact functor, and let $\mathcal{I}_F$ be an F-adapted class. The *right-derived functor* of F is given by

$$\mathsf{R}F: \mathcal{D}^+(\mathcal{A}) \to \mathcal{D}^+(\mathcal{B}), \ C^{\bullet} \mapsto F(\iota_F^{-1}(C^{\bullet})).$$

Definition 2.38 Let $F: \mathcal{A} \to \mathcal{B}$ be a left-exact functor. By taking cohomology, we get induced functors

$$\mathsf{R}^i F := \mathcal{H}^i(\mathsf{R}F): \mathcal{D}^+(\mathcal{A}) \xrightarrow{\mathsf{R}F} \mathcal{D}^+(\mathcal{B}) \xrightarrow{\mathcal{H}^i} \mathcal{B}$$

which are called *i-th right-derived functors* of F.

Moreover, we can precompose $\mathsf{R}^i F$ with $\mathcal{A} \to \mathcal{D}^+(\mathcal{A})$ and get induced functors on the abelian level, which we will denote by the same symbol.

Note that the last step in the definition of $\mathsf{R}^i F$ is taking cohomology of the complex, in particular, $\mathsf{R}^i F(A) = \mathsf{R}^0 F(A[i])$. One can check that $\mathsf{R}^0 F$ and F are naturally isomorphic.

Remark 2.39 The definition of the right-derived functor $\mathsf{R}F$ is based on the choice of an F-adapted class. But one can show that different choices yield isomorphic derived functors, as the derived functor can be characterised by a universal property. See [3, §III.6.7] for more details on this.

Example 2.40 Let $\mathcal{A}$ be an abelian category with enough injectives. One common way to define $\operatorname{Ext}^i_{\mathcal{A}}(C, D)$ is by using an injective resolution $I^\bullet$ of D:

$$\operatorname{Ext}^i_{\mathcal{A}}(C, D) := \mathcal{H}^i \operatorname{Hom}_{\mathcal{A}}(C, I^\bullet)$$

Hence we see that $\operatorname{Ext}^i_{\mathcal{A}}(C, -)$ is the i-th right-derived functor of $\operatorname{Hom}_{\mathcal{A}}(C, -)$. Moreover, we get that

$$\operatorname{Ext}^i_{\mathcal{A}}(C, D) = \operatorname{Hom}_{\mathcal{D}(\mathcal{A})}(C, D[i]).$$

Remark 2.41 Let $F : \mathcal{A} \to \mathcal{B}$ be a right-exact functor. Since we have already given the definition of a right derived functor, we leave the proper definition of a *left-derived functor* $\mathsf{L}F$ and i-th left-derived functor $\mathsf{L}_i F$ as an exercise to the reader.

The following theorem finally tells us that (right-) derived functors preserve exactness in the derived sense. As usual, there is also an analogous statement for left-derived functors.

Theorem 2.42 *Let $F : \mathcal{A} \to \mathcal{B}$ be a left-exact functor between abelian categories, such that there is an F-adapted class in $\mathcal{A}$. Then the right-derived functor $\mathsf{R}F : \mathcal{D}^+(\mathcal{A}) \to \mathcal{D}^+(\mathcal{B})$ is an exact functor.*

Proof The essential steps are

- $\iota_F : \operatorname{Hot}^+(\mathcal{I}_F)[qis^{-1}] \to \mathcal{D}^+(\mathcal{A})$ is exact and therefore ι_F^{-1} as well;
- the functor $\operatorname{Hot}^+(\mathcal{I}_F)[qis^{-1}] \to \mathcal{D}^+(\mathcal{B})$, $I^\bullet \to F(I^\bullet)$ is also exact, which follows from the adaptedness of $\mathcal{I}_F$.

□

Note that the combination of Theorem 2.42 and Remark 2.35 gives again long exact sequences, i.e. for an exact triangle $C^\bullet \to D^\bullet \to E^\bullet \to C^\bullet[1]$ and a right-derived functor $\mathsf{R}F$, there is the long exact sequence

$$\cdots \to \mathsf{R}^i F(C^\bullet) \to \mathsf{R}^i F(D^\bullet) \to \mathsf{R}^i F(E^\bullet) \to \mathsf{R}^{i+1} F(C^\bullet) \to \cdots$$

We close this section with a very important exact functor.

Definition 2.43 Let $\mathcal{A}$ be a $\mathbb{k}$-linear category such that $\mathrm{Hom}(A, B)$ is finite-dimensional for any two objects $A, B \in \mathcal{A}$. An auto-equivalence $S \colon \mathcal{A} \to \mathcal{A}$ is called *Serre functor* of $\mathcal{A}$ if for all $A, B \in \mathcal{A}$ there is an isomorphism

$$\eta_{A,B} \colon \mathrm{Hom}(A, B) \to \mathrm{Hom}(B, SA)^{\vee}$$

of $\mathbb{k}$-vector spaces, which is functorial in both A and B.

We note that without the assumption on the dimension of the Hom-spaces, one runs into problems (as $V^{\vee\vee} \not\cong V$ if V is an infinite-dimensional vector space).

Proposition 2.44 *Let $\mathcal{A}$ be a $\mathbb{k}$-linear triangulated category with Serre functor S. Then S is an exact functor.*

3   The Derived Category of Coherent Sheaves

From now on, we will specialise and consider the abelian category $\mathrm{coh}(X)$ of coherent sheaves on a noetherian scheme X over a field $\mathbb{k}$. We denote the derived category of $\mathrm{coh}(X)$ by $\mathcal{D}^b(X) := \mathcal{D}^b(\mathrm{coh}(X))$.

3.1   Deriving Left-Exact Functors

In this section, we discuss the most prominent left-exact functors in algebraic geometry: Hom, push-forward and sheaf $\mathcal{H}om$.

Theorem 3.1 *Let X be a noetherian scheme. Then there are enough injective objects in the category of quasi-coherent sheaves $\mathrm{Qcoh}(X)$.*

We remark that injective sheaves are hardly finitely generated. Actually, for our applications this is only a minor technical issue.

Proposition 3.2 *Let X be a noetherian scheme. Then the inclusion functor*

$$\mathcal{D}^b(X) \to \mathcal{D}^b(\mathrm{Qcoh}(X))$$

induces an equivalence of $\mathcal{D}^b(X)$ with $\mathcal{D}^b_{\mathrm{coh}}(\mathrm{Qcoh}(X))$, the derived category of complexes of quasi-coherent sheaves with bounded cohomology.

The following proposition will allow us to restrict derived functors to the bounded derived category of coherent sheaves.

Proposition 3.3 *Let X and Y be schemes and $F \colon \mathrm{Qcoh}(X) \to \mathrm{Qcoh}(Y)$ be a left-exact functor. Assume that there is an F-adapted class in $\mathrm{Qcoh}(X)$.*

If for any $\mathcal{F} \in \mathrm{coh}(X)$ holds $RF(\mathcal{F}) \in \mathcal{D}^b(Y)$, then the right-derived functor of F restricts to

$$RF: \mathcal{D}^b(X) \to \mathcal{D}^b(Y).$$

We have formulated this proposition using an adapted class (even though there are enough injective sheaves), because we will also use the dual statement for right-exact functors.

Proof Let $\mathcal{F}^\bullet \in \mathcal{D}^b(X)$ be a bounded complex with $\mathcal{F}^i = 0$ for $i > n$. Choose some adapted resolution $\mathcal{F}^\bullet \to I^\bullet = [\cdots 0 \to I^m \to I^{m+1} \to \cdots]$, which might be in $\mathcal{D}^+(X)$. Then its truncation complex

$$I^{\leq n}: \quad [\cdots \to 0 \to I^m \to \cdots \to I^{n-1} \xrightarrow{d^n} I^n \to \ker(d^n) \to 0 \to \cdots]$$

is still quasi-isomorphic to $\mathcal{F}^\bullet$, but $\ker(d^n) \in \mathrm{coh}(X)$ will not be F-acyclic in general. Nevertheless, by assumption $RF(\ker(d^n)) \in \mathcal{D}^b(Y)$, so we can replace $\ker(d^n)$ by some bounded adapted resolution $J^\bullet$. One can check that $I^{\leq n}$ and $J^\bullet$ fit together to form a bounded adapted resolution of $F^\bullet$. $\qquad\square$

Inner Homomorphisms Given a quasi-coherent sheaf $\mathcal{F} \in \mathrm{Qcoh}(X)$ on a noetherian scheme X over a field $\Bbbk$, there is the left-exact functor

$$\mathrm{Hom}(\mathcal{F}, -): \mathrm{Qcoh}(X) \to \Bbbk \mathrm{Mod} = \mathrm{Qcoh}(\mathrm{Spec}\,\Bbbk).$$

Since X is noetherian, there are enough injectives in $\mathrm{Qcoh}(X)$ and we get

$$R\,\mathrm{Hom}(\mathcal{F}, -): \mathcal{D}^+(\mathrm{Qcoh}(X)) \to \mathcal{D}^+(\Bbbk\,\mathrm{Mod})$$

Actually, this functor can be extended to complexes $\mathcal{F}^\bullet \in \mathrm{Com}^-(\mathrm{Qcoh}(X))$.

Proposition 3.4 *Let X be a smooth and proper variety and $\mathcal{F}^\bullet \in \mathcal{D}^b(X)$. Then $R\mathrm{Hom}(\mathcal{F}^\bullet, -)$ restricts to a functor*

$$R\mathrm{Hom}(\mathcal{F}^\bullet, -): \mathcal{D}^b(X) \to \mathcal{D}^b(\Bbbk\,\mathrm{mod}).$$

This proposition follows from $R\,\mathrm{Hom}(\mathcal{F}^\bullet, -) = R\Gamma \circ R\mathcal{H}om(\mathcal{F}^\bullet, -)$, see Proposition 3.19, and the corresponding statements for $R\Gamma$ and $R\mathcal{H}om$.

Example 3.5 Let C be a projective curve with a singular point x. Then for the skyscraper sheaf $\Bbbk(x)$ one can show that $\mathrm{Ext}^i(\Bbbk(x), \Bbbk(x)) \neq 0$ for all $i > 0$. In particular, the image of $R\,\mathrm{Hom}(\Bbbk(x), -)$ is not contained in $\mathcal{D}^b(\Bbbk\,\mathrm{mod})$.

Push-Forward Let $f: X \to Y$ be a morphism of noetherian schemes, which induces the left-exact *push-forward* functor (or *direct image*)

$$f_*: \mathrm{Qcoh}(X) \to \mathrm{Qcoh}(Y).$$

As X is noetherian, $\text{Qcoh}(X)$ has enough injectives, so f_* gives the right-derived functor

$$\mathsf{R}f_* \colon \mathcal{D}^+(\text{Qcoh}(X)) \to \mathcal{D}^+(\text{Qcoh}(Y)).$$

The $\mathsf{R}^i f_* = \mathcal{H}^i \mathsf{R}f_*$ are also known as *higher direct images* of f.

For a noetherian scheme X over a field $\Bbbk$, the push-forward of the structure map $\pi \colon X \to \text{Spec}\,\Bbbk$ is taking global sections, so $\Gamma = \pi_*$ in this case. Moreover, for a sheaf $\mathcal{F}$ we find that its cohomology groups are therefore $H^i(X, \mathcal{F}) = \mathsf{R}^i \pi_*(\mathcal{F})$.

Proposition 3.6 *Let* $f \colon X \to Y$ *be a morphism of noetherian schemes and* $\mathcal{F} \in \text{Qcoh}(X)$. *Then the derived push-forward restricts to*

$$\mathsf{R}f_* \colon \mathcal{D}^b(\text{Qcoh}(X)) \to \mathcal{D}^b(\text{Qcoh}(Y)).$$

If f *is in addition a proper morphism, then* $\mathsf{R}f_*$ *restricts further to*

$$\mathsf{R}f_* \colon \mathcal{D}^b(X) \to \mathcal{D}^b(Y).$$

Proof Using Proposition 3.3, the statements can be reduced to the following (deep) theorems:

- if $\mathcal{F} \in \text{Qcoh}(X)$ then $\mathsf{R}^i f_* \mathcal{F}$ X are trivial for $i > \dim(X)$;
- if f is proper and $\mathcal{F}$ coherent, then all $\mathsf{R}^i f_* \mathcal{F}$ are coherent.

□

Local Homomorphisms Let X be a noetherian scheme and $\mathcal{F} \in \text{Qcoh}(X)$. Then there is the left-exact functor

$$\mathcal{H}om(\mathcal{F}, -) \colon \text{Qcoh}(X) \to \text{Qcoh}(X)$$

which induces the derived functor

$$\mathsf{R}\,\mathcal{H}om(\mathcal{F}, -) \colon \mathcal{D}^+(\text{Qcoh}(X)) \to \mathcal{D}^+(\text{Qcoh}(X)).$$

Like in the case of Hom, the sheaf $\mathcal{F}$ can be replaced by a bounded above complex.

Proposition 3.7 *Let* X *be a smooth and proper variety and* $\mathcal{F}^\bullet \in \mathcal{D}^b(X)$. *Then* $\mathsf{R}\,\mathcal{H}om(\mathcal{F}^\bullet, -)$ *restricts to a functor*

$$\mathsf{R}\,\mathcal{H}om(\mathcal{F}^\bullet, -) \colon \mathcal{D}^b(X) \to \mathcal{D}^b(X).$$

Definition 3.8 Let X be a smooth projective variety and $\mathcal{F}^\bullet \in \mathcal{D}^b(X)$. Then the *dual* of $\mathcal{F}^\bullet$ is $\mathcal{F}^{\bullet\vee} := \mathsf{R}\,\mathcal{H}om(\mathcal{F}^\bullet, \mathcal{O}_X) \in \mathcal{D}^b(X)$.

3.2   Deriving Right-Exact Functors

In this section, we discussion of the most prominent right-exact functors: tensor and pull-back.

Remark 3.9 Let $X = \mathbb{P}^1$ be the projective line over an infinite field. Then there are no (non-zero) projective objects in $\mathrm{coh}(X)$ or $\mathrm{Qcoh}(X)$, see [5, Ex. III.6.2].

Tensor Product This lack of projective objects implies that we still have to work in order to derive tensor product and pull-back.

Theorem 3.10 *Let X be a scheme and $\mathcal{F}$ a quasi-coherent sheaf. Then the flat sheaves in $\mathrm{Qcoh}(X)$ form an adapted class for the left-exact functor $F = \mathcal{F} \otimes -: \mathrm{Qcoh}(X) \to \mathrm{Qcoh}(X)$.*

Proof We check the two properties in the definition of adaptedness. Let $\mathcal{F}^\bullet$ be an acyclic complex of quasi-coherent sheaves. Then for a flat sheaf $\mathcal{E}$, the complex $\mathcal{F}^\bullet \otimes \mathcal{E}$ is still acyclic by definition of flatness.

Let $\mathcal{G} \in \mathrm{Qcoh}(X)$. We use that arbitrary direct sums of flat sheaves are flat and that $\mathcal{O}_U \in \mathrm{Qcoh}(X)$ is flat for any open $U \subset X$. With this it is easy to build a surjection $\bigoplus_i \mathcal{O}_{U_i} \twoheadrightarrow \mathcal{G}$ by choosing (local) generators of $\mathcal{G}$ as a $\mathcal{O}_X$-module. $\square$

Remark 3.11 If X is a noetherian scheme and $\mathcal{F}$ a coherent sheaf on X, then $\mathcal{F}$ is flat if and only if it is locally free.

In particular, tensoring with a locally free coherent sheaf yields an exact functor. So with no need to derive, we arrive at the description of the Serre functor in the smooth case.

Theorem 3.12 *Let X be a smooth projective variety. Then the exact functor*

$$\mathcal{D}^b(X) \to \mathcal{D}^b(X), \ \mathcal{F}^\bullet \mapsto \mathcal{F}^\bullet \otimes \omega_X[\dim(X)]$$

is a Serre functor of $\mathcal{D}^b(X)$.

Projective objects in abelian categories are characterised by a lifting property dual to the one of Proposition 2.12. For an adapted class, like locally free sheaves for the tensor product, such a lifting does not exist in general.

Example 3.13 Consider $X = \mathbb{P}^1$. The structure sheaf $\mathcal{O}$ is already locally free, but tensoring the Euler sequence with $\mathcal{O}(-2)$ yields another locally free resolution:

$$
\begin{array}{ccccccccc}
P^\bullet & & 0 & \longrightarrow & \mathcal{O}(-2) & \longrightarrow & \mathcal{O}(-1)^{\oplus 2} & \longrightarrow & 0 \\
& {\scriptstyle f}\downarrow & & & & & \downarrow & & \\
\mathcal{O} & & & & 0 & \longrightarrow & \mathcal{O} & \longrightarrow & 0
\end{array}
$$

Note that f is a quasi-isomorphism, as $\mathcal{H}^{\bullet}(P^{\bullet}) = \mathcal{O}$. The (dual) lifting property of Proposition 2.12 would ask for a map g in the converse direction:

$$
\begin{array}{ccc}
\mathcal{O} & \xrightarrow{\ \mathrm{id}\ } & \mathcal{O} \\
{\scriptstyle g}\big\downarrow & & \big\| \\
P^{\bullet} & \xrightarrow{\ f\ } & \mathcal{O}
\end{array}
$$

But such a g cannot exist, since $\mathrm{Hom}(\mathcal{O}, \mathcal{O}(-1)) = H^0(\mathbb{P}^1, \mathcal{O}(-1)) = 0$.

Proposition 3.14 *Let X be a scheme and $\mathcal{F}^{\bullet} \in \mathrm{Com}^-(\mathrm{coh}(X))$. Then the right-exact functor $\mathcal{F}^{\bullet} \otimes -$ induces the left-derived functor*

$$
\mathcal{F}^{\bullet} \overset{L}{\otimes} - : \mathcal{D}^-(X) \to \mathcal{D}^-(X).
$$

If additionally, X is smooth and $\mathcal{F}^{\bullet} \in \mathcal{D}^b(X)$, then this functor restricts to

$$
\mathcal{F}^{\bullet} \overset{L}{\otimes} - : \mathcal{D}^b(X) \to \mathcal{D}^b(X).
$$

Proof The last statement can be shown using the analogue of Proposition 3.3 and the theorem that for smooth varieties any $\mathcal{F}^{\bullet} \in \mathcal{D}^b(X)$ is quasi-isomorphic to a bounded complex of locally free sheaves of length at most $\dim(X)$. □

Remark 3.15 The $(-i)$-th derived functor of the tensor product is denoted by

$$
\mathrm{Tor}_i(\mathcal{F}^{\bullet}, -) := \mathcal{H}^{-i}(\mathcal{F}^{\bullet} \overset{L}{\otimes} -).
$$

Pull-Back Let $f : X \to Y$ be a morphism of noetherian schemes. Note that the *pull-back* (or *inverse image*) f^* is the composition of the exact functor f^{-1} with the tensor product $\mathcal{O}_X \otimes_{f^{-1}\mathcal{O}_Y} -$.

From this, using flat sheaves as an adapted class, we get the left-derived functor

$$
\mathsf{L}f^* : \mathcal{D}^-(\mathrm{Qcoh}(Y)) \to \mathcal{D}^-(\mathrm{Qcoh}(X))
$$

which is the composition of f^{-1} and $\mathcal{O}_X \overset{L}{\otimes}_{f^{-1}\mathcal{O}_Y} -$.

Proposition 3.16 *Let $f : X \to Y$ be a morphism of noetherian schemes with Y smooth. Then $\mathsf{L}f^*$ restricts to*

$$
\mathsf{L}f^* : \mathcal{D}^b(Y) \to \mathcal{D}^b(X).
$$

Proof As for the tensor product, smoothness of Y implies that we can replace a bounded complex of coherent sheaves by a bounded complex of locally free sheaves. □

3.3   Compatibilities

In this section, we will speak a bit about the interaction between the above introduced derived functors: adjunction of pull-back and push-forward, projection formula and flat base change.

Remark 3.17 The crucial technical tool for this section is hidden in Remark 2.39: the right- (or left-) derived functor associated to a left- (or right-) exact functor is essentially unique.

In particular, an equality of functors on an adapted class extends to an equality of derived functors.

As a first application of this remark, we see that the adjunction of f^* and f_* on the abelian level extends to derived categories.

Proposition 3.18 *Let* $f \colon X \to Y$ *be proper morphism of smooth varieties. Then* Lf^* *and* Rf_* *form a pair of adjoint functors, i.e. there is an isomorphism functorial in both arguments*

$$\mathrm{Hom}_{\mathcal{D}^b(X)}(Lf^*\mathcal{F}, \mathcal{G}) \xrightarrow{\sim} \mathrm{Hom}_{\mathcal{D}^b(Y)}(\mathcal{F}, Rf_*\mathcal{G}).$$

Another equality of abelian functors is $\Gamma \circ \mathcal{H}om = \mathrm{Hom}$.

Proposition 3.19 *Let* X *be a smooth projective variety and* $\mathcal{F}^{\bullet} \in \mathcal{D}^b(X)$. *Then* $R\Gamma \circ R\mathcal{H}om(\mathcal{F}^{\bullet}, -) = R\mathrm{Hom}(\mathcal{F}^{\bullet}, -)$.

Proof Hidden in this statement is the equality $R(\Gamma \circ \mathcal{H}om(\mathcal{F}^{\bullet}, -)) = R\Gamma \circ R\mathcal{H}om(\mathcal{F}^{\bullet}, -)$, which follows from the fact that $\mathcal{H}om(\mathcal{F}^{\bullet}, -)$ maps injective sheaves to Γ-acyclic ones. □

The next proposition is the so-called *projection formula* which, on the abelian level of coherent sheaves, holds for locally free sheaves $\mathcal{F}$.

Proposition 3.20 *Let* $f \colon X \to Y$ *be proper morphism of smooth varieties and* $\mathcal{E}^{\bullet} \in \mathcal{D}^b(X)$, $\mathcal{F}^{\bullet} \in \mathcal{D}^b(Y)$. *Then there is a natural isomorphism*

$$Rf_*(\mathcal{E}^{\bullet}) \overset{L}{\otimes} \mathcal{F}^{\bullet} \xrightarrow{\sim} Rf_*(\mathcal{E}^{\bullet} \overset{L}{\otimes} Lf^*(\mathcal{F}^{\bullet})).$$

Finally, there is also the *flat base change*. For this, note that pull-backs along flat morphisms do not need to be derived.

Proposition 3.21 *Let* $u \colon X \to Z$ *be a flat morphism and* $f \colon Y \to Z$ *a proper morphism of smooth varieties. Consider the fibre product*

$$
\begin{array}{ccc}
X \underset{Z}{\times} Y & \xrightarrow{\;v\;} & Y \\
{\scriptstyle g}\downarrow & & \downarrow{\scriptstyle f} \\
X & \xrightarrow{\;u\;} & Z
\end{array}
$$

Then for $\mathcal{F}^{\bullet} \in \mathcal{D}^b(Y)$ there is a natural isomorphism

$$u^* Rf_*(\mathcal{F}^{\bullet}) \xrightarrow{\sim} Rg_* v^*(\mathcal{F}^{\bullet})$$

and in particular, $u^* R^i f_*(\mathcal{F}^{\bullet}) \cong R^i g_* v^*(\mathcal{F}^{\bullet})$.

3.4 Fourier-Mukai Transforms

We introduce an important class of exact functors between derived categories of coherent sheaves.

Throughout this section, let X and Y be smooth projective varieties. Moreover, we denote the two projections from their product by:

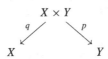

Definition 3.22 Let $\mathcal{P}^{\bullet} \in \mathcal{D}^b(X \times Y)$ then the induced exact functor

$$\Phi_{\mathcal{P}^{\bullet}} : \mathcal{D}^b(X) \to \mathcal{D}^b(Y), \ \mathcal{E}^{\bullet} \mapsto Rp_*(q^*\mathcal{E}^{\bullet} \overset{L}{\otimes} \mathcal{P}^{\bullet})$$

is called the *Fourier-Mukai transform* with *Fourier-Mukai kernel $\mathcal{P}^{\bullet}$*.

Note that in the above definition, there is no need to derive q^*, as a projection is flat. Moreover, a Fourier-Mukai kernel can be used to define also a Fourier-Mukai transform in the converse direction. To stress the direction, we sometimes write $\Phi_{\mathcal{P}^{\bullet}}^{X \to Y}$.

For a Fourier-Mukai kernel $\mathcal{P}^{\bullet} \in \mathcal{D}^b(X \times Y)$, its *left and right adjoint Fourier-Mukai kernels* in $\mathcal{D}^b(X \times Y)$ are

$$\mathcal{P}_L^{\bullet} := \mathcal{P}^{\bullet \vee} \otimes p^* \omega_Y[\dim(Y)], \quad \mathcal{P}_R^{\bullet} := \mathcal{P}^{\bullet \vee} \otimes q^* \omega_X[\dim(X)].$$

This notation is justified by the following statement.

Proposition 3.23 *Let $\mathcal{P}^{\bullet} \in \mathcal{D}^b(X \times Y)$, then the Fourier-Mukai transforms $\Phi_{\mathcal{P}_L^{\bullet}}^{Y \to X}$ and $\Phi_{\mathcal{P}_R^{\bullet}}^{Y \to X}$ are left and right adjoint to $\Phi_{\mathcal{P}^{\bullet}}^{X \to Y}$.*

Finally, one can ask, whether a given exact functor $F : \mathcal{D}^b(X) \to \mathcal{D}^b(Y)$ is *of Fourier-Mukai type*, i.e. can be written as a Fourier-Mukai transform with some kernel. The central result to this question is the following theorem by Orlov.

Theorem 3.24 *Let* $F: \mathcal{D}^b(X) \to \mathcal{D}^b(Y)$ *be an exact fully faithful functor and assume that* X *and* Y *are smooth projective varieties. Then there is a* $\mathcal{P}^{\bullet} \in \mathcal{D}^b(X \times Y)$ *such that* $F \cong \Phi_{\mathcal{P}^{\bullet}}$. *Moreover,* $\mathcal{P}^{\bullet}$ *is unique up to isomorphism.*

Remark 3.25 The original statement also assumes the existence of a left adjoint. Based on work of Bondal and van den Bergh, the existence of both adjoints is automatic. Moreover, over an algebraically closed field of characteristic zero, a non-zero exact full functor $F: \mathcal{D}^b(X) \to \mathcal{D}^b(Y)$ between smooth projective varieties is already faithful. Details on both can be found in the survey [2, Prop. 3.5 & Thm. 3.14].

For the following, let $\iota: X \xrightarrow{\sim} \Delta \subset X \times X$ denote the inclusion of the diagonal Δ, in particular, $\iota_* \mathcal{O}_X = \mathcal{O}_\Delta$.

Examples 3.26 Let $f: X \to Y$ be a morphism between smooth projective varieties and denote by Γ_f its graph in $X \times Y$. Then push-forward and pull-back are of Fourier-Mukai type:

$$\mathsf{R}f_* \cong \Phi_{\mathcal{O}_{\Gamma_f}}^{X \to Y}, \quad \mathsf{L}f^* \cong \Phi_{\mathcal{O}_{\Gamma_f}}^{Y \to X}.$$

Notable special cases are

- $\mathrm{id} = \Phi_{\mathcal{O}_\Delta}$, where Δ is the diagonal;
- $H^*(X, -) = \Phi_{\mathcal{O}_X}$, using that $X \cong X \times \mathrm{Spec}(\Bbbk)$ and $H^*(X, -) = \mathsf{R}\pi_*$ for $\pi: X \to \mathrm{Spec}(\Bbbk)$.

The shift functor is of Fourier-Mukai type using the kernel $\mathcal{O}_\Delta[1]$.

The tensor product $\mathcal{F}^{\bullet} \overset{\mathsf{L}}{\otimes} -$ is of Fourier-Mukai type, using the kernel $\iota_*(\mathcal{F}^{\bullet})$ with $\iota: X \hookrightarrow X \times X$. By Proposition 3.23, also $\mathcal{H}om(\mathcal{F}^{\bullet}, -)$ is of Fourier-Mukai type, as it is the left adjoint of the tensor product.

Proposition 3.27 *The composition of two functors of Fourier-Mukai type is again of Fourier-Mukai type.*

Example 3.28 Let X be a smooth projective variety. Then the Serre functor $S = - \otimes \omega_X[\dim(X)]$ of $\mathcal{D}^b(X)$ is of Fourier-Mukai type.

With this Proposition 3.27 by Mukai, we can use the above functors as building blocks, yielding a vast array of functors of Fourier-Mukai type. It is probably fair to say that all geometrically meaningful functors are of Fourier-Mukai type. For further discussion see the survey [2].

4 The Derived Torelli Theorem for K3 Surfaces

This last section serves as a bridge to [11]: we discuss the auto-equivalences of the derived category of K3 surfaces. In this section, the ground field will be $\mathbb{C}$.

For basic facts about K3 surfaces and its Hodge theory needed here, already the recap in [7, §10.1] is enough. We will only recall the global Torelli theorem. For this let X be a K3 surface. The standard polarised Hodge structure on the second cohomology $H^2(X, \mathbb{C})$, which uses the intersection pairing, can be restricted to the integral cohomology:

$$H^2(X, \mathbb{Z}) = H^{2,0}(X, \mathbb{Z}) \oplus H^{1,1}(X, \mathbb{Z}) \oplus H^{0,2}(X, \mathbb{Z}).$$

Note that a smooth rational curve $C \subset X$ becomes a (-2)-class $[C]$ inside $H^{1,1}(X, \mathbb{Z}) \subset H^2(X, \mathbb{Z})$. In particular, the associated reflection

$$s_{[C]} \colon H^2(X, \mathbb{C}) \to H^2(X, \mathbb{C}), \ \alpha \mapsto \alpha + (\alpha, [C])[C]$$

is a Hodge isometry, i.e. $s_{[C]}$ respects the intersection pairing and the decomposition. Moreover, this reflection restricts to an (integral) Hodge isometry $s_{[C]} \colon H^2(X, \mathbb{Z}) \to H^2(X, \mathbb{Z})$.

Theorem 4.1 (Torelli) *Let X and Y be two K3 surfaces. Then there is an isomorphism $f \colon X \xrightarrow{\sim} Y$ if and only if there exists a Hodge isometry $\phi \colon H^2(X, \mathbb{Z}) \xrightarrow{\sim} H^2(Y, \mathbb{Z})$.*

In this case, there are smooth rational curves $C_1, \ldots, C_m$ on X such that

$$\phi = \pm s_{[C_1]} \circ \cdots \circ s_{[C_m]} \circ f_*.$$

Derived Torelli In the following, we will see that the above statement is the cohomological "shadow" of a statement involving the respective derived categories.

Let $f \colon X \to Y$ be a morphism of K3 surfaces. On the level of rational cohomology, f induces a ring homomorphism

$$f^* \colon H^*(Y, \mathbb{Q}) \to H^*(X, \mathbb{Q}),$$

the *cohomological pull-back*. Using Poincaré duality, the *cohomological push-forward*

$$f_* \colon H^*(X, \mathbb{Q}) \to H^*(Y, \mathbb{Q})$$

can be defined as the dual map to f^*. Given a class $\alpha \in H^*(X \times Y, \mathbb{Q})$ the *cohomological Fourier-Mukai transform* with kernel α is

$$\Phi_\alpha^H \colon H^*(X, \mathbb{Q}) \to H^*(Y, \mathbb{Q}), \ \beta \mapsto p_*(\alpha.q^*(\beta)).$$

Definition 4.2 Let X be an algebraic K3 surface. Then the *Mukai vector* of $E^{\bullet} \in \mathcal{D}^b(X)$ is defined as

$$v(E^{\bullet}) := (\mathrm{rk}(E^{\bullet}), c_1(E^{\bullet}), \mathrm{rk}(E^{\bullet}) + c_1^2(E^{\bullet})/2 - c_2(E^{\bullet})).$$

Moreover, for $\alpha = (\alpha_0, \alpha_1, \alpha_2)$ and $\beta = (\beta_0, \beta_1, \beta_2)$ with $\alpha_k, \beta_k \in H^{2k}(X, \mathbb{Q})$, the *Mukai pairing* is given as

$$\langle \alpha, \beta \rangle := \alpha_1.\beta_1 - \alpha_0.\beta_2 - \alpha_2.\beta_0.$$

Remark 4.3 Up to a sign, the Mukai pairing can be seen as a cohomological shadow of the *Euler characteristic* χ. To be precise, for $E^{\bullet}, F^{\bullet} \in \mathcal{D}^b(X)$ holds

$$-\langle v(E^{\bullet}), v(F^{\bullet}) \rangle = \chi(E^{\bullet}, F^{\bullet}) := \sum (-1)^k \dim \mathsf{R}^k \operatorname{Hom}(E^{\bullet}, F^{\bullet}).$$

This follows quite immediate from the Hirzebruch-Riemann-Roch formula.

The definition of the Mukai pairing is made in such a way, that the pairing extends the intersection pairing on $H^2(X, \mathbb{Z})$. Even more, we can extend the integral Hodge structure on $H^2(X, \mathbb{Z})$ by setting

$$\tilde{H}^{2,0}(X, \mathbb{Z}) := H^{2,0}(X, \mathbb{Z}),$$

$$\tilde{H}^{1,1}(X, \mathbb{Z}) := H^0(X, \mathbb{Z}) \oplus H^{1,1}(X, \mathbb{Z}) \oplus H^4(X, \mathbb{Z}),$$

$$\tilde{H}^{0,2}(X, \mathbb{Z}) := H^{0,2}(X, \mathbb{Z}).$$

This gives an integral weight-two Hodge structure on $H^*(X, \mathbb{Z})$, which is polarised by the Mukai pairing. In the following, we will denote this polarised Hodge structure by $\tilde{H}(X, \mathbb{Z})$, which is called the *Mukai lattice*.

Theorem 4.4 ([13]) *Let* $\Phi \colon \mathcal{D}^b(X) \xrightarrow{\sim} \mathcal{D}^b(Y)$ *be an equivalence of the derived categories of two algebraic K3 surfaces. Then this induces a map on cohomology which defines a Hodge isometry*

$$\Phi^H \colon \tilde{H}(X, \mathbb{Z}) \xrightarrow{\sim} \tilde{H}(Y, \mathbb{Z}).$$

Proof By Theorem 3.24, Φ can be written uniquely as a Fourier-Mukai transform with Fourier-Mukai kernel $P^{\bullet} \in \mathcal{D}^b(X \times Y)$. As the key ingredient, Mukai showed that $v(P^{\bullet}) \in \tilde{H}^{1,1}(X, \mathbb{Z})$. As a consequence, $\Phi^H := \Phi^H_{v(P^{\bullet})} \colon H^*(X, \mathbb{Q}) \to H^*(X, \mathbb{Q})$ can be restricted to the integral part. Finally, as an application of the Grothendieck-Riemann-Roch formula, one obtains that

$$v(\Phi(E^{\bullet})) = \Phi^H(v(E^{\bullet})).$$

$\square$

Remark 4.5 If $\Phi \colon \mathcal{D}^b(X) \to \mathcal{D}^b(Y)$ is an equivalence between derived categories of arbitrary smooth projective varieties, then there is a natural pairing on cohomology such that the induced $\Phi^H \colon H^*(X, \mathbb{Q}) \to H^*(X, \mathbb{Q})$ is an isometry, see [7, §5.2]. Note that in the general situation, Φ^H will not restrict to the integral part.

Corollary 4.6 *Let X be an algebraic K3 surface. Then there is a homomorphism of groups*

$$\varpi : \text{Aut}(\mathcal{D}^b(X)) \to \text{O}(\tilde{H}(X, \mathbb{Z})), \ \Phi \mapsto \Phi^H,$$

where $\text{Aut}(\mathcal{D}^b(X))$ *is the group of auto-equivalences of* $\mathcal{D}^b(X)$ *and* $\text{O}(\tilde{H}(X, \mathbb{Z}))$ *denotes the group of Hodge isometries of the Mukai lattice.*

Orlov strengthened the Mukai's result to the so-called derived Torelli Theorem.

Theorem 4.7 ([14]) *Two algebraic K3 surfaces X and Y have equivalent derived categories if and only if there exists a Hodge isometry of their Mukai lattices.*

Proof The strategy of the proof is to reduce to the case that an isometry $\phi : \tilde{H}(X, \mathbb{Z}) \to \tilde{H}(Y, \mathbb{Z})$ preserves H^2. This reduction is built on results about moduli spaces of sheaves on K3 surfaces, see [7, §10.3] for an overview. As soon as ϕ preserves H^2, by Theorem 4.1 such an isometry is of the form

$$\phi = \pm s_{[C_1]} \circ \cdots \circ s_{[C_m]} \circ f_*.$$

for some $f : X \xrightarrow{\sim} Y$. In particular, $\mathcal{D}^b(X) \cong \mathcal{D}^b(Y)$.   □

The relationship of auto-equivalences and Hodge isometries was clarified further by Hosono et al. [6], Ploog [15] and Huybrechts et al. [8].

The central observation is that the Mukai lattice $\tilde{H}(X, \mathbb{Z})$ has signature $(4, 20)$ and that there is a *natural orientation* of the positive directions. Given an ample class $\alpha \in H^{1,1}(X)$ and a generator $\sigma \in H^{2,0}(X)$, the four classes

$$\Re(\exp(i\alpha)) = 1 - \alpha^2/2, \ \Im(\exp(i\alpha)) = \alpha, \ \Re(\sigma), \ \Im(\sigma)$$

define an orientation which is independent of the choices of α and σ. We denote by $\text{O}^+(\tilde{H}(X, \mathbb{Z}))$ the Hodge isometries which preserve this orientation.

Proposition 4.8 *Let X be an algebraic K3 surface and* $\Phi \in \text{Aut}(\mathcal{D}^b(X))$. *Then* Φ^H *preserves the natural orientation, i.e.* $\Phi^H \in \text{O}^+(\tilde{H}(X, \mathbb{Z}))$. *Conversely, for any* $\psi \in \text{O}(\tilde{H}(X, \mathbb{Z}))$ *there is a* $\Psi \in \text{Aut}(\mathcal{D}^b(X))$ *with*

$$\Psi^H = \psi \circ (\pm \text{id}_{H^2}),$$

in particular, $\text{O}^+(\tilde{H}(X, \mathbb{Z})) \subset \text{O}(\tilde{H}(X, \mathbb{Z}))$ *has index two.*

Spherical Twists As a corollary of Proposition 4.8 we obtain the following short exact sequence for an algebraic K3 surface X:

$$0 \to \ker(\varpi) \to \text{Aut}(\mathcal{D}^b(X)) \xrightarrow{\varpi} \text{O}^+(\tilde{H}(X, \mathbb{Z})) \to 0.$$

One may wonder which elements lie in the kernel of ϖ, i.e. auto-equivalences that act as the identity on cohomology. Or one might ask whether the reflections $s_{[C]}$ for smooth rational curves $C \subset X$ can be lifted to auto-equivalences of $\mathcal{D}^b(X)$. Both questions lead to the notion of a spherical twist. We recall the central properties, for further details see [7, §8.1].

Definition 4.9 Let X be a smooth projective variety. An object $E^{\bullet} \in \mathcal{D}^b(X)$ is called *spherical* if

- $E^{\bullet}$ is a *spherelike* object, i.e.

$$\mathrm{Hom}^*(E^{\bullet}, E^{\bullet}) := \bigoplus_k \mathrm{Hom}(E^{\bullet}, E^{\bullet}[k])[-k] \cong \mathbb{C}[t]/t^2;$$

- and $E^{\bullet}$ is a *Calabi-Yau* object, i.e. $E^{\bullet} \otimes \omega_X \cong E^{\bullet}$.

Remark 4.10 The graded vector space $\mathrm{Hom}^*(E^{\bullet}, E^{\bullet})$ is the cohomology of the complex $\mathrm{R}\,\mathrm{Hom}(E^{\bullet}, E^{\bullet}) \in \mathcal{D}^b(\mathbb{C}\,\mathrm{mod})$, and actually quasi-isomorphic to it. Note that $\mathrm{Hom}^*(E^{\bullet}, E^{\bullet})$ becomes a $\mathbb{C}$-algebra with the Yoneda product. So the first property asks that there is an (up to scalar) unique self-extension of $E^{\bullet}$ that squares to zero. By the second property this extension has to be of degree $\dim(X)$.

Theorem 4.11 *Let $E^{\bullet}$ be a spherical object in $\mathcal{D}^b(X)$. Then there is an auto-equivalence $T_{E^{\bullet}}$ of $\mathcal{D}^b(X)$ which fits into an exact triangle of functors:*

$$\mathrm{Hom}^*(E^{\bullet}, -) \otimes E^{\bullet} \xrightarrow{ev} \mathrm{id} \to T_{E^{\bullet}} \to \mathrm{Hom}^*(E^{\bullet}, -) \otimes E^{\bullet}[1],$$

which is called the spherical twist *along $E^{\bullet}$.*

Remark 4.12 The first arrow in the above triangle is the *evaluation map* ev which comes from the adjunction of Hom and $\otimes$. This triangle of functors cannot serve as a definition of $T_{E^{\bullet}}$, as cones are not functorial. But ev induces a morphism between the respective Fourier-Mukai kernels, which allows to define $T_{E^{\bullet}}$ as a Fourier-Mukai transform to the cone of this morphism.

The triangle of functors above allows to deduce easily two important properties of a spherical twist $T_{E^{\bullet}}$:

- $T_{E^{\bullet}}(E^{\bullet}) = E^{\bullet}[1 - \dim(X)]$ and
- $T_{E^{\bullet}}(F^{\bullet}) = F^{\bullet}$ for $F^{\bullet}$ with $\mathrm{Hom}^*(E^{\bullet}, F^{\bullet}) = 0$.

On the level of cohomology, these properties become

- $T_{E^{\bullet}}^H(v(E^{\bullet})) = (-1)^{1-\dim(X)} v(E^{\bullet})$ and
- $T_{E^{\bullet}}^H(\alpha) = \alpha$ for α with $\langle v(E^{\bullet}), \alpha \rangle = 0$.

In particular, $T_{E^{\bullet}}^H$ is already completely determined: it is the reflection along $v(E^{\bullet})^{\perp}$ if $\dim(X)$ is even and the identity if $\dim(X)$ is odd.

Corollary 4.13 *For an algebraic K3 surface X holds:*

- *If $E^\bullet$ is a spherical object, then $T^2_{E^\bullet}$ is a non-trivial element of $\ker(\varpi)$.*
- *If $C \subset X$ is a smooth rational curve, then $\mathcal{O}_C(-1)$ is a spherical object with $T^H_{\mathcal{O}_C(-1)} = s_{[C]}$.*

Proof The first part follows from the observation that $T^H_{E^\bullet}$ is a reflection. For the second part, one can check that all $\mathcal{O}_C(k)$ are spherical objects for $k \in \mathbb{Z}$. But only for $k = -1$, one obtains that $v(\mathcal{O}_C(-1)) = [C]$. □

So in the presence of smooth rational curves on an algebraic K3 surface, we obtain elements in $\ker(\varpi)$. The question about the structure of $\ker(\varpi)$ in general is hard, so far only the case of Picard rank 1 is solved.

Theorem 4.14 ([1]) *Let X be an algebraic K3 surface of Picard rank 1. Then $\ker(\varpi)$ is the product of $\mathbb{Z} \cdot [2]$ and the free group generated by T^2_V with V running over all spherical vector bundles on X.*

Acknowledgements The author thanks Klaus Altmann, Andreas Krug, Ciaran Meachan, David Ploog and Paolo Stellari for comments and suggestions.

References

1. A. Bayer, T. Bridgeland, Derived automorphism groups of K3 surfaces of Picard rank 1. Duke Math. J. **166**, 75–124 (2017), also http://arxiv.org/abs/1310.8266
2. A. Canonaco, P. Stellari, Fourier-Mukai functors: a survey, in *Derived Categories in Algebraic Geometry – Tokyo 2011*, ed. by Y. Kawamata. European Mathematical Society (2013), also http://arxiv.org/abs/1109.3083
3. S. Gelfand, Y. Manin, *Methods of Homological Algebra* (Springer, Berlin, 2003)
4. R. Hartshorne, *Residues and Duality* (Springer, Berlin, 1966)
5. R. Hartshorne, *Algebraic Geometry* (Springer, Berlin, 1977)
6. S. Hosono, B.H. Lian, K. Oguiso, S.-T. Yau, Autoequivalences of derived category of a K3 surface and monodromy transformations. J. Algebraic Geom. **13**, 513–545 (2004), also http://arxiv.org/abs/math/0201047
7. D. Huybrechts, *Fourier-Mukai Transforms in Algebraic Geometry* (Oxford University, Oxford, 2006)
8. D. Huybrechts, E. Macrì, P. Stellari, Derived equivalences of K3 surfaces and orientation. Duke Math. J. **149**, 461–507 (2009), also http://arxiv.org/abs/0710.1645
9. M. Kashiwara, P. Schapira, *Sheaves on Manifolds* (Springer, Berlin, 1990)
10. J. Lipman, Notes on derived functors and Grothendieck duality, in *Foundations of Grothendieck Duality for Diagrams of Schemes*, ed. by J. Lipman, M. Hashimoto (Springer, Berlin, 2009)
11. E. Macrì, P. Stellari, Lectures on non-commutative K3 surfaces, Bridgeland stability, and moduli spaces, in *Birational Geometry of Hypersurfaces*, ed. by A. Hochenegger et al. Lecture Notes of the Unione Matematica Italiana, vol. 26 (Springer, Cham, 2019). https://doi.org/10.1007/978-3-030-18638-8_6
12. J.P. May, The additivity of traces in triangulated categories. Adv. Math. **163**, 34–73 (2001)

13. S. Mukai, On the moduli space of bundles on K3 surfaces I, in *Vector Bundles on Algebraic Varieties* (Oxford University, Oxford, 1987)
14. D. Orlov, Equivalences of derived categories and K3 surfaces. J. Math. Sci. **84**, 1361–1381 (1997), also http://arxiv.org/abs/alg-geom/9606006
15. D. Ploog, Groups of autoequivalences of derived categories of smooth projective varieties, PhD thesis, Berlin, 2005

LECTURE NOTES OF THE UNIONE
MATEMATICA ITALIANA

Editor in Chief: Ciro Ciliberto and Susanna Terracini

Editorial Policy

1. The UMI Lecture Notes aim to report new developments in all areas of mathematics and their applications - quickly, informally and at a high level. Mathematical texts analysing new developments in modelling and numerical simulation are also welcome.

2. Manuscripts should be submitted to
   Redazione Lecture Notes U.M.I.
   umi@dm.unibo.it
   and possibly to one of the editors of the Board informing, in this case, the Redazione about the submission. In general, manuscripts will be sent out to external referees for evaluation. If a decision cannot yet be reached on the basis of the first 2 reports, further referees may be contacted. The author will be informed of this. A final decision to publish can be made only on the basis of the complete manuscript, however a refereeing process leading to a preliminary decision can be based on a prefinal or incomplete manuscript. The strict minimum amount of material that will be considered should include a detailed outline describing the planned contents of each chapter, a bibliography and several sample chapters.

3. Manuscripts should in general be submitted in English. Final manuscripts should contain at least 100 pages of mathematical text and should always include

   – a table of contents;
   – an informative introduction, with adequate motivation and perhaps some historical remarks: it should be accessible to a
     reader not intimately familiar with the topic treated;
   – a subject index: as a rule this is genuinely helpful for the reader.

4. For evaluation purposes, please submit manuscripts in electronic form, preferably as pdf- or zipped ps-files. Authors are asked, if their manuscript is accepted for publication, to use the LaTeX2e style files available from Springer's web-server at
   ftp://ftp.springer.de/pub/tex/latex/svmonot1/ for monographs
   and at
   ftp://ftp.springer.de/pub/tex/latex/svmultt1/ for multi-authored volumes

5. Authors receive a total of 50 free copies of their volume, but no royalties. They are entitled to a discount of 33.3% on the price of Springer books purchased for their personal use, if ordering directly from Springer.

6. Commitment to publish is made by letter of intent rather than by signing a formal contract. Springer-Verlag secures the copyright for each volume. Authors are free to reuse material contained in their LNM volumes in later publications: A brief written (or e-mail) request for formal permission is sufficient.

Printed in the United States
By Bookmasters